Participation and Learning

Alan Reid • Bjarne Bruun Jensen • Jutta Nikel
Venka Simovska
Editors

Participation and Learning

Perspectives on Education and the Environment, Health and Sustainability

Springer

Alan Reid
University of Bath
UK

Jutta Nikel
University of Bath
UK

Bjarne Bruun Jensen
Danish School of Education
University of Aarhus
Copenhagen Denmark

Venka Simovska
Danish School of Education
University of Aarhus
Copenhagen Denmark

ISBN 978-1-4020-6415-9 e-ISBN 978-1-4020-6416-6

Library of Congress Control Number: 2007933829

Printed on acid-free paper.

9 8 7 6 5 4 3 2 1

springer.com

Preface

The *Research in Participatory Education Network* (RIPEN) was initiated by the Research Programme for Environmental and Health Education at the Danish School of Education, University of Aarhus, in 2003. It embraces a broad spectrum of researchers, scholars, students, and practitioners of participatory education, working in or from Europe, North America, Africa, and Australasia. Given the international scope of the network and the range of interests it now has, as initiators and early participants in the network the editorial team invited RIPEN to discuss what a critical perspective on participatory approaches to education might mean for education and the environment, health and sustainability, and how network members might research and substantiate their claims and arguments. Following the introductory chapter on the scope of this collection, 19 chapters illustrate the contributors' responses to that invitation.

Our focus on critical perspectives was prompted by earlier work by Majid Rahnema in Wolfgang Sachs's (1992), *Development Dictionary*. Critiquing concepts of participation in a volume that set out to stimulate cultural, historical, and anthropological debate on the key concepts of development, Rahnema (p. 126) wrote:

> Participation, which is also a form of intervention, is too serious and ambivalent a matter to be taken lightly, or reduced to an amoeba word lacking any precise meaning, or a slogan, or fetish, or for that matter, only an instrument or methodology. Reduced to such trivialities, not only does it cease to be a boon, but it runs the risk of acting as a deceptive myth or a dangerous tool for manipulation. To understand the many dimensions of participation, one needs to enquire seriously into all its roots and ramifications, these going deep into the heart of human relationships and the socio-cultural realities conditioning them.

The broader critical work of which this was a part demonstrates how easily a potentially radical, fundamental social concept can become sterile and obsolescent, through its co-option by powerful interests or its reduction to a meaningless buzzword. Recent edited collections on participation in the field of development – most notably, those edited by Cooke and Kothari (2001) *Participation: The New Tyranny*? and by Hickey and Mohan (2004) *Participation: From Tyranny to Transformation*? – are provocative collections on these issues, and were instrumental in stimulating the members of RIPEN to debate the status and jargon of participation in the policy statements, practices, ethics, research, and discourses of education and the environment, health, and sustainability. For example, with Cooke and

Kothari (2001) identifying 'overarching and fundamental problems with participatory approaches to development' (p. 13), that include (p. 14) …

- The naivety of assumptions about the authenticity of motivations and behaviour in participatory processes
- How the language of empowerment masks a real concern for managerialist effectiveness
- The quasi-religious associations of participatory rhetoric and practice
- How an emphasis on the micro level of intervention can obscure, and indeed sustain, broader macro-level inequalities and injustice

…, in this collection, we address whether these also obtain in education, or if education in both formal and lifelong learning settings either faces or raises a different set of problems?

Network meetings exploring such issues took place at the University of Bath's Centre for Research in Education and the Environment (UK), at the Graduate Center of the City University of New York with members of the Children's Environments Research Group (USA), and in Copenhagen with members of the Research Programme for Environmental and Health Education at the Danish School of Education (Denmark). The meetings presented members and invited contributors to RIPEN with opportunities to discuss themes and develop frameworks and agendas that would advance our shared deliberations. A key event was a collaborative writing workshop in Copenhagen in 2005, at which draft chapters were prepared and shaped to produce the core of the book. As a whole, the collection represents our first joint attempt to share more broadly our initial ideas and reflections on the debates about participation as they relate to education, the environment, health, and sustainability.

In *Participation and Learning*, readers will find case studies and commentaries on participatory work in the field of education that we hope will variously inspire practice, encourage reflection, and stimulate further critique and development in relation to theoretical, institutional, and methodological challenges. Pulling out some of the key themes in the chapters illustrates the range of subject areas that are addressed in the collection: children's participation in community development; participation within and as a result of environmental learning; teachers' and learners' participation in organising environmental, health or sustainability education; youth participation in school curriculum development; young people's participation in meaning-making about health and sustainability; youth participation in and through school councils and whole-school approaches; facilitators' role and dilemmas in fostering public participation; researching conceptions of citizen participation in teaching materials; and many others.

As editors of the volume, we welcome this diversity of material, but have also sought to find coherence and cohesion in the arguments. As we discuss in the Introduction, this collection takes up a series of key debates in the area of participatory approaches to education and explores them from multiple and diverse perspectives. However, what makes this book unique is that we engage these issues from within the areas of environmental, sustainability, and health education, grappling with the tensions within these areas, in order to add to and complicate the discussions on participation in wider fields.

Thus, the chapters might be read as encouraging readers to consider the unique and shared assumptions amongst contributors about, for example, human agency and the roles, liberties, and responsibilities of teachers and learners in education, as well as the value of current and possible heuristics, frameworks and models of participation for teaching and learning in general, and specifically in relation to the environment, health, and sustainability. There are contributions that illustrate distinct perspectives on procedural and emancipatory interests in participatory education (for example, various viewpoints on the requirements and efficiencies of using participatory approaches to promote particular outcomes such as participatory democracy, e.g. bottom-up versus top-down approaches), as well as others on the purposes and processes associated with active forms of citizenship, social capital, empowerment, social (behaviour?) change, civic engagement, and local planning.

Several chapters address such matters from theoretical and analytical perspectives; others mainly focus on commentaries and accounts of empirical content or context. Some illustrate how participatory education might require greater consideration of role of empathy and guidance on the part of participants and facilitators; others suggest a deeper level of attentiveness to the authenticity and relationality of the levels of involvement in participatory work, particularly if its personal, cultural, and linguistic appropriateness is to be the focus of evaluation and debate. Moreover, while some authors demonstrate considerable and long-term engagement with the key issues and their challenges and ramifications, other contributions bring new work to the debate and represent thinking in progress.

As a result, recognising that the volume will engage readers in diverse ways, we invite the reader to consider shifting the metaphor of engagement from that of a *lens* to that of *mediation*: that is, to read the various chapters of the book not with the principal goal of understanding what it might *look like* to experience, live, and research participation from the reader's perspective alone, but to engage with and respond to the material in terms of what doing and researching participatory education *involves*. This is, in effect, an invitation to participate more fully in meeting the goals of both the network and the editors in producing this collection: we invite you to not only analyse the arguments and claims as an outsider, but to foster a critical empathy with the authors and their work from an insider's perspective.

The value of doing so is underlined by the fact that contributors draw on a wide range of theoretical and methodological perspectives and positions on participation, including: ethnographic, socio-cultural, post-structural, feminist, pragmatic, critical, postpositivist, conceptual, deliberative, amongst others. Some of these coincide conceptually and pragmatically; others are in tension. For an edited collection, such a situation highlights the need for authors to be clear about the particular meanings and practices of the participation they discuss, and so each chapter starts with an introduction that should help readers locate and navigate the diversity of perspectives and material contained within the collection's pages.

What is distinctive to this volume, and hence its subtitle, is the focus on discussing participation in terms of the sense, health, and well-being of self, place, and community. While some of this work may be more commonplace in the fields of development studies, health promotion, or environmental psychology, as yet little has been written on these topics in the educational field, particularly at their intersections with pedagogy, sustainability, health, and the environment.

Our intention then is that this collection will foster critical awareness and debate about participatory approaches in these areas, through a focus on the following key questions:

- What counts as participation in education?
- What are the preconditions to participatory forms of education?
- What processes might such participatory education involve or require?
- Are the outcomes or consequences that it might lead to always worthwhile?

It is with such tasks in mind that we hope you find *Participation and Learning* a stimulating and useful resource for exploring and examining these challenges, and for rethinking the possibilities for and linkages between participation, environmental education, health education, and education for sustainable development.

Finally, we would like to acknowledge and thank the following people in supporting, encouraging, or helping inspire the shape and production of this collection: Clare-Marie Beaman, Warren Critchlow, Justin Dillon, Tracey Halliday, Seyoung Hwang, Bob Jickling, Junko Katayama, Elin Kelsey, Mary Ellen Lewis, Cecilia Lundholm, Marcia McKenzie, Susanne Müller, Sue Oakley, Mark Rickinson, Connie Russell, Kelly Teamey, Judith Terpos, Asimina Vergou, and Tamara Welschot.

Alan Reid, Bjarne Bruun Jensen, Jutta Nikel, and Venka Simovska

References

Cooke, B. and Kothari, U. (eds) (2001) *Participation: The New Tyranny?* London: Zed Books.

Hickey, S. and Mohan, G. (eds) (2004) *Participation: From Tyranny to Transformation? Exploring New Approaches to Participation in Development*. London: Zed Books.

Rahnema, M. (1992) Participation. In: Sachs, W. (ed) *The Development Dictionary*. London: Zed Books, pp. 116–131.

Contents

Contributors

Mary Jeanne Barrett, College of Education, University of Saskatchewan, Saskatchewan, S7N 0X0, Canada, Tel.: + 1 306 585 4536, Fax: + 1 306 585 4880, E-mail: mj.barrett@usask.ca, Web site: www.usask.ca/education/people/barrett.htm

Mary Jeanne (M.J.) Barrett is completing her doctorate, which interrogates ways in which dominant discourses of teaching, learning and research may make it difficult to traverse the human/nature divide. She is at the College of Education, University of Saskatchewan, Saskatchewan, Canada.

Robert Barratt, School of Education, Bath Spa University, Newton Park, Newton St Loe, Bath, BA2 9BN, UK, Tel.: + 441225 875549, Fax: + 44 1225 875499, E-mail: r.barratt@bathspa.ac.uk

Robert Barratt is Assistant Dean of Education, Bath Spa University. He has conducted a range of geographical and environmental research relating to young people and the environment. He is interested in researching young people's access to the environment, outdoor learning and the associated challenge of risk and safety management. His research is child centred and involves young people in participatory projects.

Elisabeth Barratt Hacking, Centre for Research in Education and the Environment (CREE), Department of Education, University of Bath, Bath, BA2 7AY, UK, Tel.: + 44 1225 386768, Fax: + 44 1225 386113, E-mail: edsecbh@bath.ac.uk, Web site: www.bath.ac.uk/education/profiles/ecbhacking.htm

Elisabeth Barratt Hacking is a lecturer in education at the University of Bath where she is a member of the Centre for Research in Education and the Environment. Her research interests are in the fields of geographical and environmental education. She is particularly interested in how children's local environment and community perspectives can be used by schools and local communities. Her research has been undertaken in school settings using participatory research approaches with children and their teachers.

Søren Breiting, DPU, Research Programme for Environmental and Health Education, Danish School of Education, University of Aarhus, Tuborgvej 164, DK-2400 Copenhagen NV, Denmark, Tel.: + 45 8888 9134, Fax: + 45 8888 9705, E-mail: breiting@dpu.dk, Web site: www.dpu.dk/about/breiting

Søren Breiting is an associate professor at the Danish School of Education, Department of Curriculum Research, in Copenhagen. He works in the research and development of interdisciplinary issue-orientated teaching to develop theory and practice of democratic environmental education and education for sustainable development. As one of the founders of the Research Centre for Environmental and Health Education, he has a special interest in environmental education and education for sustainable development in the general school system and has served as an advisor in Thailand, Hungary, and Namibia besides his involvement in Denmark and other Nordic countries.

Monica Carlsson, DPU, Research Programme for Environmental and Health Education, Danish School of Education, University of Aarhus, Tuborgvej 164, DK-2400 Copenhagen NV, Denmark, Tel.: + 45 8888 9122, Fax: + 45 8888 9705, E-mail: monica@dpu.dk, Web site: www.dpu.dk/about/monica

Monica Carlsson is an associate professor at the Department of Curriculum Research and a member of the Research Programme for Environmental and Health Education – both at the Danish School of Education (DSE). She coordinates and teaches on DSE's M.Ed. programme in Health Promotion and Education, and teaches educational evaluation on several other master programmes at DSE. Her research interests focus on curriculum development and evaluation of educational approaches within health education and environmental education, based on the concepts of action competence, pupil participation, and school-community collaboration.

Louise Chawla, College of Architecture & Planning, University of Colorado, Campus Box 314, Boulder, Co 80309-0314, USA, Tel.: +1 303 492 5228, E-mail: louise.chawla@colorado.edu

Louise Chawla holds degrees in developmental psychology and environmental psychology, and is the international coordinator of UNESCO's Growing Up in Cities project, which engages low-income urban adolescents in evaluating and improving their local communities. She is a professor at Whitney Young College, an interdisciplinary honours programme at Kentucky State University. Dr. Chawla also serves as an adjunct professor in the doctoral programme in Environmental Studies at Antioch New England Graduate School in New Hampshire.

Janet Dyment, Centre for Human Movement, University of Tasmania, Locked Bag 1330, Launceston, Tasmania 7250, Australia, Tel.: + 61 3 6324 3510, Fax: + 61 3 6324 3679, E-mail: Janet.Dyment@utas.edu.au, Web site: fcms.its.utas.edu.au/educ/educ/pagedetails.asp?lpersonId=2892

Janet Dyment has worked extensively with Canadian-based environmental organisation Evergreen to research the impacts of school-ground greening on school communities across Canada. Currently, she is investigating the relationship between school-ground greening and physical activity. She has been involved in advocating for green school grounds, through her work with schools, school boards, and governments. She is currently a Senior Lecturer in Outdoor Education in the Centre for Human Movement, Faculty of Education at the University of Tasmania. Prior to this appointment, she taught in the School of Outdoor Recreation, Parks, and Tourism at Lakehead University.

Stephen Gough, Centre for Research in Education and the Environment (CREE), Department of Education, University of Bath, Bath, BA2 7AY, UK, Tel.: + 44 1225 383919, Fax: + 44 1225 386113, E-mail: s.r.gough@bath.ac.uk, Web site: www.bath.ac.uk/ education/profiles/srgough

Steve Gough is Deputy Director of the Centre for Research in Education and the Environment at the University of Bath. His special area of interest is the role of learning in the management (whether successful or otherwise) of the relationship between society and nature. He has conducted research funded by a wide variety of organisations including the UK Economic and Social Research Council, the MacArthur Foundation, the World Wildlife Fund, The UK Department for International Development (DFiD), and the UK National Health Service. He serves on the international editorial boards of Environmental Education Research and the Journal of Education for Sustainable Development.

Paul Hart, Faculty of Education, University of Regina, Regina, Saskatchewan, S4S 0A2, Canada, Tel.: + 1 306 585 4626, Fax: + 1 306 585 4880, E-mail: hartpa@uregina.ca, Web site: education.uregina.ca/index.php?q=faculty.html&type=faculty&uid=35

Paul Hart is a Professor of Science and Environmental Education at the University of Regina, Canada where he teaches both undergraduate and graduate students. He has published widely in the areas of science and environmental education. He is an Executive Editor of the Journal of Environmental Education *and a Consulting Editor for other journals including* Environmental Education Research *and the* Canadian Journal of Environmental Education. *His research interests include the genealogical roots of teacher's thinking, children's ideas about the environment, and connections between environmental education and social learning.*

Roger A. Hart, Children's Environments Research Group, Ph.D. Program in Environmental Psychology/Center for Human Environments, Graduate Center, The City University of New York, 365 5th Avenue, New York, NY10016–4309, USA, Tel.: + 1 212 817 1887, Fax: + 1 212 817 3355, E-mail: roghart@gmail.com, Web site: www.cerg1.org

Roger Hart has a BA from Hull University in England and a Ph.D. from Clark University in Massachusetts, USA. His research has focused on children's relationship to the physical environment. In recent years, he has also been more broadly concerned with research and theory on how to foster the greater social participation of young people in their communities as a way of better fulfilling their rights and of building participatory democracy.

Debbie Heck, Griffith University, Gold Coast Campus, G06 3.56 PMB 50, Gold Coast Mail Centre, Queensland 9726, Australia, Tel.: + 61 7 555 28013, E-mail: d.heck@griffith.edu.au, Web site: www3.griffith.edu.au/01/griffithprofiles/content_profile.php?id=73333237323535

Debbie Heck is a Lecturer in the Faculty of Education at Griffith University where she is a member of the Centre for Innovative Conservation Strategies. Her work focuses on improving policy and practice in sustainability education and science education contexts. This includes discourse analysis of policy and programmes developed for environmental education and sustainability education in Australian schools, and also involves research with teachers in their science classrooms to further develop science education pedagogy and practice.

Bjarne Bruun Jensen, DPU, Research Programme for Environmental and Health Education, Danish School of Education, University of Aarhus, Tuborgvej 164, DK-2400 Copenhagen NV, Denmark, Tel.: + 45 8888 9138, Fax: + 45 8888 9729, E-mail: bjbj@dpu.dk, Web site: www.dpu.dk/about/bjbj

Bjarne Bruun Jensen is Professor in the Department of Curriculum Research at the Danish School of Education and directs the Research Programme on Environmental and Health Education. His research interests are in action-oriented health and *environmental education; pupils' concepts of health, inequality in* health and *action for health, and children as catalysts of social change. He is member of the editorial board of* Environmental Education Research *and* Health Promotion International, *and National Coordinator for the Danish network of health-promoting schools. He also works as a consultant for WHO within the fields of Health Education and Health-Promoting Schools.*

Marianne Krasny, Department of Natural Resources, Cornell University, Ithaca, NY 14853, USA, Tel.: + 1 607 255 2827, E-mail: mek2@cornell.edu, Web site: www.dnr.cornell.edu/mek2/index.php

Marianne Krasny is Professor and Chair in the Department of Natural Resources at Cornell University, specialising in environmental, science, and sustainability education in urban and other settings in the USA and internationally. She conducts outreach and research focusing on the interface between science education, civic participation, cultural understanding, and the environment, and on social learning among community-based educators. Dr. Krasny provides leadership for a diverse group of graduate and undergraduate students, all of whom integrate research and practice to address the following question: How can we create experiences for youth and adults that lead to science learning, environmental and cultural understanding, and more vital democratic communities?

Jeppe Læssøe, DPU, Research Programme for Environmental and Health Education, Danish School of Education, University of Aarhus Tuborgvej 164, DK-2400 Copenhagen NV, Denmark, Tel.: + 45 8888 9041, Fax: + 45 8888 9729, E-mail: jepl@dpu.dk, Web site: www.dpu.dk/about/jepl

Jeppe Læssøe is a senior lecturer affiliated to the Research Programme for Environmental and Health Education at the Danish School of Ecucation. He has an MA in psychology and a Ph.D. in communication studies. He has been involved in a range of interdisciplinary research projects and evaluations, including: (a) the history, knowledge interests, and strategies of the environmental movement, (b) participatory and action-oriented adult education related to risk issues and local sustainable development planning, and (c) the sociocultural dynamics behind changes in modern everyday life and material consumption. His research focus during recent years has been on the strategies, roles, and competences of professional mediators involved in sustainable development issues.

Heila Lotz-Sisitka, Environmental Education and Sustainability Unit, Education Department, Rhodes University, P.O. Box 94, Grahamstown 6139, South Africa, Tel.: + 27 46 603 8390, Fax: + 27 46 636 1495, E-mail: H.Lotz@ru.ac.za, Web site: campus.ru.ac.za/eesu

Heila Lotz-Sisitka is the Murray and Roberts Chair of Environmental Education at Rhodes University. She is actively involved in national policy development and recently coordinated the research component of the South African Department of Education's National Environmental Education Programme. She is also actively involved in development of regional strategies and policy for the United Nations Decade of Education for Sustainable Development in southern Africa. She is the editor of the Southern African Journal of Environmental Education *and serves on the council of the Environmental Education Association of Southern Africa. She currently coordinates the Masters and Ph.D. programmes in Environmental Education.*

Jutta Nikel, Centre for Research in Education and the Environment (CREE), Department of Education, University of Bath, Bath, BA2 7AY, UK, Tel.: + 44 1225 385622, Fax: + 44 1225 386113, E-mail: edpjn@bath.ac.uk, Web site: www.bath.ac.uk/ education/profiles/jnikel

Jutta Nikel is a postdoctoral research fellow at the Centre for Research in Education and the Environment, at the University of Bath. Her research interests focus on empirical and theoretical research regarding environmental education and education for sustainable development, and teachers' thinking and practice in these areas. Her recent work for the Royal Academy of Engineering focused on how sustainability issues are integrated into undergraduate engineering education. Her current work examines the contribution of Education for Sustainable Development for improving quality in education in low-income countries.

Rob O'Donoghue, Environmental Education and Sustainability Unit, Education Department, Rhodes University, P.O. Box 94, Grahamstown 6139, South Africa, Tel.: + 27 46 603 8390, Fax: + 27 46 636 1495, E-mail: r.odonoghue@ru.ac.za, Web site: campus.ru.ac.za/eesu

Rob O'Donoghue is Director of the Gold Fields Environmental Education Service Centre. He is actively involved in the development of practical tools for environmental learning and in research to clarify environmental education processes in school, environmental education centre, and community contexts. He has supported work on indigenous environmental knowledge and has given attention to situated learning actions that draw on intergenerational ways of knowing to engage environment concerns and the question of sustainability. He is currently supporting the C.A.P.E. Project (Cape Action for People and Environment) and has reviewed biodiversity education materials and the way that biodiversity conservation is being addressed in southern Africa.

Alan Reid, Centre for Research in Education and the Environment (CREE), Department of Education, University of Bath, Bath, BA2 7AY, UK, Tel.: + 44 1225 386294, Fax: + 44 1225 386113, E-mail: a.d.reid@bath.ac.uk, Web site: www.bath.ac.uk/education/profiles/adreid

Alan Reid is a senior lecturer in education and member of the Centre for Research in Education and the Environment, University of Bath. He is the editor of Environmental Education Research, *and coordinates doctoral programmes in research methods in education and management, and teaches on Masters programmes in environmental education, technologies and learning, and qualitative research approaches. His research interests focus on teachers' thinking and practice in environmental education, and policy-related and philosophical issues in environmental education theory and practice.*

John Robinson, The Education and Social Research Institute, Manchester Metropolitan University, 799 Wilmslow Road, Didsbury, Manchester, M20 2RR, UK, Tel.: + 44 161 247 5193/5210, Fax: + 44 161 247 6981, E-mail: j.f.robinson@mmu.ac.uk, Web site: www.esri.mmu.ac.uk/resstaff/profile.php?surname=Robinson&%20name=John

John Robinson is Director for Research Development in the Centre for Urban Education, Manchester Metropolitan University. His research interests focus on education for sustainable futures and education for social and environmental justice. He is particularly interested in an assets approach to educational provision. He is co-editor of Global Citizenship and Environmental Ethics: Probing the

Boundaries/At the Interface *(Rodopi, 2004) and* School Development Through Whole School Approaches: Critical Case Studies in Sustainability Education for Active Global Citizenship *(Trentham Books, 2005).*

Dawn Sanders, Learning Department, The Natural History Museum Cromwell Road, London, SW7 5BD, UK, Tel.: + 44 20 7942 5313, Fax: + 44 20 7942 5175, E-mail: dawngarden@hotmail.com, Web site: www.nhm.ac.uk

Dawn Sanders leads the science educator team at the Natural History Museum London. Her doctoral study considered botanic gardens as environments for learning (Sussex University). Dawn's current research interests focus on historical aspects of botanical education and women botanists. She is particularly interested in collaborative forms of working and writing, and is a keen volunteer at the South London Botanical Institute.

Karsten Schnack, DPU, Research Programme for Environmental and Health Education, Danish School of Education, University of Aarhus, Tuborgvej 164, DK-2400 Copenhagen NV, Denmark, Tel.: + 45 8888 9404, Fax: + 45 8888 9705, E-mail: schnack@dpu.dk, Web site: www.dpu.dk/about/schnack

Karsten Schnack is a professor in curriculum studies and educational theory at the Danish School of Education. He is co-founder and an active member of the Research Programme for Environmental Education and Health Education, and engaged also in critical analysis of Education for Sustainable Development. Based on both educational philosophy and action research and together with colleagues, he has developed conceptions of action competence, experiential learning, notions of environmental issues as societal problems constituted by conflicting interests, and participation and co-determination as keys to democratic education. He was head of the 'Didaktik-program' (programme for curriculum and educational theory) at the Royal Danish School of Educational Studies.

Tania Schusler, Department of Natural Resources, Cornell University, Fernow Hall, Ithaca, NY 14853–3001, USA, Tel.: + 1 607 272–2292, Fax: + 1 607 272–7088, E-mail: tms23@cornell.edu, Web site: n/a

Tania Schusler, environmental issues educator with Cornell Cooperative Extension, recently completed her doctorate in the Department of Natural Resources at Cornell University. Her research interest is in community participation in environmental management, and her dissertation specifically explored youth–adult partnerships creating positive environmental change. This focus grew out of her own experience as a young person participating in environmental action and as a National Science Foundation Graduate Teaching Fellow in K-12 Education guiding high school students in a community-based environmental project. She facilitates community education around sustainability issues in Tompkins County, New York, and previously managed a statewide volunteer programme for The Nature Conservancy.

William Scott, Centre for Research in Education and the Environment (CREE), Department of Education, University of Bath, Bath, BA2 7AY, UK, Tel.: + 44 1225 386648, Fax: + 44 1225 386113, E-mail: w.a.h.scott@bath.ac.uk, Web site: www.bath.ac.uk/education/profiles/wahscott

William Scott is a Professor of Education at the University of Bath where he directs the Centre for Research in Education and the Environment. He was a founding editor of Environmental Education Research, *and is a Fellow of the Royal Society of Arts. He is particularly interested in the role of learning within sustainable development, in the contributions that institutions can make to this, and in the problems of researching the effectiveness of such activities. He has conducted a range of research, development and evaluation studies on behalf of government, industry, NGOs, and other agencies in this country and abroad, most significantly with WWF.*

Tony Shallcross, Institute of Education, Manchester Metropolitan University, Crewe Green Road, Crewe, CW1 5DU, UK, Tel.: + 44 161 247 2010/5279/5030, Fax: + 44 161 247 6370, E-mail: A.G.Shallcross@mmu.ac.uk, Web site: www.esri.mmu.ac.uk/resstaff/profile.php?surname=Shallcross&%20name=Tony

Tony Shallcross is the leader of International Programmes, at the Institute of Education, Manchester Metropolitan University, Cheshire (Crewe). He directs the EC-funded Sustainability Education in European Primary Schools Project (SEEPS), which supports school development of whole-school approaches to sustainability education. With John Robinson, he is joint editor of a special issue of the Journal of Education for Teaching on the UN Decade of ESD. *His research interests include pupil agency in influencing more sustainable community actions, whole-school approaches as embryonic communities of practice, and multi-modal analysis in environmental education research.*

Venka Simovska, DPU, Research Programme for Environmental and Health Education, Danish School of Education, University of Aarhus, Tuborgvej 164, DK-2400 Copenhagen NV, Denmark, Tel.: + 45 8888 9116, Fax: + 45 8888 9705, E-mail: vs@dpu.dk, Web site: www.dpu.dk/about/vs

Venka Simovska, Ph.D., is assistant professor in health education and health promotion at the Research Programme for Environmental and Health Education, the Danish School of Education. Venka has extensive experience in school-based projects and research, including the European Network of Health Promoting Schools. Her research interests revolve around issues relating to health, young people, and teaching/learning processes in the area of health education and promotion. In particular, her research focuses on student participation and empowerment in learning.

Paul Vare, Learning South West, Bishops Hull House, Bishops Hull, Taunton, Somerset, TA51EP, UK, Tel.: + 44 1823 335491, Fax: + 44 1823 323388, E-mail: Learning4L@aol.com, Web site: www.learning-southwest.org.uk

Paul Vare is a senior manager at Learning South West, a voluntary membership organisation for learning providers in post-compulsory education. His background in education for sustainable development informs his work in 'mainstream' activity, such as supporting learning within small businesses. Paul represented European ECO Forum (an NGO coalition) at the United Nations Economic Commission for Europe (UNECE) where he helped draft their ESD Strategy and later joined the UNECE Expert Group on ESD indicators. He currently chairs the Learning for Sustainability Coalition in South West England through which he coordinates the development of a 'sustainable schools dating agency'.

Chapter 1
Participation and Learning: Developing Perspectives on Education and the Environment, Health and Sustainability

Alan Reid[1], **Bjarne Bruun Jensen**[2], **Jutta Nikel**[3], **and Venka Simovska**[4]

Keywords participation, education, environment, health, sustainability

1.1 Introduction

Participation and Learning emerged out of a loose set of interests and events that have brought the collection's contributors together at a series of formal conferences and informal workshops since 2003 as the *Research in Participatory Education Network* (RIPEN). RIPEN's work proceeds from the view that current discourses on participatory approaches to education have become increasingly diverse and contested in both theoretical and practical terms, as assumptions and activities have been analysed and tested in general education as well as in the contributors' fields of interest and expertise. While this has resulted in a range of tensions and challenges for practitioners and researchers, it is exactly in exploring and unravelling the methodological and pedagogical knots that have emerged, that we believe the potential value of participatory approaches to education is found.

As noted in the Preface, the contributors to this collection are scholars, practitioners, and researchers who share a common interest in understanding what works *in* and *as* participatory education in both formal and lifelong learning settings. The book is grounded in wide-ranging discussions of the features and operations of various participatory approaches to education that focus on environmental, health,

[1]Centre for Research in Education and the Environment (CREE), Department of Education, University of Bath, Bath, BA2 7AY, UK, a.d.reid@bath.ac.uk

[2]DPU, Research Programme for Environmental and Health Education, Danish School of Education, University of Aarhus, Tuborgvej 164, DK-2400 Copenhagen NV, Denmark, bjbj@dpu.dk

[3]Centre for Research in Education and the Environment (CREE), Department of Education, University of Bath, Bath, BA2 7AY, UK, edpjn@bath.ac.uk

[4]DPU, Research Programme for Environmental and Health Education, Danish School of Education, University of Aarhus, Tuborgvej 164, DK-2400 Copenhagen NV, Denmark, vs@dpu.dk

A. Reid et al. (eds.), *Participation and Learning,* 1–18.

and sustainability-related themes and challenges, and chapters draw on a broad base of experience and research data to illustrate a range of critical and self-critical reflections on these themes.

Hart (Chapter 2), amongst others, argues that participatory approaches to education are best viewed as tools for developing and sharing knowledge, skills, and experiences that ideally lead to cognitive gains, action competence, and community building (see also Chapter 9 by Læssøe, Chapter 15 by Dyment, and Chapter 17 by Schusler and Krasny). Another commonly voiced expectation is that participatory approaches afford the co-determination of educational processes and outcomes through the sustained social engagement and interaction of teachers and learners in planning and negotiating the focus and modes of their learning and teaching (see Schnack, Chapter 11). Closely associated with empowerment, these kinds of activities often promote bottom-up over top-down processes, and a distrust of external or state-imposed ways of doing things, particularly those that have historically excluded or ignored the marginal, powerless, or weaker members of society. Participatory approaches then can offer participants redress and opportunities on a variety of fronts; most positively, with citizens both young and old exercising their democratic rights to participate in civil society, and in decision-making and actions that promote justice, equality, and well-being for all.

1.2 Complicating the Field

Mindful of these potential contributions, we would like to take the opportunity this collection presents to highlight some of the productive tensions that can constitute such work.

To begin, we recognise that there are a number of concerns around the rhetoric of participation being used to legitimise particular educational practices to the exclusion of alternatives. In this collection, this is exemplified by the emergence of a 'culture of participatory workshops' in education and training, and the marginalisation of critical perspectives on the focus and content of professional development initiatives (see Lotz-Sisitika and O'Donoghue, Chapter 7). Related to this is unease about participatory practices that assume a lack of capacity, motivation, or engagement represents 'deficits' in learners, in contrast to alternatives in learning and social theory which suggest more productive starting points in operationalising participatory imperatives in education (e g. situative and constructivist perspectives, as discussed by Reid and Nikel, Chapter 3). Other concerns have been expressed about the ways that children's participation is managed (often only by adults, rather than with or by youth themselves, see Barratt and Barratt Hacking, Chapter 18); and about participatory projects that proceed as though importing a solution to a problem that the expert or outsider has diagnosed is the most appropriate *modus vivendi*, even if the participant or insider might not experience this as culturally appropriate or valuable (see Vare, Chapter 8). At the other extreme are situations in which it is assumed that the less interference by professionals or outsiders and the more autonomy and participation of the target group, the better.

With these concerns being taken up within the theorising, planning, and evaluation of environmental, sustainability, and health education, new debates have emerged. They include whether to assess participation primarily in terms of the membership of a project (e.g. access-focused participation), and how to evaluate the degrees and kinds of participation in selecting a topic, investigating themes, taking action, assessing processes and outcomes, and so forth (e.g. process-focused participation) (Jensen 2000). As earlier work in the Research Programme for Environmental and Health Education at the Danish School of Education has shown, it cannot be assumed that the contributions from participating in different phases of an environmental, health, or sustainability education project are of equal value in terms of learning outcomes, particularly if in action-orientated initiatives, the actions are to some extent prescribed, such as in raising levels of health, improving the environment, or demonstrating that sustainable development is taking place (see Simovska, Chapter 4). Setting aside the methodological challenges of demonstrating the outputs and outcomes of such learning for one moment (though see Scott and Gough 2003:31–43 for a discussion), while at some levels participation might be quite genuine and real, if the action is 'necessary' owing to donor, funder, or other politically grounded requirements, one must ask whether at that point participation is anything other than tokenistic?

This problematical situation reminds us to consider the risks associated with any project that valorises or privileges some constituencies in participatory approaches over others, even if 'local' or 'indigenous' forms of knowledge are at play (see Hart, Chapter 2; Reid and Nikel, Chapter 3; Vare, Chapter 8). For example, Bühler (2002) argues that, be it about non-participation, the 'participation deficit', or the processes or goals of participation, the notion that whatever local people or children have to say is valid can be as patronising as its opposite. Rather, the challenge is, being able to work through where overdetermination by particular voices and positions ends, and genuinely participatory co-determination begins.

Hart in Chapters 12 and 14, and Carlsson and Sanders in Chapter 20 accentuate the challenge, arguing that some participatory approaches have been built on naïve understandings. Most prominent amongst these is that structural barriers are automatically overcome by increasing stakeholder participation or the participation of the marginalised. Power is often described in such circumstances in unidimensional terms: as the powerful versus the powerless, with reallocation of power as the primary objective of participatory approaches. However, shifting the relative proportions of power-holding and power-sharing may do little more than reinvent or reinforce these barriers. As Heck (Chapter 16) suggests, there can be persistent gaps between a rhetoric of democracy and its practice in education, and given the various instances of patronage and paternalism in the localised governance and management of participatory processes, it may well be the case that a participatory approach is not necessarily intrinsically better than previous or alternative forms of leadership in attempting to foster inclusionary practices or processes (see Vare, Chapter 8; and Læssøe, Chapter 9, on different interpretations of democracy and intervention in participatory projects).

Given these tensions, a key issue for participatory approaches to education is the degree to which facilitators of participation and participants can both cede control and offer transparency in their working arrangements and practices in participatory teaching and learning situations (see Breiting, Chapter 10; and Schnack, Chapter 11). Understanding social power and cultural capital as circulating, rather than as simply tied to the pedagogical, political, or economic structures of educational systems, alerts us to the possibility that control can (continue to) be exercised (held or withheld) in less obviously apparent ways in participatory approaches to learning and teaching. For example, in response to the apparent diversity and inequality of the target group for the participatory work, some local knowledge and expertise might be excluded as much as others is included within the project framework or by the facilitators of participation (see, e.g. Shallcross and Robinson, Chapter 19, on the 'truths' and 'untruths' of 'communities of practice' in participatory approaches to environmental and sustainability education).

Similar outcomes and hence, tensions, can arise when the negotiation and discussion of alternative goals and processes in participation focuses on securing consensus rather than on pursuing dialogue about the project's terms of reference and its ongoing degree of fixity. Typically, this might occur when participation becomes the 'recipe' of the facilitator, rather than a stimulus to developing or reconfiguring processes, situations, and relationships in teaching and learning, with one eye on local circumstances and the other on transformational possibilities (Simovska, Chapter 4). Indeed, questions about the sources, circulation, and exercise of power in participatory approaches arise quite readily when the actions and outcomes are not 'benevolent' but rather lead to hostility and conflict – be it cognitive, among peers or stakeholders, or with the wider community (Scott and Gough, Chapter 5).

Acknowledging that inequalities and discord might increase during participation or as a result of it, suggests that practitioners and researchers should take account of the range of motivations and efforts of the members of the target group, and their various capacities to participate and learn, individually and together. In addition, addressing the potential extension of inequity and dispute during or after participation becomes doubly important when the lines between the 'objects' and 'subjects' of participation become blurred, as when participatory approaches and techniques become institutionalised within educational practices and systems, or when the 'researched' become the 'researcher', as in participatory action research. According to Lotz-Sisitka and O'Donoghue (Chapter 7), such participatory approaches are now prevalent and an embedded 'apparatus' of teaching and learning in South Africa, as well as in research and professional development. The situation gives prominence to some 'subjectivities of participation' at the expense of others, most notably in terms of the differing degrees of participant, researcher, or practitioner passivity and activity expected and practised, among a diverse range of stakeholders. Barrett (Chapter 13) also explores this issue, considering the marginalisation of the subaltern in participatory approaches to education, that is, what might be done *about* or perhaps better, *by*, those who are rendered without agency in participatory processes by their social status. The issue and its difficulties raise a key question for participatory approaches: what are the theoretical and practical grounds for resisting the structures of dominance of 'more acceptable' mainstream positions and discourses about participatory education?

In response, and as these initial comments suggest, even as pedagogical tools it is our view that we should not regard participatory approaches as either neutral or automatically beneficial to any form of education that focuses on environmental, health, and sustainability-related themes and issues. Participatory approaches mobilise particular world views and visions of society and its interactions while obviating others, perhaps most strikingly when non-participation is not deemed an acceptable option. Attending to such broader questions of the purpose and context for their use foregrounds the growing need for pedagogical and research work that explores and potentially resolves the tensions between, on the one hand, the so-called 'tyranny of participatory methods', and on the other, the 'tyranny of cultural appropriateness', in choosing and implementing a participatory approach successfully (cf. Chawla, Chapter 6; and Lotz-Sisitka and O'Donoghue, Chapter 7). In the following section, we introduce how the contributors have sought to address such matters before concluding the chapter with reflections on this particular landscape of research and commentary.

1.3 Overview of the Chapters

The basis for the grouping and ordering of the material in this book is the nature of a chapter's contribution to the overall themes, tensions, and questions that underpin the collection. Across the course of *Participation and Learning*, chapters shift in focus from providing commentary on conceptual and analytical themes and models, to descriptive and interpretive reports based on empirical research and case studies. Chapters 2–5 discuss current approaches and evaluative tools for participatory practice and offer new heuristics and frameworks for developing and critiquing participatory education. Chapters 6–11 focus on examining examples of educational and institutional practices and discourses of participation, and identify issues for further debate. Chapters 12–16 concentrate on researching participatory education and participatory research methods with educators, learners, and curriculum materials. Chapters 17–20 focus on case examples (mainly in the area of children and youth participation in curriculum development and school and community development) and offer critical perspectives on participatory education and the community, science, civics, school councils, and institutional development.

In Chapter 2, Roger Hart, a researcher synonymous with work on participatory approaches with children, reflects on the genesis and reception of the 'Ladder of Participation' (Hart 1992). Since its development in the 1980s, the Ladder (Figure 2.1) has become a well-used model in planning, discussing, and evaluating approaches to child and youth participation in community-based projects. The Ladder illustrates different degrees and categories of participatory approaches, some of which Hart considers to be tokenistic, decorative, or manipulative forms of participation (the lowest rungs). Hart traces the model's development and early success to its ability to bridge rather disparate conceptual work and discussion in the area about children's involvement in the initiation of participatory projects, as well as consultation, decision-making, and directing the work with and without adult involvement. He then reviews the positive debates the model has engendered in education and community work, as well as what he sees as the misinterpretations and subsequent misuse of the Ladder, as, for example, in cases where there has been disregard for its metaphorical underpinnings or where work has taken place without due regard to, or with faulty assumptions about, the power and rights of children living and learning in diverse educational and social contexts. Drawing on examples from community development initiatives in low-income countries, Hart recommends further discussion of participatory approaches and the development and synthesis of new models that address current issues in youth participation work in these places. These include further research and development on children's informal learning in their communities and their play with peers, addressing cultural difference in community participation settings, and working in and across contexts with children living within individualist or collectivist subcultures and societies.

Alan Reid and Jutta Nikel (Chapter 3) highlight the necessity of taking a critical perspective on conceptions and practices of participatory approaches in the field of education, through a discussion of three major perspectives on learning: behaviourist, cognitive, and situative, focusing on how these perspectives inform alternative (and perhaps, competing) framings of participation in environmental learning. They also suggest an evaluative framework of questions to help interrogate ideas about, and examples of, participation in environmental learning. Their questions include: *Who defines what is called participation? What degree of freedom does the participant have to participate? What are the criteria for being a participant? Participation in what? How important is the participants' participation within the complete process? And, how is the participation justified, if at all?* Reid and Nikel illustrate the ways in which such a heuristic framework can be helpful in thinking about the phenomena of participatory education and for ordering the diverse material on participation, revealing both patterns and tensions. Through a close reading of German policy documents concerned with participation and education for sustainable development, they argue attention be paid to the theoretical, ideological, and pedagogic positions from which questions are asked about participatory education. In so doing, key issues emerge for participatory approaches regarding: the practicalities of engaging in participatory education; the conceptualisation of participation in terms of epistemological, pedagogical, and ideological theories; and the understandings of reality, power, and discourse structures in environmental education, health education, and sustainability education.

In Chapter 4, Venka Simovska introduces a model that differentiates between two different qualities of participation, the 'token' and the 'genuine'. Simovska argues that what counts as 'genuine' or 'token' student participation can be distinguished by: (a) the focus of the learning activities in which the student participate, (b) the expected outcomes of those activities, and (c) the target of change for the participation. Her model builds on two sets of theoretical concepts, drawing on work on a democratic approach to health-promoting schools, and a sociocultural perspective on learning. Taken together, Simovska highlights how learning involves a shared process of seeking and constructing meanings about socially important concepts and practices, such as health and health promotion. The 'genuine' and 'token' qualities of participation inform the main analytical framework in a case study of *Young Minds*, a cross-cultural project involving schools from countries belonging to the European Network of Health Promoting Schools. Students' involvement in teaching and learning can often be dominated by a behaviouristic paradigm, focusing primarily on individual students and the modification of their personal lifestyles. However, in drawing on the model, case study and analytical framework, Simovska argues that student participation can be a broader transformative process. For example, taking action should be a deliberate part of learning about health and can be an important indicator of transformation, particularly when participatory approaches encourage wider dialogue and reflexivity about what constitutes health promotion and health education in schools, the community and society.

In the first part of Chapter 5, William Scott and Steve Gough highlight the variety of forms in which environmental learning is promoted and initiated by teachers, environmental interpreters, field study officers, conservation/heritage scientists, environmental activists, sociobiologists, Gaianists, and educational researchers. They explore the sources of this diversity in relation to the widely differing assumptions that people who promote or encourage environmental learning hold about learners and purpose and process in participation, including the aims of the learning of participants and any social action they might take. In other words, these groups have different concerns and interests in promoting participatory forms of environmental learning. Scott and Gough differentiate their analysis over nine categories of interest (ranging from 'those interested in sharing the joy and fulfilment derived from nature', to 'those advocating/promoting individual behaviour change', to 'those interested in the study of environmental learning' amongst others), and they also relate the categories to corresponding modes of participation in learning. The second part of their chapter explores what this might mean in an organisational context (via a case report of a membership environmental non-governmental organisation (NGO) with a remit to enhance biodiversity), and the chapter concludes with a discussion of the tensions that can arise when the interests, values, goals, and organisational imperatives of educators and participants collide.

In Chapter 6, Louise Chawla discusses young people's (mostly voluntary) participation in local environmental initiatives and the precursors to participatory activity understood as children's engagement with the world and their interest in making a mark on that world. Chawla takes up two questions essential to action-oriented participatory environmental education: *What experiences prepare children to be aware of their environment and to take action on its behalf?* And, *how can communities support children's environmental learning and action?* The chapter begins by introducing her conceptual framework, which draws on an ecological approach to psychology (as developed by James and Eleanor Gibson) and elements of social constructivism. Chawla argues for conceptualising environmental learning and participation from an ecological psychology perspective since it can generate new perspectives on children's agency, and the environmental contexts of action, by placing children and environment in a shared sphere of interest. Key concepts at work here include action (as a means of staying in touch with the environment's significant properties), the functional significance of environmental aspects, the properties of objects, affordance (Gibson 1979), and behaviour settings (Barker 1968). Participation is central, as 'people flourish more fully when they have a rich range of opportunities to realise their capabilities, and their capabilities include seeing the environment accurately and knowing how to take effective action in response (p. 101). The chapter also discusses four conditions for supporting the development of children's environmental awareness and competence and their implications for participatory forms of environmental education: (a) affordances that promote discovery and responsive person/environment relationships, (b) access and mobility to engage with those affordances, (c) perceptual learning to notice and value the environment, and (d) opportunities to take responsible roles in the community. These conditions are illustrated with examples from research on the significant life experiences of adults committed to environmental education and action, showing how their concern can be understood within the framework and in terms of the four conditions, and how this might refocus current educational initiatives with young people.

The goal of Chapter 7, by Heila Lotz-Sisitka and Rob O'Donoghue, is to contextualise contemporary idealisations of participatory education and training within wider 'socio-political projects'. The appearance and uptake of a participation discourse in South African society has elevated citizen participation to mandatory status, particularly in promoting the rights and responsibilities of individuals in a democracy. While this might be viewed as social discourse caught up in a global drive for participation, spearheaded for example by UN-related agencies and other donor organisations, Lotz-Sisitka and O'Donoghue analyse the institutional context for social and pedagogical participation within the overarching intentions of expanding the liberation of the oppressed and democratising social life in a post-apartheid state. Their chapter offers both a theoretical exploration and data-driven investigation of how participatory education has more recently developed as an idealised and 'techniqued logic of practice', and how and why this has become increasingly self-referential, chiefly within its operations as an imperative for training and education within a South African environmental education context. Their conclusions illustrate how the emergence of participatory education can be historically and sociologically understood, as both a central feature of, and a contested terrain of ambivalence within, the developing landscape of environmental education theory and practice.

In Chapter 8, Paul Vare reviews participation in the context of his involvement in developing, running, and evaluating rural communication programmes throughout Africa. The main feature of Chapter 8 is how Vare brings his practical experience in these programmes into conversation with concepts of participation and learning, from the perspective of how a 'practitioner might look at theory'. The chapter starts with an autobiographical account of his experiences in creating communication programmes within sustainable development projects in rural Africa (mainly in Uganda). He then charts the evolution of a participatory approach within the projects before reflecting on the way people learn in such settings. Vare summarises his views in this regard as learning being ‘a complex process of dialogue’, relying primarily on known and trusted sources. He is also quick to acknowledge that accounts of the projects largely used self-reported empirical data and they lacked theoretical underpinnings. Consequently, the second part of the chapter explores his response to these shortcomings, within the twin context of investigating how one might: (a) build a durable legacy in terms of learning, and (b) secure further donor funding. Starting from an analysis of different approaches to learning, Vare considers the relevance of such concepts as the ‘zone of proximal development’ and ‘legitimate peripheral participation’ to the project experiences. He continues by tackling possible limitations in these approaches and considering the value of cultural–historical activity theory (CHAT). In conclusion, Vare argues that the principles of situated learning and activity theory resonate strongly with real-world examples of participatory ‘education for sustainable development’ (ESD), and as well as raising critical issues in planning, managing, and reporting on participatory education, there might be direct benefits to learners and funders if CHAT and ESD were to become more closely aligned.

Chapter 9 by Jeppe Læssøe contributes to the debate about how public participative processes for sustainable development are planned conceptually and carried out. His particular focus is the emergence of a new kind of professional agent – the mediator – in Danish civil society, who is charged with implementing such participative processes. Læssøe argues for a transgression of narrow and static fixations on top-down versus bottom-up approaches to public or civic participation, as these have led some environmental sociologists to turn their criticism of the former into an idealisation of the latter. Moreover, when confronted with the quest for sustainable development, an emancipatory conception of participatory forms of learning can neglect the tensions that can emerge between participation as a defence of democratic rights (‘listening to the voices of people’), and participation as a tool for promoting learning processes that aim to replace limited project framework interests with a broader and collective responsibility for securing a sustainable future. With this in mind, Læssøe takes a closer look at the relationships between the key concepts – participation and sustainable development – analysing dominant conceptions of their relationship as they have developed in the political discourse and in environmental sociology since the 1970s. His discussion focuses on participation as part of a social emancipation process, the professionalisation of public debate about environmental issues, and issues raised by the mandatory participation of citizens with the overarching aim of reaching consensus. Alongside such developments has been the emergence of the ‘mediator’, a role that can be differentiated into ‘networkers’, ‘interpreters’, and ‘facilitators’. Drawing on examples from Denmark, Læssøe outlines four dilemmas with which these ‘change agents’ have had to cope and the qualitative differences between them. These are summarised as ‘Populism versus paternalism’, ‘Local settings versus global scope’, ‘Environment-centredness versus cultural orientation’, and ‘Independence versus involvement’. Læssøe concludes by arguing that instead of simply opposing top-down approaches with an idealised bottom-up process, the focus should now be on the high-level education of mediators, and how this qualifies and enables them to cope with the dilemmas their tasks and roles demand in both fostering and harnessing participation in sustainable development and social learning.

In Chapter 10, Søren Breiting provides an alternative perspective to the preceding chapters on what we might mean and understand as 'genuine participation'. He reconfigures existing conceptualisations of the 'ownership of participation' to unpack the specific qualities that make participation genuine, drawing on his professional experience in educational development initiatives in a number of countries around the world (mainly Thailand, Namibia, and Denmark) over the last decade. Breiting makes the case that the level of 'mental ownership' that a participatory initiative is able to generate among participants, corresponds with the experienced quality of the participatory approach. He then traces how his conceptions of mental ownership in education have evolved over recent years, starting from an environmental education project in Namibia and how this has influenced a large-scale curriculum and professional development project in Thailand, by highlighting the need for participants to be able to find their 'fingerprint' in the final outcome, or receive some form of recognition for their contribution to the participatory process. Breiting also argues that generating mental ownership through participatory approaches to professional development and curriculum development processes improves the quality of innovations in education, and requires attention to a range of issues raised by democratic views of education, cooperation, and empowerment. Doing so repositions arguments about mental ownership towards engaging wider debates on the need to democratise participatory forms of environmental education and ESD, and the need to foster deliberations about educational goals for young people and adults in terms of societal adaptation or emancipation (e.g. Hellesnes 1982).

Chapter 11 by Karsten Schnack also addresses the crucial question of how to foster deeper and more meaningful participation in education. Schnack's chapter pursues its line of investigation through educational philosophy, exploring how practice grounded in a humanistic view of education or driven by a shared ideal of improving the world in which we live can make a difference to participatory education. He then unpacks the relevance of general education to 'adjectival educations' (such as environmental education, health education, and ESD), and vice versa, arguing that adjectival educations still promote efforts to educate pupils in a broader, humanistic sense in the face of increasing vocationalisation and instrumentalism of general and adjectival education policy. The main part of the chapter explores the implications of a Klafkian understanding of the key didactics concept, *Bildung*, in curriculum and school examples from Denmark. Schnack illustrates the challenges and tensions that surround attempts to increase student participation through legislation, alongside issues related to enforcement and resistance in schools within a democratic educational system, for general education, environmental education, and ESD. The chapter concludes with a discussion of research implications in relation to studies of decision-making *about* and *in* the classroom (e.g. in terms of what takes place there, and the phases and phasing of teaching and learning processes), coupled with the need for research that focuses on the tensions between, on the one hand, standardised and objectified curricula and an intensified culture of educational assessment, and on the other, participatory, open-ended approaches to education.

In Chapter 12, Paul Hart examines why teachers come to participate in environmental education at all, in contrast to say, participation in conservation, social justice initiatives, or other matters of concern. Central to his perspective is the need to probe our understandings of agency, identity building, discursive practices, teachers' stories, and the impact of teaching, as well as the relevance of such themes to explaining teachers' practices and their drivers. Hart's research interests focus on the assumptions and predispositions that underlie teachers' notions of what counts as participatory learning within the genealogies of the contexts for their environmental education activities. This involves examining teachers' stories of their actions within their personal and professional theories of that context. The first part of the chapter outlines changes within the last two decades to understandings of narrative and epistemology to research on teacher thinking. Hart argues that story-based research methodologies can attempt to explore teachers' theory/practice connections as genealogical tracings of those pedagogical ideas that appear to have helped them reflect on the origins of their participatory practices in environmental education, as sociocultural dimensions of their own learning (see Gutiérrez and Rogoff 2003). The second part of the chapter focuses on the notion of 'telling identities' and how these stories, 'even if individually told, are products of collective story telling' (Sfard and Prusak 2005:10). Excerpts from an ongoing project illustrate how a university professor and three former students' interactions generated lasting meaning for each of their lives as teachers. The chapter ends by emphasising the importance of active participation for both surfacing and understanding deeply held values, as well as for inquiring into and understanding teachers' thinking and practice from within this perspective.

Whilst beginning with an acknowledgment of the advantages of participatory approaches to research and pedagogy in initiating change and providing space for marginalised voices, Mary Jeanne Barrett (Chapter 13) goes on to argue that despite the best of intentions to promote open and collaborative processes, participatory approaches can 'still impose agendas that support particular versions of what is appropriate thought, behaviour and action' (p. 212) and do not 'give open access to non-dominant and counter-hegemonic discourses' (p. 214). Barrett draws on feminist post-structuralist analysis (of power) to support her arguments, applying this to a critical examination of students' accounts of participatory encounters in classroom interactions. Drawing on key ideas related to identity, discourse, power, hegemony, and subject positioning, alongside two studies that draw on feminist and post-structural theories (Barron 1995; McKenzie 2004), Barrett argues for a different focus in interpreting teacher and pupil difficulties in taking up particular subject positions around environmental issues and activism. Rather than pointing to individual failings, lack of interest, or inability or discouragement to taking up unpopular positionings of the self or individual, Barrett discusses these phenomena in terms of: (a) the ways in which the subject positions available to students and teachers primarily reflect the discourses in the school and the broader community and not those of the individual, and (b) how through speech and action, teachers and research can circumscribe and (re)position students' lives and subjectivities. Thus, Barrett argues, what is marked as normal or aberrant in relation to participation clearly favours the conservative and safe, rather than the transgressive or radical, in terms of one's contributions to and grounds for environmental activism through participatory environmental learning.

Paul Hart picks up on some of these themes in Chapter 14 in this collection, again through exploring why educators come to participate in environmental education. While Hart continues to trace the ideas, structures, or events that work as precursors or barriers to participatory activity, Chapter 14 is more a methodological deliberation than a report of findings, in that Hart arranges his material around the question of *what it means to inquire* into the precursors to participation. Concepts such as agency, identity, and self-consciousness are again important here, particularly in terms of how we come to understand participatory education relationally, in the sense that participation is formed in and through relationship. Thus, Hart considers how 'strong narratives of participation' with people and places may also constitute an invitation to participate in an ecological consciousness vicariously. But the work of research in this area does not end there. Rather, Hart illustrates a shift from narrative-based (autobiographic work) to genealogical inquiry. This foregrounds the need to examine notions of self, identity, and agency in the inquiry process, and necessitates researchers finding ways to support and interact with educators in going beneath the surface of anecdote to examine motives, implications, and connections, whilst also encouraging individuals to look at themselves as agents and how they are formed as subjects – both as teachers and inquirers (Meadmore *et al.* 2000). The chapter concludes by addressing issues of general methodological interest such as authentic representation of experience and the interpretation and critical appraisal of authenticity, with Hart arguing the need to maintain rather than suspend suspicion of the self-evidency of self-expression, and the changes and challenges to the role of the researcher in such work.

In Chapter 15, Janet Dyment focuses on the nature and context of children's participation within school-ground greening initiatives. Interest in these initiatives is growing given the rise of 'whole school approaches' to participation (see also Chapter 19), and the hope that such initiatives allow young people to acquire environment-based skills, including those of participation, that can be extended or transferred to other contexts for democracy and civic participation. Dyment presents the key findings of a study of youth participation in Canadian school-ground greening initiatives, focusing on student involvement in problem identification, visioning and planning, actual greening, and maintenance. The chapter contains Dyment's methodological reflections on the research process, the turning points in her thinking and learning about participation and methodology, and the strength and limitations of the chosen mixed-method approach to the research. This involves engaging the broader debate about whether participatory action research should be seen as *one* or the *only* way to research participation, and besides traditionally grounded modes of inquiry, the chapter also considers the possible contributions of post-positivist approaches to understandings the role and identity of research about participation, and disseminating research findings to diverse audiences for research.

The focus of Chapter 16 by Debbie Heck is not yet typical of research on participation. Heck asks how the various forms and features of individual and civic participation are presented in curriculum resources, and in the context of Australian citizenship materials, the types of citizenship advocated within the forms of citizenship education they seek to promote or hamper. While the chapter provides an overview of different approaches to citizenship education, such as a 'legal status' view of citizenship or a 'public practice' perspective, it mostly focuses on critical discourse analysis as a research methodology appropriate to this focus for research about participation. Heck recounts how she aimed to: (a) deconstruct the core concepts and senses of citizenship implied by the texts (federally distributed citizenship material considered for national distribution), (b) identify how to reconstruct the process of the text development, dissemination, and consumption, and (c) analyse the view of participation and the power relationships that influenced the type of participation evident in the materials. Key challenges in such work include developing a theoretical frame which deliberately operates from within a critical paradigm, at the same time as developing a rigorous, non-partisan approach that would be acceptable to bureaucrat funders who tend to favour quantitative forms of research. Heck's preference for Fairclough's (1992) approach to critical discourse analysis is discussed, with particular consideration given to rigour and credibility in research. The chapter concludes with recommendations for researchers using critical discourse analysis to explore participation, and identifies possible future directions for research regarding curriculum materials and participatory educational programmes, such as the match between documents and the curriculum-in-use; educators' interpretations of the curriculum documents and dominant perspectives on participation; and cross-national, comparative work to examine similar or different social processes and understandings of participation within socioecological and sustainability processes.

In Chapter 17, Tania Schusler and Marianne Krasny review youth participation in local environmental action initiatives. The authors argue that science education and civics education can be integrated meaningfully in such projects as they create opportunities for young people to participate in local decision-making and action in relation to environmental issues. The key premise here is that, if the projects are appropriately organised, young people can develop their understanding of environmental science and political processes, and develop skills in scientific enquiry and civic engagement, and – most importantly, for Schusler and Krasny – these can all be experienced and learned at the same time. Drawing on the literatures of civic engagement, science education, and youth development, they suggest six guiding principles for youth participation in local environmental action. These include addressing issues such as adult perceptions of young people: (a) exercising democracy, (b) engaging in deliberate action, (c) generating scientific knowledge, and (d) undertaking critical reflection on processes and outcomes. The six principles are discussed within a comparative analysis of three environmental action programmes in the USA: the Earth Force programme, the Seneca Falls Landfill Project programme, and the Garden Mosaic Programme. The chapter concludes with suggestions for further work on science, civics, and participatory education; the adequacy and scope of the principles of youth environmental action; and how the practices, contexts, and characteristics of educators create opportunities for youth participation in local environmental action.

Chapter 18 by Robert Barratt and Elisabeth Barratt-Hacking focuses on children's participation in curriculum development. The chapter introduces a school-based participatory research project undertaken in England, which inquired into how children's everyday experiences could be defined, understood, and shaped by a school curriculum that foregrounds local community action and community development. The underlying assumptions of the project team highlight the importance of children being able to make sense of the relationship between home, school, and the local environment for identity-formation processes. As part of this, children's confidence and security at a local scale is seen as critical to developing broader understandings of, for example, their place, role, and contribution in the wider community, now, and for the future. Conceptually, the project is strongly informed by Baacke's model of four 'ecological zones' and by Neale's differentiation between children's participation in society as either 'welfare dependants' or 'young citizens'. In the school, learning was contextualised in relation to children's community experience, but at the same time, it aimed at initiating actions towards community development. Consequently, the project was designed to entail 'researching collaboratively with children' (Garbarino *et al.* 1989), using a grounded theory approach to the data analysis to ensure awareness of 'pupil voice' in the outcomes. The authors conclude from the project that the children often lacked a forum in the school and in the community for participating in curriculum decision-making and community development. The children also recommend the establishment of a 'school environment curriculum council', and outlined the features of the changed role of the school in light of fundamental prerequisites for their participation in curriculum and community development, based on the outcomes of the data analysis and its processes. These include: creating structures and processes within school, which recognise children's contributions and supports children's decision-making; appreciating that children want to bring about community change; and recognising the apparent discord between children's personal aspirations and their opportunity to effect local action.

Chapter 19 is a commentary on whole-school approaches and their intrinsic association with participatory education. Drawing on ideas about 'whole school action-focused learning' (Lave and Wenger 1991) and 'communities of practice' (Wenger 1999), Tony Shallcross and John Robinson argue for whole-school approaches on the basis that they enable an action focus to environmental education and influence the lifestyles of children, adults, and the community. They suggest that the main advantage that a situated, whole-school action-focused learning approach offers over behaviourist or cognitive/constructivist approaches to learning is that they not only generate sustainable actions by individuals, but also support changes towards sustainable lifestyles at a broader, societal scale. The authors set out the underlying concepts and theoretical frameworks for their work and then present examples and evidence about the impact of whole-school approaches from a pan-European project, *Sustainability Education in European Primary Schools* (SEEPS), funded by the

European Commission. The chapter then raises the question of how to analyse and evaluate such approaches. Here, Shallcross and Robinson are particularly concerned with how to deal with: (a) the possibility of results revealing a high discrepancy between existing practice and the desired, ideal state, (b) the lack of research and evaluation into the impact and processes of whole-school approaches, and (c) associated research ethics. Like Barratt and Barratt Hacking, Shallcross and Robinson also raise concerns about the quality and quantity of student 'voices' and about student contributions to research and conceptual development in this area; that is how the developments of our understandings of participation might better address the perspectives and insights of participants themselves.

Finally, in Chapter 20, Monica Carlsson and Dawn Sanders examine pupil participation in collaborative environmental education projects between schools and external organisations. Their two case examples are the Danish Eco School project and the school grounds development projects (Learning Through Landscapes) in England. Carlsson and Sanders begin their chapter by discussing pupil participation and non-participation within school councils and environmental councils, from a perspective that explicitly addresses issues of power. They define collaboration as a process that involves the sharing of an area or field of action – an arena *of* and *for* power (based on the work of Katzenelson 1994 and Fink 1989). This leads to questioning the roles of pupils, teachers, and outside actors in collaborative projects, and the different, and sometimes inconsistent, understandings of school councils as arenas for collaboration and participation. Regarding the latter point, the authors draw on Micheletti's work on 'everyday makers' to highlight that the processes and purposes of collaborative projects run via such councils raise a series of key issues. These include: whether the collaborative projects aim to produce individualised collective action, individualised political participation, or students' ownership and engagement; and, working across these categories, whether the projects favour liberal or direct representational notions of democracy.

1.5 Concluding Comments

Given our expressed intention that this collection fosters critical awareness and debate about participatory approaches in environmental, health, and sustainability education, we conclude this chapter by selecting four themes for further work and consideration, drawing across our grouping of the aforementioned chapters in the book.

First, in probing what counts as participation in education and the preconditions to participatory forms of education, a recurring theme across the chapters that suggests a focus for further work, is the range and scope of views, understandings and assumptions about *childhood, youth*, and *adulthood*, and how at different life stages, one is understood to participate in social life. Recent childhood research (Christensen and James 2000), for example, with its postulate that 'children are social actors, with a part to play in their own representation', stands to enrich our historical and contemporary appreciation of participatory education in these areas, and in particular, what it means to be a participant in different contexts, settings, and life trajectories, in that like Barratt and Barratt Hacking (Chapter 18), and Shallcross and Robinson (Chapter 19), it draws attention to the 'student voice movement' (see Ruddock and McIntyre 2007). This movement blurs the positioning of children as passive objects of research with that of active subjects. Children can and are expected to speak for themselves and report valid views and experiences about teaching and learning reliably, they are capable of engaging in research conversations with adult researchers, and they have the right to do so (see Farrell 2005). In the context of this collection, the movement also highlights the challenge to researchers and practitioners of participatory education of addressing power relations and differentials, rather than assuming they can be eliminated, particularly if researching the theory and practice of participatory approaches to education becomes a part of the ongoing, day-to-day encounters of teachers and learners in practising participatory forms of teaching and learning.

Second, again on the preconditions to participatory approaches to education but also in relation to the processes that they might involve or require, in order to enable and allow genuine participation, it is clear that what is required is more than just the people and the task to be in place. Many of the chapters call into question the role of the facilitator in the *mediation* of participation, particularly when the participatory process is novel for the participants but not for the facilitator as this can exacerbate rather than reduce the power gradient. Similar outcomes may occur when the prior experience and capabilities of the project's initiators are discounted in order to create the impression of a 'level playing field' (see Vare, Chapter 8), or past approaches to participation are preferred over the possibility of devising new ones that are perhaps better aligned to current capabilities or future needs of the participants, or that offer new routes for social, ecological, or personal transformation. A key challenge here is to better appreciate and understand the benefits and drawbacks of the time and effort put into clarifying and working with an existing or newly organised and structured protocol for participatory engagement, how much can be presumed on the part of participants and facilitators, and how often this all needs to be reviewed or even abandoned, given the exigencies of local circumstances and developments.

Third, while the contributors clearly emphasise that working within the participatory task is in itself a learning challenge, the broader question this feeds into is, are the outcomes or consequences that it might lead to always worthwhile? More specifically, we can ask how does a participatory approach enhance teaching and learning directly rather than incidentally, and relatedly, why do people 'do participation' in its various forms, and at its various stages? Working across the case studies and examples, a clear research challenge is to quantify and qualify the *reasoning and rationales for participating in participatory approaches*, such as in relation to perceptions and experiences of ownership, and the role of well-being-related, ethical, or pragmatically based logics of practice for participatory environmental education, health education, or ESD. An additional challenge is to supplement this by tracing out the lines and bodies of evidence that link rationales with outcomes, as Roger Hart seeks in Chapter 2, and Reid and Nikel, in Chapter 3.

A fourth theme for further work and consideration is to ask, why do teachers and learners engage in diverse forms of *joint participatory actions*, some of which support and others challenge the status quo, and to consider what their effects, strengths, and drawbacks are within and for education, particularly in the case of trans-institutional work, and across formal and lifelong learning settings. These kinds of inquiry might be pursued in the context of an analysis of how the diverse lifeworlds of students and teachers (e.g. their homelife and worklife) are shaped and constituted, alongside an analysis of the inscriptions in discourse (be that in education or participation) that focus participation on voluntaristic or personalistic responses to the issues, rather than, say, situating them in relation to broader cultural, historical, or economic contexts, arrangements and trajectories (e.g. as these relate to the boundaries and dynamics of family, community, nation, and so forth). Thus, while we can recognise that much participation takes place within the context of responding to programmes and initiatives originating from outside schools (e.g. as a result of the work of NGOs or government agencies), because participation invariably involves working *for* some goals for teaching and learning and *against* others, a key area for future research is to understand the framings, linkages, and dynamics of locally initiated, 'autonomous' participatory projects, in contrast to those that are either voluntarily taken part in, or legislated for by others working in or through the education field.

To conclude and reflect back on this introduction, these four areas suggest to us a wider need in education to engage with the metaquestion of where our examples and inquiries come from regarding participatory education, why they are currently constituted so, and what alternatives or new horizons should be pursued in relation to the environment, health, and sustainability. A key test for networks and interests such as those of RIPEN is to be able to continue the conversation and debate about participatory approaches, in order to further uncover and analyse what experience, scholarship, and research have to say about these issues, and develop better, genuine, and more sustainable forms of participation. Put otherwise, and to reframe the core focus of the tensions and challenges that underpin this book, our key questions for participatory education could now be expressed as:

- Why should participation continue to count in education?
- How will we know what processes participatory education should involve or require?
- How can the preconditions to participatory forms of education be enabled?
- How will worthwhile outcomes or consequences of participation be ensured?

References

Barker, R. (1968) *Ecological Psychology*. Stanford, CA: Stanford University Press.

Barron, D. (1995) Gendering environmental education reform: identifying the constitutive power of environmental discourses. *Australian Journal of Environmental Education*, 11: 107–120.

Bühler, U. (2002) Participation 'with Justice and Dignity': beyond the 'New Tyranny'. *Conflict and Development*, 1: 1–16.

Christensen, P. and James, A. (eds) (2000) *Research with Children*. London: RoutledgeFalmer.

Fairclough, N. (1992) *Discourse and Social Change*. Cambridge: Polity Press.

Farrell, A. (ed) (2005) *Ethical Research with Children*. Maidenhead, Birkshire, UK: Open University Press.

Fink, H. (1989) *Arenaer – om politik og iscenesættelse*. Copenhagen: Universitetsforlaget.

Garbarino, J., Stott, F., *et al.* (1989) *What Children Can Tell Us*. San Francisco, CA: Jossey-Bass.

Gibson, J.J. (1979) *The Ecological Approach to Visual Perception*. Boston, MA: Houghton Mifflin.

Gutiérrez, C. and Rogoff, B. (2003) Cultural ways of learning: individual traits and repertoire of practice. *Educational Researcher*, 32, 5: 19–25.

Hart, R. (1992) *Children's Participation: From Tokenism to Citizenship*. Florence: International Child Development Centre/UNICEF.

Hellesnes, J. (1982) Education and the concept of critique. In: Bredo, E. and Feinberg, W. (eds), *Knowledge and Values in Social and Educational Research*. Philadelphia, PA: Temple University Press, pp. 355–368.

Jensen, B.B. (2000) Participation, commitment and knowledge as components of pupils' action competence, In: Jensen, B.B., Schnack, K., and Simovska, V. (eds), *Critical Environmental and Health Education – Research Issues and Challenges*. Copenhagen: Danish University of Education, pp. 219–238.

Katzenelson, B. (1994) Homo socius. *Socialpsykologisk Grundbog*. Copenhagen: Gyldendal.

Lave, J. and Wenger, E. (1991) *Situated Learning: Legitimate Peripheral Participation*. Cambridge: Cambridge University Press.

McKenzie, M. (2004) The 'Willful contradiction' of poststructural socio-ecological education. *Canadian Journal of Environmental Education*, 9: 177–190.

Meadmore, D., Hatcher, C., and McWilliam, E. (2000) Getting tense about genealogy. *Qualitative Studies in Education*, 13, 5: 463–476.

Ruddock, J. and McIntyre, D. (2007) *Improving Learning Through Consulting Pupils*. London: Routledge.

Scott, W. and Gough, S. (2003) *Sustainable Development and Learning: Framing the Issues*. London: RoutledgeFalmer.

Sfard, A. and Prusak, A. (2005) Telling identities: in search of an analytic tool for investigating learning as a culturally shaped activity. *Educational Researcher*, 34, 4: 14–23.

Wenger, E. (1999) *Communities of Practice: Learning, Meaning and Identity*. Cambridge: Cambridge University Press.

Chapter 2
Stepping Back from 'The Ladder': Reflections on a Model of Participatory Work with Children

Roger A. Hart

Keywords participation, youth, children, models, citizenship

2.1 Introduction

It is with some reticence that I write these comments on the 'the ladder of children's participation' for this metaphor was introduced by me long ago in order to problematise an issue that now has a significant body of practice and critical reflection. But my colleagues, the editors of this volume, suggested that because the ladder is still used a great deal as a model it might be useful to stand back and make a few comments about the ways it has been interpreted. The ladder probably drew so much attention because when it was first published in *Children's Participation: From Tokenism to Citizenship* by UNICEF in 1992, there was very little written of a conceptual nature on the theme of children's participation in their programmes, projects, or organisations. The book was simply meant to stimulate a dialogue on a theme that needed to be addressed critically. But many people have chosen to use the ladder as a comprehensive tool for measuring their work with children rather than as a jumping-off point for their own reflections.

This chapter tries to correct some of the misinterpretations of the ladder of children's participation but, more importantly, it addresses the debates that the model has engendered. I hope in this way to contribute to the ongoing discussions and to encourage the development of new models. First, I discuss what types of participation are, and are not, addressed by the model. Next, in order to clarify my purpose, I recount how I came to write about the subject and to use the ladder as a metaphor. This leads to a review of some of the misunderstandings about the ladder model. I then discuss some of the more important issues that have been raised by others before adding some concerns of my own. The chapter concludes with a request for further critical reflection and the generation of new models.

Children's Environments Research Group, Ph.D. Program in Environmental Psychology/Center for Human Environments, Graduate Center, The City University of New York, 365 5th Avenue, New York, NY10016–4309, USA, roghart@gmail.com

A. Reid et al. (eds.), *Participation and Learning*, 19–31.

2.2 The Need to Think More Broadly About How Children Participate in Society

It is important to begin by noting that the ladder of participation addresses only a rather narrow range of ways that most children in the world participate in their communities. It focuses on programmes or projects rather than on children's everyday informal participation in their communities and it is largely limited to describing the varying roles adults play in relation to children's participation. This is because the ladder was drawn from the perspective of an author living and working in the 'Minority World' where, at this point in history, children's participation in meaningful community activity is largely limited to formal programming of their activities by adults. For the past one hundred years or so most children in North America and Europe (i.e. the minority world) have been segregated from everyday participatory activities with adults in their communities. The creation of schools, play and recreation programmes, and, most recently, 'after-school programmes', alongside the retreat of children into the private spaces of their homes, have all resulted in the removal of children from informal participation with adults in their communities. They are also segregated for much of the time from children of other age groups than their own. In these parts of the world children spend less and less time in activities where they learn though informal participation to take on roles of greater competence in their communities. In fact, most people who work with children in participatory ways speak about children's participation as though it were a new thing. The emphasis on formal participation and verbal dialogue in these countries of the minority world is new but we must also be aware of what has been lost in these cultures. Children all over the world traditionally learned with adults through a kind of 'apprenticeship', or participation, in work. Gradually adults conceded greater and greater degrees of opportunity for a child to take on responsibilities. This is still common in countries of the majority world but it is so greatly reduced in minority world countries that many seem to have forgotten its importance.

Children also build competencies from their participation in play or work with one another, often without adults. While this also continues to occur in many countries it is a kind of 'participation' that has been increasingly replaced by the segregated world of formal participation in projects and programmes with adults and so it seems to go unrecognised by the writers on 'children's participation'. We would do well to try to integrate our thinking on children's formal participation with what is known of children's informal participation and culture-building through play with their peers (e.g. Hart 1979; Corsaro 2003). We should therefore be equally concerned with how adults differently 'set the stage' for children to self-organise such as how they arrange public settings for play and recreation. In *Children's Participation* (Hart 1992), I noted the high degrees of group participation that children can often achieve in their early years in their socio-dramatic play with one another. Fortunately, theory and research in developmental psychology is now increasingly recognising the social context of children's development and that children learn through their interaction with more experienced members of their community, including other children (Vygotsky 1978; Rogoff 1990; Valsiner 2000). Consequently, when children of mixed ages and different abilities play together there can be greater opportunities for them to learn through participating in activities with one another than can come from the normal school structures where children are typically graded by age and ability into classes. The term *scaffold* helps illuminate the kind of structures that are provided to a learning child by the social world (Gauvain 2001), and in some ways a scaffold may be a more suitable model than a ladder for much of what we are discussing because it implies multiple routes to growth. Also, whereas the ladder metaphor is usually used to characterise only child–adult relations, the scaffold metaphor can be thought of as a mutually reinforcing structure where all people, including adults and children of different abilities, help each other in their different climbing goals (Hart 2005).

2.3 Genesis of the Model

The ladder model was first published in the *Childhood City Newsletter* (1980). A quarterly newsletter on children's environments, it had a small circulation and was produced by a small group working at the City University of New York, with the close collaboration of Robin Moore in North Carolina. Robin and I had both struggled to find effective ways to involve children and youth in community environmental research, planning, and design. Designers were increasingly trying to involve children in design but this was sometimes being done in rather tokenistic ways. We knew from experience that it was a challenge to find effective ways to work with young people and we felt it would be a good idea to raise the level of dialogue on this issue by bringing together the ideas of as many people as we could find to discuss it. We produced three sequential issues on the theme of participation as the North American contribution of the International Association for the Child's Right to Play (IPA) to the international Year of the Child. As part of this series, I sketched out the ladder of children's participation and Selim Iltus, a doctoral student who was then working as an assistant on the newsletter, turned it into an attractive graphic form. But it was not until it was published in a little book for UNICEF in 1992 that it caught the interest of a large number of child advocates and others who work closely with young people, and became translated into many languages. The United Nations *Convention on the Rights of the Child* (CRC) had recently been launched and UNICEF, along with many international non-governmental organisations (NGOs), were struggling to interpret what this visionary document meant for their work. The parts of the Convention that seemed to be the most difficult for them to interpret were those that concerned the participation of children. Because of this, James Himes, the Director of the UNICEF International Centre for Child Development in Florence, Italy, felt it would be good if there were a book on the subject (Hart 1992).

The ladder was simply offered as a schema to help bring a critical perspective to a subject that at that time altogether lacked one. There was no intention for it to serve as any kind of comprehensive evaluative tool. Sherry Arnstein (1979) had

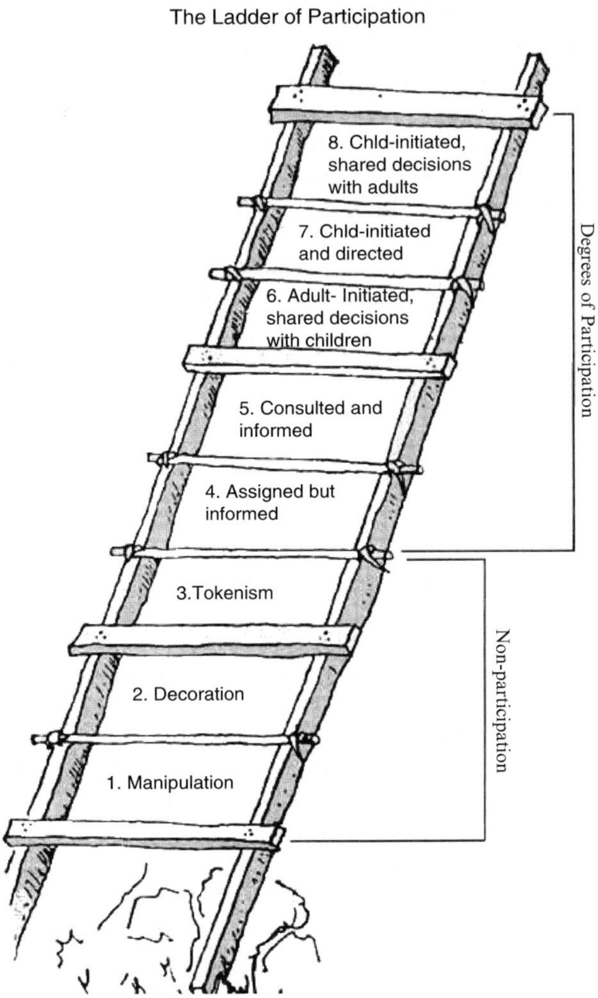

Figure 2.1 Eight levels of young people's participation in projects (the ladder metaphor is borrowed from the well-known essay on adult participation by Arnstein (1969), the categories are new) (Hart 1992:8.)

used the ladder metaphor for her own writing on participatory planning with adults and I found it useful to play with this metaphor in organising my own thinking on the subject of children's participation. The ladder model has since been extended and modified in many fascinating ways. The most beneficial quality of the model has probably been its utility for helping different professional groups and institutions to rethink how they work with young people: youth workers, television and radio directors, scout leaders, play workers, street workers, health professionals, and even some schoolteachers. Its simplicity of form and clarity of goals enable them to find a language to look at their current ways of working systemically, and in so doing, come up with something more complex and useful to their particular context. I enthusiastically embrace these many applications and extensions of the model but I will not focus on them here for the editors have asked instead that this chapter focus on the debates that the ladder metaphor has engendered. So the next part of the chapter necessarily offers some corrections to how the ladder has been interpreted. Like any metaphor, the use of the ladder involves distorting reality as one tries to reduce the complexity of the issues and so these comments are offered in the original spirit of stimulating dialogue rather than as a hardened defence of an old model.

2.4 The Tendency to Think in Terms of a Developmental Model with Sequential 'Stages' or Levels of Participation

In some ways the ladder metaphor is unfortunate for it seems to imply a necessary sequence to children's developing competence in participation (Reddy and Ratna 2002:18; Kirby and Woodhead 2003:243). This was not the intention but given the metaphor of a ladder it is not at all surprising that the model has been interpreted as stepwise climbing. In fact the ladder is primarily about the degree to which adults and institutions afford or enable children to participate. I gave the example of how when adults leave very young children alone to play in an environment that affords children a high degree of opportunity to manipulate it, they are often able to cooperate with each other at a very high level to build structures such as dams and canals in sand and water. But for most children in urbanised countries their lives are increasingly planned and managed by adults with little opportunity to initiate activities themselves. Play is the one domain where it is common for the normal patterns of adult power to be suspended for a while, although, as described earlier, this has been changing in many countries (Hart 2004).

But some people ask if there should even be different 'levels' of participation. As a result they have created different models (see, for example, the summary in Reid and Nikel 2004). Jensen has suggested that the rungs of the ladder be described as different *forms* rather than different *levels* of participation (Jensen 2000). Treseder (1997) came up with a circle model and Mannion (2003) has devised something like a fountain of participation! While I understand the desires of some authors to try to improve upon this potentially misleading aspect of the ladder, I do wish to stress that I do not consider these different forms of participation to be equal. I think of the upper rungs of the ladder as expressions of different 'degrees' of agency or participatory engagement by young people. But again, the ladder metaphor may be a problem in that it seems to some to suggest that in all cases the higher rungs of the ladder are superior to the ones beneath. This is unfortunate for the intention is to show that there are different degrees to which children are allowed, enabled, and supported to initiate their own projects and make decisions in carrying them out with others. While a child may not want at all times to be the one who initiates a project they ought to know that they have the option, and to feel that they have the confidence and competence to do so on occasion. Adult facilitators of projects should not be made to feel that they must always support their child participants to operate on the 'highest' rungs of the ladder, but they must manage to communicate to children that they have the option to operate with these 'higher' degrees of engagement. It is not appropriate that some children feel that they must always only follow the initiative of others any more than it is good for any child to feel that they should always be a leader (see, for example, the National Youth Council of Ireland and the Children's Rights Alliance 2003). The ladder should be thought of as some kind of scale of competence not performance: children should feel that they have the *competence* and confidence to engage with others in the way outlined on any of the rungs of the ladder, but they should certainly not feel that they should always be trying to *perform* in such ways. Because the types of engagement expressed in the higher rungs of the ladder are commonly denied to young people there is obviously some value in having some way of representing these 'higher', forbidden forms. But why, I am often asked, is the eighth rung higher? Surely a child in power is the highest level, they ask …!

2.5 A Preoccupation with 'Children's Power'!

One of the most surprising critiques of the model for me has been the desire of some to transform the top rung of the ladder to be 'children in charge' or children's decision-making without adults (e.g. Melton 1993). The top rung of the ladder as it was drawn in the original essay was called 'child initiate, shared decision with adults'. My purpose in creating this scheme had not been to argue naïvely that we should think of children as repressed individuals who needed to be liberated through a series of steps whereby all adult engagement was removed. My concern was rather to argue that children's potentials as citizens needs to be recognised to the fullest and, to that end, children ought to be able to participate at times at their highest possible level. The highest possible degree of citizenship in my view is when we, children or adults, not only feel that we can initiate some change ourselves but when we also recognise that it is *sometimes* appropriate to also invite others to join us because of their own rights and because it affects them too, as fellow-citizens. When people recognise the rights of others to have a voice and involve them, then this, in my mind, is *morally* superior to children being 'in-charge'. Yet others disagree for they wish to stress that liberation from adults is the end goal of their efforts (Ackermann *et al.* 2003; Ministry of Children and Families of British Colombia 2005). This issue is related to a common problem throughout the world with how children's rights have been conveyed to many parents. Children's rights to have their perspective taken into account 'in all matters that concern them' are often wrongly understood to mean that children should have the last word. The CRC recognises the need for children to also understand and respect the rights of others but this is not generally discussed. Not surprisingly Africa responded to the CRC by creating their own children's rights charter that stress children's responsibilities as well as rights.

It is important also to note here that when children participate with one another, without adults, that does not in any way change the story of power and participation. When children are on the top rung of the ladder organising a project together they still have similar struggles of leadership and opportunities for democratic participation. This is an equally important area for us to conceptualise that was not considered at all by my ladder model. Furthermore, sometimes adults set up children's and youth groups that seem to function as self-determining groups when, in fact, they have been designed to operate under adult power, with an ideology designed by them but with children as the appointed agents of influence and control over one another. They might sometimes be better called 'regime-organised'. The Hitler youth movement obviously had some of these qualities. But even the scouts, with all of the wonderful opportunities that they have offered to so many children, have an adult-imposed hierarchic structure of 'Leaders' and 'Seconds'. This creates the conditions whereby children carry out activities with one another when adults are not around in a less participatory manner than they might have. There is still a great deal to be done to creatively work out how to work with children in participatory ways that enable them to be maximally self-determining (Hart *et al.* 2001).

2.6 The Dangers of Adopting a Single Comprehensive Tool for Evaluating Projects

It is an easy step from thinking of the ladder as a developmental model to using it as a comprehensive tool to evaluate how participatory a programme is. But this was not the objective. We have so much to do to develop new appropriate tools of evaluation. It is true that the ladder has proved itself to be valuable in enabling groups to monitor the degree of agency young people have in initiating projects or activities in their programmes or projects. Its clear vocabulary and simple logic is useful in enabling people of a wide range of ages and abilities to periodically discuss issues of power in their day-to-day work with each other. Furthermore, Jensen (2000) has usefully modified the ladder to be relevant to different stages in a project rather than just its initiation. But there is clearly a need for more schemes that can help groups look systematically at their practice. Gaitan (1998), for example, has developed a model that distinguishes between the degree to which children understand their rights and capacities to participate, their opportunities to make decisions with others, and their ability to be involved in action with others. But while many group facilitators say the ladder is valuable for their work with children and youth, the emphasis should be on a *beginning* dialogue. With older children, group facilitators might even conclude that it is most effective to involve children themselves in making more appropriate schemes for evaluating their participation.

2.7 Cultural Limitations of the Model

It is most surprising to me that I could not find more cultural critiques of the ladder, particularly from Asia and Africa, for I can think of some important ones. The reason may well be that many of those who write about the issue of children's participation are themselves educated in the West and rely on Western theories of children's development which, sadly, almost completely dominates the child development literature globally. There have however been some important critiques of the Western orientation to childhood and child development in the overseas development work of international governmental and non-governmental agencies (Boyden 1990; Woodhead 1999; Boyden and Myers 1999; Hart *et al.* 2004). These parallel a small but important debate within academic psychology on the need to correct the normalising and universalising tendencies of the field of child development and to build theory that recognise the historical, cultural, and socio-political contexts of children's development (Henriques *et al.* 1984; Burman 1994). Recently, this kind of critique has been applied specifically to the universal adoption of the language of 'participation' in the development agendas of non-government agencies (Cooke and Kothari 2001). The ladder reflects the assumption that individual agency in participating with others to make decisions is a key factor in the achievement of good programmes and programmes that are designed to fulfil children's rights. But this degree of emphasis on individual agency may not be appropriate to many cultures. Allow me a brief personal digression to explain how I came to learn about the problem for I think it highlights the depth of the issues we need to address.

In 1981, the director of the participation section at UNICEF headquarters in New York contacted me because they had heard of the work of our Children's Environments Research Group on children's participation. She expressed some embarrassment that UNICEF had never thought of children's participation at that time but only about community participation *for* children. She had no reason to be embarrassed for that was true of all international NGOs working for children at that time, and most NGOs in all countries. I was in turn embarrassed that she wanted to distribute our publications to UNICEF field offices internationally for I knew that these newsletters were hopelessly biased as international commentaries – they drew exclusively from industrialised countries, primarily Europe and North America and did not include such important experiments as Tanzania's work with youth and Sri Lanka's Sarvodaya movement. As a result it was decided that I should travel overseas to see what I could learn. But after three weeks in Sri Lanka I knew that I had embarked on a course that was way beyond my abilities. It was true that children in Sri Lanka were participating in hundreds of villages to a very high level in such centrally important projects as building water wells and decorating and maintaining them. But to an outsider, raised in England and living in the USA, this was a very different kind of children's participation than I had ever seen before. In the villages I visited, all of the children were involved in community projects and from my questioning it did not seem that they were initiating or designing any projects of their own. But they did show great pride in their projects and no sense of being compelled to do them against their wishes. I tried to look to my own experiences as a child and my work with children to interpret what I was seeing but I could not. Clearly these projects were being designed by adults and being carried out by all of the children according to a plan from above. Was this not regime-organised, I asked myself, and hence was it really participative? One of my indicators of a project being participatory is that that the participants volunteered for it. But the fact that all children participated made it unlikely that these programmes were entirely voluntary – at least in my understanding of the word. After a week or so of difficulty with these questions I began to realise what many more experienced travellers could have said from the first day: this was a more collective kind of culture where children are raised from an early age to see themselves deeply as members of a community with a responsibility to the development and care of others. But while I liked this degree of collective concern I found myself asking where is the right of the individual to make their own choices? But my time was running out and I had to return with no answer to such complex questions. Fortunately the UNICEF staff understood my cultural dilemma and they made it easy for me to say how uncomfortable I felt in drawing any conclusions from what I had seen and putting these down on paper.

The contrast between *individualist* and *collectivist* cultures has long been a subject of interest amongst social scientists but has been little studied by developmental psychologists (e.g. Triandis 1995). The Western model of child development that is being sown internationally though development programmes of education and child development stresses the development of personal independence and the autonomy of individuals. But we need to ask what the implications might be for children's development within a more collectivist culture. There is of course great danger in jumping to any simple comparisons between this dualism. Some might be tempted, for example, to take the romantic view that all cultures that look collective are necessarily superior for the caring for others that they exhibit. But the work of such child advocates as the Sri Lankan lawyer, Savitri Goonesekere, has shown us that in Asian countries there can also be great value in giving children the individual right to know and to speak out in order to protect their own rights (Goonesekere 1998). For example, girls in the *Development and Education Programme for Daughters and Communities* in Northern Thailand learn to speak about their rights and act directly in their villages to rescue their peers from being taken to the city to work as prostitutes by those families who are close to sacrificing one of their girls to the city (Development and Education Programme for Daughters and Communities 2005). But the issue of how children can work with adults in different cultures to participate with others to achieve their rights remains an area where so much more exploration is needed. Fostering a debate about cultural differences in children's participation is important for anyone who is both concerned with the achievement of human rights and with maintaining the integrity of cultures. The debate needs to begin by levelling the ground: allowing local cultures to declare their own local 'bill of rights' before engaging in a dialogue with the universal and universalising CRC. The use of such simple tools as the ladder of participation must be done with care and with a great readiness to critique and invent anew.

The cultural-bias critique can even be levelled to a degree at the UN CRC for although it was ten years in its drafting by an international committee, its authors relied upon Western notions of child development and the importance of the development of individual autonomy in children (Alston 1994). Sociologists and anthropologists have discovered in recent years an interest in the study of childhood and they are understandably offering cultural critiques of psychology (Jenks 1996; James *et al.* 1998). But for many years there has been a gradual turn taking place towards the study of the cultural context of children's development in the generation of theory within the field of developmental psychology (Rogoff 1990; Cole 1996; Valsiner 2000). Much of this focuses on the cultural context of thinking but there is also a small literature that is concerned with cultural differences in children's social relations, and in particular, questions of children's independence and agency (e.g. Kagitcibasi 1996; Markus and Kitayama 1994; Murphy-Berman *et al.* 2003; Welch and Leary 1990). Sadly, in spite of these important academic developments, the old universalistic models of child development from the West already dominate schools of education and early childhood development all over the world and so there is a lot of work to be done to reverse this cultural form of imperialism. Fortunately in recent years some Western development advisers have begun to critique these old models (e.g. Woodhead 1999; Boyden 1990; Boyden and Myers 1999; Hart *et al.* 2004). I predict that these new cultural theories of child development and methods that are more grounded in the local study of child development and child-rearing will be enthusiastically embraced by non-governmental agencies and NGOs working in development programmes with children and youth all over the world because they meet with the analyses that many of those who work with children have already made for themselves.

2.8 Conclusion: The Need for New Models

Two members of the *Concerned for the Working Children* in India who work with Bhima Singha, a union of child workers in Bangalore, have developed valuable schemas for thinking about the varying roles adults play in relation to children's participation, including some valuable critiques of the ladder model (Reddy and Ratna 2002). They suggest two rungs of non-participation below the ones that I suggested: *Active resistance* is where adults actively work against children's participation because they feel that children should not be burdened with participation, or that they do not have the capacity or that they can be easily manipulated to further adult agendas. Some adults in this category actually mobilise support to lobby against children's participation. *Hindrance* is when adults block opportunities for children and discourage them from participating, intentionally or unintentionally undermining their abilities and making them feel inadequate and reluctant to participate. This second rung seems to be just a weaker form of the first one but taken together I feel that they do add to the original ladder typology – maybe these rungs of the ladder should be drawn on the diagram underground! Reddy and Ratna also introduce rungs, of the ladder called *Tolerance* and *Indulgence*. These seem to be elaborations of what I generally call 'tokenism' but while it may not be necessary to imagine these as entirely new rungs this elaboration is useful. At the top of the ladder Reddy and Ratna suggest two categories: *Children initiated and directed* and *jointly initiated and directed by children and adults*. It may be useful for some programmes to consider more rungs like this but, as I have tried to explain above, the important distinction in my mind is how the children think of themselves and the adults. I was trying to express in the ladder that the top of the ladder should not be 'children in charge' but children as citizens who think of themselves as members of a larger community that includes adults and other children who they may sometimes invite to join them.

Clearly the *Concerned for Working Children* is one of these groups that recognise that it is a struggle to find better ways of working with young people and that we must work on this with the children and youth themselves. More important than what any static model like 'the ladder' looks like is the recognition that we must be engaged in a never-ending process of working across generations to generate improved ways of adults and children working together, both on the realisation of children's rights but also on their shared involvement in the future of their communities. When I first wrote about the ladder most of my thinking came from dialogue with facilitators from NGOs working in the field. At the time they were well ahead of academia in recognising that the participatory capacities of children, and the relationships they can build with adults, had been grossly under-recognised. In recent years research and theory-building in the social sciences, and in psychology, has begun to catch up. What is now needed are programmes of collaboration between academics and those who work directly with children as well as with children and youth themselves. In particular we need to find ways of monitoring and evaluating the way that we work with children and the quality of the realisation of their participation rights. Hopefully we can avoid turning this enterprise into yet another specialised profession; we need to build monitoring and evaluation into the everyday practices of groups of young people and those who work closely with them (e.g. Sabo 2003).

I have probably not laid the ladder debate completely to rest with these words but, from my perspective, I see the ladder lying in the long grass of an orchard at the end of the season. It has served its purpose. I look forward to the next season for I know there are so many different routes up through the branches and better ways to talk about how children can climb into meaningful, and shall we say fruitful, ways of working with others.

Acknowledgments This Chapter has benefited greatly from comments by Louise Chawla, Peter Dixon, Jason Hart, Selim Iltus, Bjarne Brunne Jensen, David Kritt, Gerison Lansdowne, Kim Sabo, Jason Schwartzman, Venka Simovska, Alan Reid, Jutta Nikel, Joachim Theis, and Henk Van Beers.

References

Ackermann, L., Feeny, T., Hart. J., and Newman, J. (2003) *Understanding and Evaluating Children's Participation: A Review of Contemporary Literature*. London: PLAN UK/PLAN International. (http://www.plan-uk.org/pdfs/literaturereview.pdf)

Alston, P. (ed) (1994) *The Best Interests of the Child: Reconciling Culture and Human Rights*. Florence, Italy: International Child Development Centre, UNICEF and Oxford: Clarendon Press.

Arnstein, S.R. (1979) Eight rungs on the ladder of citizen participation. *Journal of the American Institute of Planners*, 35, 4: 216–224.

Boyden, J. (1990) Childhood and the policy makers: a comparative perspective on the globalisation of childhood. In: James, A. and Prout, A. (eds), *Constructing and Reconstructing Childhood: Contemporary Issues in the Sociological Study of Childhood*. London: Falmer, pp. 190–229.

Boyden, J. and Myers, W. (1999) *What Works for Working Children*. London: Jessica Kingsley.

Burman, E. (1994) *Deconstructing Developmental Psychology*. London: Routledge.

Childhood City Newsletter (1980) *Special Issue*: *Participation Methods*. New York: Children's Environments Research Group, Graduate School of the City University of New York.

Cole, M. (1996) *Cultural Psychology*. Cambridge, MA: Belknap Press.

Cooke, B. and Kothari, U. (eds) (2001) *Participation: The New Tyranny*? London: Zed Books.

Corsaro, W. (2003) *We're Friends Right? Inside Kids' Culture*. Washington DC: Joseph Henry Press.

Development and Education Programme for Daughters and Communities (2005) Thailand. (http://www.thailife.de/wecare/depdc)

Gaitan, A. (1998) Protaganismo Infantil, *Actas del Semianario de Participation*. Bogotá, Colombia: Regional Office of UNICEF for Latin America.

Gauvain, M. (2001) *The Social Context of Cognitive Development*. New York: Guilford Press.

Goonesekere, S. (1998) *Children, Law and Justice*. New Delhi/Thousand Oaks/London: Sage.

Hart, J. (2005) Personal commentary on draft of this chapter.

Hart, J. (2004) *Children's Participation in Humanitarian Action: Learning from Zones of Armed Conflict*. Oxford: Refugee Studies Centre, University of Oxford.

Hart, J., Newman, J. Ackermann, L., and Feeny, T. (2004) *Children Changing Their World: Understanding and Evaluating Children's Participation in Development*. London: Plan UK.

Hart, R. (1979) *Children's Experience of Place*. New York: Irvington (Wiley).

Hart, R. (1992) *Children's Participation: From Tokenism to Citizenship*. UNICEF Innocenti Essays, No. 4, Florence, Italy: International Child Development Centre of UNICEF.

Hart, R., Rajbhandari, J., and Khatiwada, C. (2001) *The Children's Clubs of Nepal: Evaluation of a Democratic Experiment and the Associated Methodological Video Mirrors of Ourselves: Tools of Democratic Self Reflection for Groups of Children and Youth*. Kathmandu, Nepal: Save the Children Alliance and the Children's Environments Research Group. (www.cerg1.org)

Henriques, J., Hollway, W., Urwin, C., Venn, C., and Walkerdine, V. (1984) *Changing the Subject: Psychology, Social Regulation and Subjectivity*. London: Methuen.

James, A., Jenks, C., and Prout, A. (1998) *Theorising Childhood*. Cambridge: Polity Press.

Jenks, C. (1996) *Childhood*. London: Routledge.

Jensen, B.B. (2000) Participation, commitment and knowledge as components of pupils' action competence. In: Jensen, B.B., Schnack, K., and Simovska, V. (eds), *Critical Environmental and Health Education: Research Issues and Challenges*. Copenhagen: Danish University of Education, pp. 219–238.

Kagitcibasi, C. (1996) *Family and Human Development Across Cultures: A View from the Other Side*. London: Erlbaum.

Kirby, P. and Woodhead, M. (2003) Children's participation in society. In: Montgomery, H., Burr, R., and Woodhead, M. (eds), *Changing Childhoods: Local and Global*. Chichester, UK: Wiley, pp. 233–284.

Mannion, G. (2003) Children's participation in school grounds development: Creating a place for education that promotes children's social inclusion. *International Journal of Inclusive Education*, 7, 2: 175–192.

Markus, H. and Kitayama, S. (1994) *Emotion and Culture: Empirical Studies of Mutual Influence*. Washington, DC: American Psychological Association.

Melton, G. (1993) Review of 'children's participation: from tokenism to citizenship'. *The International Journal of Children's Rights*, 1: 263–266.

Ministry for Children and Families of British Colombia, Canada (2005) *The Youth Participation Strategy. Degrees of Involvement – The Ladder of Participation*. Vancouver, BC: Ministry for Children and Families and the McCreary Centre Society. (http://www.mcs.bc.ca/yps/howyps.htm)

Murphy-Berman, V., Berman, J.J. *et al*. (2003) *Cross-cultural Differences in Perspectives on the Self*. Lincoln, NE: University of Nebraska Press.

National Youth Council of Ireland and the Children's Rights Alliance (2003) *Hearing Young Voices – Guidelines for Consulting Children and Young People in Relation to Developing Public Policy and Services in Ireland*. Dublin: Open Your Eyes to Child Poverty Initiative. (http://www.youth.ie/research/content.html)

Reddy, N. and Ratna, K. (2002) *A Journey in Children's Participation*. Vimanapura: The Concerned for Working Children. (http://www.workingchild.org/prota9.htm)

Reid, A. and Nikel, J. (2004) Towards evaluating the theory and practice of participation in environmental education? Paper presented at the Annual Conference of the *American Educational Research Association*, San Diego, CA, April 12–16.

Rogoff, B. (1990) *Apprenticeship in Thinking: Cognitive Development in Sociocultural Activity*. New York: Oxford University Press.

Sabo, K. (ed) (2003) *Youth Participatory Evaluation. New Directions for Evaluation*. San Francisco, CA: Jossey-Bass.

Treseder, P. (1997) *Empowering Children and Young People Training Manual: Promoting Involvement in Decision-making*. London: Save the Children-UK.

Triandis, H.C. (1995) *Individualism and Collectivism*. Boulder, CO: Westview Press.

Valsiner, J. (2000) *Culture and Human Development*. Thousand Oaks, CA: Sage.

Vygotsky, L.S. (1978) *Mind and Society*. Cambridge, MA: Harvard University Press.

Welch, C.E.L. and Leary, V.A. (1990) *Asian Perspectives on Human Rights*. Boulder, CO: Westview Press.

Woodhead, M. (1999) Combating child labour: listen to what the children say. *Childhood*, 6, 1: 27–49.

Chapter 3
Differentiating and Evaluating Conceptions and Examples of Participation in Environment-Related Learning

Alan Reid[1] **and Jutta Nikel**[2]

Keywords learning, development, participation, learning theory, evaluation

3.1 Introduction

This chapter draws together a range of critical perspectives on the concepts and practices of participation. Starting with recent debates about the tyrannical and transformational possibilities of participatory approaches in the field of development, we explore echoes of these critiques in education, with a focus on learning and teaching about the environment and sustainable development. The chapter illustrates how three major perspectives on participatory learning – behaviourist, cognitive, and situative – help differentiate current understandings and can inform individualised and shared expectations of participatory approaches to environment-related learning. The chapter also sets out a series of questions to aid critical investigation of examples of participatory forms of environment-related learning, outlining an evaluative framework that highlights three key dimensions to participatory activities – practice, theory, and meta-theory. We conclude the chapter with an extended example of the application of the framework, and discuss a range of issues for participatory work in Education for Sustainable Development (ESD), the development of learner competences, and citizen engagement with participatory learning that aims to foster wider and deeper participation in civil society.

3.2 Being Critical: the Development Context

Rahnema's (1992) review of participatory approaches to development is a landmark in the attempts to foster a critical appraisal of the concept and practice of participation. Noting that the concept is linguistically rooted in the notion of 'taking part'

[1]Centre for Research in Education and the Environment (CREE), Department of Education, University of Bath, Bath, BA2 7AY, UK, a.d.reid@bath.ac.uk

[2]Centre for Research in Education and the Environment (CREE), Department of Education, University of Bath, Bath, BA2 7AY, UK, edpjn@bath.ac.uk

A. Reid et al. (eds.), *Participation and Learning*, 32–60.

with others or 'having a share' in something with others (*Oxford English Dictionary*), Rahnema argues that the term has become a part of modern jargon and can now be used to 'support the most fanciful constructions' (p. 116).

Rahnema sketches the history of participatory approaches to development and how the term has come to be both increasingly fashionable and commonly associated with attempts to promote social change through such processes as 'popular participation', 'empowerment', 'dialogical intervention', and 'conscientization'. In a more recent analysis, Hickey and Mohan (2004:3) argue that at the heart of such participatory approaches lies the 'promise of empowerment and transformative development for marginal people', and to many observers, it is the ubiquity of the terminology of participation within and across policies, preferred approaches to development, and evaluation techniques, that has become most striking.

For development, key concepts and expressions of participation have tended to be those linked with Participatory Rural Appraisal (PRA), Participatory Action Research (PAR), and with PRA and PAR theorists like Orlando Fals-Borda, Anisur Rahman, and Robert Chambers, but as Rahmena (1992), Williams (1983), and Cornwall and Brock (2005) show, their cultural and conceptual roots go much deeper. In education there are similar links, particularly to PAR, and most typically to Paulo Freire and Ivan Illich in the wider educational discourse, while in health- and environment-related learning, as illustrated by Roger Hart in Chapter 2 of this volume, discussion often return to the *Ladder of Children's Participation* (e.g. Mordock and Krasny 2001).

In a variety of ways, these approaches and their advocates have helped encourage a broader critical awareness of participation. With a socially critical perspective in mind, they have highlighted that participatory processes and situations related to decision-making *about, in*, and *for* social change are replete with power issues and subject to hegemonic ideological and cultural forces. For example, from a Gramscian standpoint (Gramsci 1971:412–413), the notion of hegemony calls for educators to attend to the prevailing 'common sense' formed in culture and diffused by civic institutions (like schools), and which informs those values, customs, and spiritual ideals that induce 'spontaneous' consent to the status quo, through their various channels of 'persuasion' and 'propaganda'. In such circumstances, both ensuring and securing ongoing stakeholder participation may represent a powerful, concrete response to address ideological and political domination in society. However, as Rahnema (amongst others) has argued, there is also the distinct possibility – and a long history – of practice not matching the theory of participation, with approaches and participants failing to realise the counterhegemonic potential of participatory processes.

This gap, between the discourse and experience of participation, is a key starting point for critique of participation and participatory approaches. Attempts to explain the existence (and persistence) of gaps through inquiries as to whether current theorisations and praxis genuinely aim at and achieve socially critical objectives or participant ownership (see Cornwall 2002; Cooke and Kothari 2001), have dominated recent debates in the field of development. For the purposes of our discussion, we briefly focus on three inter-related themes in this recent debate and critique: terminology, history, and practice.

First, Williams (1983) warns of the potential loss of meaning and the risks associated with our forgetfulness about the cultural origins of the terminology, describing contemporary 'keywords' such as 'participation' and 'empowerment' as 'warmly persuasive' or feel-good words. Emptying the meaning of the concept, plasticizing it, or making it a 'floating signifier', are key issues in addressing the hegemonic role of such concepts (Laclau 1990). For example, in a recent UNRISD publication, Cornwall and Brock (2005:4) describe 'participation' and 'empowerment' as part of larger 'chains of equivalence', those long list of key terms that tend to be bunched together, such as, participation, empowerment, poverty reduction and then partnership, governance, accountability, and so forth (ibid.):

> [T]he more words that become part of the chain, the more that meaning resides in the connections between them. Pared down to the elements that would permit coherence, the terms that form part of today's development jargon are reduced to monochrome.

While typologising the concept and practice also risks making participatory approaches appear either black or white, the emergence of a range of ways in which to categorise them also serves to illustrate how the rhetoric may not necessarily correlate with reality. How participation is lived out reveals a diversity of forms and formulations: as *transitive* or *intransitive* participation; as *moral, amoral*, or *immoral* participation; as *forced* or *free* participation; or, as *manipulative* or *spontaneous* participation (Rahnema 1992). For each subcategory, who is 'targeted' to take part in such participation raises direct questions about how a related 'keyword', such as democracy (see Chapter 11 by Schnack, this volume), is also enacted:

- Is it enough to limit participation to those who are invited and amenable to participation?
- Should it be those who in some sense 'need' to participate?
- In what sense must the participants represent all sectors (and not just stakeholders) in society?

While such distinctions help map out the intended and actual uses of participatory terminology, they also reveal something more in terms of the lived experiences of the concept-in-use. Rahnema (1992), for example, has been particularly quick to tie this potential diversity of experience and practice of participation to the unmatched interest governments and development institutions have taken in the concept since the 1970s (see pp. 117–120). Regarding this 'co-option of participation', Rahnema observes:

1. The concept is no longer perceived as a threat.
2. Participation has become a politically attractive slogan.
3. Participation has become, economically, an appealing proposition.
4. Participation is now perceived as an instrument for greater effectiveness as well as a new source of investment.
5. Participation is becoming a good fund-raising device.
6. An expanded concept of participation could help the private sector to be directly involved in the development business.

Elaborating and substantiating such observations, Cornwall and Brock (2005) outline a 'genealogy of participation' in development, paying particular attention to the switch that occurred in the 1980s, from a more 'people-centred' notion to one fitting into the neoliberal 'regime', when, 'community participation became a channel through which popular participation began to be operationalized' (p. 7). According to Cornwall and Brock, the late 1990s saw participation assume the primary interpretation of being about providing mechanisms through which policy objectives could be realised, yet where: 'conflict and power are as absent from this world as they are from the world we are offered in today's development policies' (p. 9).

Such developments have resulted in participation becoming mainstream rhetoric, often with positive connotations for the public, politicians, and economists alike, despite the fact that interests and incentives for using participation and promoting it are often poles apart. This has prompted development critics such as Gustavo Esteva to identify a wide range of negative examples of the strategic use of participatory processes by planners, experts, and economists (e.g. in World Bank-funded structural adjustment programmes and those inspired by the Poverty Reduction Strategy Paper in the late 1990s, where it is argued participation tends to be by invitation only and the voices of the poor remain marginalised). There are also case studies and reviews of the weaknesses in the grounding, theorisation and operationalisation of participatory concepts by community activists (see Escobar 1995; Nederveen 2001; Chambers 2004).

For Esteva, key weaknesses in participatory praxis have included: the pitfalls, shortcomings, and blind spots of empowerment strategies as operated by some donor or NGO initiatives; the false promises or subtle manipulations offered by some 'concientization' or 'change' agents working from 'progressive historical-cultural realities'; and the professionalisation of grass roots activities, particularly through upscaling and mainstreaming participation in development work (see Esteva 1985, as an example, and Chapter 7 by Lotz-Sisitka and O'Donoghue, and Chapter 8 by Vare, this volume). Indeed, Rahnema (1992:126), in voicing concern about both the potential or systemic counterproductiveness of some participatory processes and concepts, has helped redirect attention away from simply acknowledging shortcomings towards promoting the ongoing need for critical appraisal of the term:

> Participation, which is also a form of intervention, is too serious and ambivalent a matter to be taken lightly, or reduced to an amoeba word lacking any precise meaning, or a slogan, or fetish, or for that matter, only an instrument or methodology. Reduced to such trivialities, not only does it cease to be a boon, but it runs the risk of acting as a deceptive myth or a dangerous tool for manipulation. To understand the many dimensions of participation, one needs to enquire seriously into all its roots and ramifications, these going deep into the heart of human relationships and the socio-cultural realities conditioning them.

Such a standpoint is important for tracing the roots of what some might regard as the backlash towards participation in the field of development studies over recent years (typified perhaps by Cooke and Kothari 2001). Indeed, recent responses to the critical appraisal of participation now attempt to work constructively with such critique (see, for example, Hickey and Mohan 2004) by explicitly addressing how 'politics matters', and by emphasising the need to understand:

- The ways in which participation relates to existing power structures and political systems
- How participation works with and for the 'poorest of the poor'
- How it can shatter an all to comfortable 'myth of community', that is, when community also assumes the status of a 'feel good' word and is left unproblematised and unexamined in participatory approaches and appraisal (see, for example, Chapter 8 by Vare, this volume)

Yet while we should recognise that these lines of critique have tended to both emanate from and been contextualised within the development field, as a chapter in a book about learning and participation, we must also consider their relevance and ramifications for participation in environment-related educational contexts (see also Chapters 7 and 8).

3.3 Participation and Environmental Education

Carlos Seré, in his position as Regional Director for Latin America and the Caribbean, of Canada's International Development Research Centre, once noted (IDRC 1998; unpaginated): 'Development isn't just a technical issue, it's a social issue. Sustainability can only be established with the participation of all.'

While we would support such a view, participation is clearly neither a given nor unambiguous in development or sustainability – for the individual, or society at large. Some participate more than others, and some participate – for want of better words – more effectively and efficiently. Important questions for educators then are, how do people learn to participate, and relatedly, why?

We can start addressing these questions by noting that the emergence of the discourse on participation in education resonates with broader historical shifts in Western understandings of childhood, the main locus for educational efforts. Broadly speaking, since the 19th century understandings have been shaped by such social ideologies as protectionism towards children, and in the 20th century, by biological and psychological models of developmentalism in childhood. The latter highlights the fact that children tend to be treated as 'human becomings' (Farrell 2005:6) rather than fully human (beings). Consequently, while children may develop and share many of the qualities of adulthood, by the very nature of the way these terms have come to be understood, the full humanity of children is regarded as not yet having been reached or established – childhood is organic in essence, it is about growth and maturation (Matthews *et al.* 1999). Thus both metaphorically and literally, children will inevitably 'fall short' in having full and equal rights of participation in decision-making about education, the environment, health, and development, or for that matter, research (see, for example, Chapter 2 by Hart, and Chapter 18 by Barratt and Barratt Hacking, this volume).

An alternative perspective on children that has received more attention since the 1980s and 1990s, is grounded in the then newly emerging fields of the sociology and psychology of childhood (see Roberts 2001; Shier 2001; Danby and Farrell 2004). Strong emphasis is placed on viewing childhood as a socio-culturally constructed category (Tobin 1995; James *et al.* 1998), and thus attention is drawn to the legal, social, economic, and political dimensions to childhood and the naming and framing of the category, 'child'. Thus for Farrell (2005:6), children are to be regarded as 'competent participants in their everyday worlds' and quite 'capable of participation in or withdrawal from' organised activities, such as learning, teaching, and inquiry – a stance that echoes many of the sentiments of the UN Convention on the Rights of the Child (CRC) (see Chapter 2).

In the field of environmental education, alongside key concepts such as interest, awareness and sensitivity, knowledge and understanding, attitudes and values, and skills, participation remains enshrined as a key objective (and approach) for learning. The understandings of environmental education set out in key foundational documents for the field include statements to this effect: the Belgrade Charter argues that students should be provided 'with opportunities for active participation in all levels of activities to solve environmental problems' (UNESCO–UNEP 1976). The Tbilisi Declaration is another major UNESCO–UNEP reference point for the field, and it states that participation is a component of the key objectives of environmental education: *to help provide social groups and individuals with an opportunity to be actively involved at all levels in working toward resolution of environmental problems* (UNESCO–UNEP 1978). Even though there is no formal recognition of the aforementioned shifts in the understandings of childhood in wider society, broadly supportive outlooks to those outlined by Farrell can be found in the UNESCO–UNEP documentation, with both the Tbilisi and Belgrade documents (and subsequent revisions and restatements at Moscow 1987, and Thessaloniki 1997) providing a benchmark for many subsequent definitional statements.

The UNESCO–UNEP perspectives on participation have reverberated throughout the national standards and guidelines of many Western environmental education organisations, like those of the North American Association for Environmental Education (NAAEE). In Iozzi *et al.*'s (1990) *Assessment of Learning Outcomes in Environmental Education*, for example, published by UNESCO and endorsed by NAAEE in its *Excellence in Environmental Education* guidelines for learning (NAAEE 2004), 'active participation' is endorsed as a key component of environmental education when it comes to promoting responsible environmental behaviours. Here, participation is regarded as helping solve problems and resolve issues by engaging learners in 'environmentally sound consumer purchasing, methods for conserving resources, assisting with the enforcement of environmental regulations, using personal and interpersonal means to encourage environmentally sound practices, and encouraging environmentally sound policies and legislative initiatives.' The sense here is that while a learner has the immediate horizon as a child, there is the more distant one as an adult to bear in mind; put differently, participation is something that happens across the lifespan and should not be limited to either childhood or adulthood, hence the value of lifelong environmental education.

Underpinning such a position is a core value within much of environmental education: recognising and advocating ongoing personal acceptance – now and for the future – of the need for a sustainable lifestyle and a commitment to participation and change (see also Chapter 17 by Shusler and Krasny, this volume). Indeed, the core value is echoed and broadened in international statements on sustainable development and ESD, as in *Agenda 21*, Chapter 23, where it is argued that: 'One of the fundamental prerequisites for the achievement of sustainable development is broad public participation in decision-making.'

As Hart has outlined in Chapter 2, environmental and health educators have felt a need for models or frameworks to help practitioners reflect on and develop their practice when initiating, planning, and evaluating participatory approaches and programmes with learners. Arnstein's (1979) *Ladder of Participation*, or more commonly Hart's (1992) adaptation of this model (Figure 2.1), has met this need well, and the Ladder has become a widely established and frequently used tool in environmental and health education and other youth-, community-, and education-oriented sectors.

However, the general usefulness of such models, as in development, has received critique in recent years, and from a variety of perspectives. Key themes within this critique can reflect debates around the aforementioned shift in understandings of childhood and the capabilities of children, as well as critical perspectives on how underlying hegemonic structures are (best) addressed (see Chapter 2 on responses to this).

In more detail, criticism has highlighted the significance of the counter-intuitive notion of 'non-participation', which emphasises the principle of creating or considering the value of situations where participants can make a choice about whether they wish to participate or not in the first place, or perhaps more importantly, why they might continue to do so (Treseder 1997; Farrell 2005). This suggests that 'non-participation' may well be a valuable and legitimate option, no matter whether participants are children or adults, and even if the participatory project is widely regarded as a 'good thing'. Second, arguing from a more pragmatic viewpoint, and as Hart discusses in Chapter 2, theoretical and experiential critique has questioned whether activities on the 'lower rungs' of the ladder should be considered of less value than those described on the higher rungs, and whether it is always best to aim for the highest level of participation possible. Thus Lardner (2001) is one of the critics who argue for a contextual set of expectations and evaluations of participatory approaches and processes: that different levels of participation are appropriate to different circumstances (see also Treseder 1997; Jensen 2000). Third, in the tradition of interpreting the very notion of participation as an endeavour for increasing the level of empowerment of 'marginal' peoples (used here in the broadest sense but in this instance, a notion that might also include school students or children), critics have sought to create awareness of or 'unmask' cases, examples or situations where 'participation' is commended, but in terms of empowerment or transformation, it has not (yet?) materialised. These 'tyrannical' expressions of participation are typified by the 'manipulative', 'decorative', and 'tokenistic' forms of participation – the 'lower rungs', illustrated and discussed at length by Mannion (2003).

3.4 Participation and Learning

So far we have considered participation within the context of debate and critique in development and about the models of participatory situations or initiatives with children, but there is another major focus for discussion that should not go ignored: namely, the various understandings of *learning* available in participation discourses and their role in conceptualising and differentiating conceptions and practices of participation. To illustrate the range of possibilities here, the increased emphasis on participation in educational projects and processes are considered in relation to various traditions in learning theory. Table 3.1 summarises an international review by Greeno *et al.* (1996) on the key characteristics of three mainstream perspectives on learning: the behaviourist, cognitive, and situative. We use this table to highlight:

- Differences in possible understandings of participatory learning in the light of their underlying epistemology and the source(s) of their constitutive concepts
- Alternative conceptions and understandings of 'knowing' and some examples of understandings of learning within each perspective
- The diversity in the constructions of the motivation for and transfer of learning, and hence, of motivation for and transfer of learning in participation, and finally
- Responses to matters of accountability and assessment in participatory learning

Are the differences in perspective irreconcilable when it comes to participation? Not really, according to Greeno *et al.* (1996:24), who go on to argue that: 'All of the psychological perspectives on learning school subjects assert that learning requires the active participation of students achieved via extrinsic motivation, intrinsic motivation or engaged participation.' Moreover, seen within the context of education and its institutions (typically schools, but increasingly non-formal and informal settings for lifelong learning), each of the three perspectives suggests a deliberate linking of the quantity and quality of learning outcomes to a focus on the individual and the 'internal' rather than to the 'external' of the learner. That is, the principal focus for understanding and engaging in participatory learning processes should no longer be solely that of the learner's environment and observable behaviours in education but the core challenges of fostering sustained and deeper levels of learner motivation and engagement.

Nonetheless, digging a little deeper into the material summarised in Table 3.1, the answer is a little more complicated than our first reading. For example, from a cognitive perspective, learning is understood to be an active, constructive,

Table 3.1 Key marker characteristics of alternative perspectives on learning. (Based on Greeno *et al*. 1996.)

Perspective	Behaviourist	Cognitive	Situative
Epistemology	Empiricism	Rationalism	Sociohistoricism/ Pragmatism
Traditions and source of concepts contributing	Associationism	Gestalt psychology	'Lave and Wenger'
	Behaviourism Connectionism	Constructivism Symbolic information processing Zone of Proximal Development	Communities of Practice Legitimate peripheral participation
Knowing as…	having associations affecting behaviour	conceptual development cognitive abilities personally meaningful	distributed and embodied through community practices
Learning as…	an organised accumulation of associations and components of skills	understanding of concepts and theories in different subject matter domains, and general cognitive abilities	becoming more adept at participating in distributed cognitive systems, focusing on engagement that maintains the person's interpersonal relations and identity in communities in which the person participates
Learning and Transfer	• Acquiring and applying associations • Behavioural and attitudinal change	• Acquiring and using conceptual and cognitive structures	• Initiation and induction • Becoming attuned to constraints and affordances through participation • Shared repertoire between the community
Motivation and Engagement	Extrinsic motivation	Intrinsic motivation	Engaged participation Legitimation
Focus of Accountability and Assessment	External	Individual	Community

Note: This Table is intended to be illustrative rather than exhaustive.

cumulative, and goal-oriented process. The learner is positioned as the key agent rather than a passive recipient in a teaching situation. It is a conception of learning supported by and which extends humanistic and Enlightenment-style arguments for fostering a high level of involvement of children and youth in the construction of their learning and decision-making about it (see Chapter 11 by Schnack, this volume). However, as in the development field, it also serves to draw attention towards economic and efficiency-style arguments about how best to achieve learning, i.e. inauthentic, tokenistic forms of participation in learning will not suffice, according to these criteria. Thus, the learner's active participation is not just a desirable emancipatory boon; rather, as in development, it is viewed as a necessity for learning to occur, to last, and be both effective and efficient. Put starkly, rote learning has no place in participatory learning.

Situative perspectives on learning, in contrast, relocate the focus on the individual by emphasising the communal and relational aspects of the individual's participation in learning. Thus, while cognitive theories have emphasised the active involvement of the individual, from a situative perspective we must consider active involvement in terms of 'participation in a community'. Typically, this is termed a community of practice or action (see, for example, Chapter 19 by Shallcross and Robinson, this volume), and it is widely regarded as a necessity for learning processes and outcomes to be marked by the same outcomes listed earlier, that is, without an active engagement with/in a community over the longer term, participatory learning risks becoming disembedded and incoherent, in relation to the learners, the processes, and the outcomes.

Table 3.2 presents distinctive positions on the focus and motivation for participating, but it also serves to illustrate some of the common ground that emerges across them. A key feature of the table is that each perspective shares the position that one's capacity to participate is learnt, constructed and dynamic – and thus can be enhanced (rather than being regarded as something that it is, for example, largely inherited, fixed, or stable). A key implication of this view, as with the shift in conceptions of childhood and development, is that schools, teachers, learners, and communities, can make a profound difference, positive and/or negative, to an individual's capacity to participate in formal, informal, and non-formal learning contexts (for further discussion, see Arnot and Reay 2001; Fielding 2001; Kirby 2001; and for specific examples related to environment-related learning, Hart 1997; Hart 2000; Brierley *et al.* 2002).

Indeed, as Greeno *et al.* (1996:16) note:

> All three [perspectives] … have contributed, and continue to contribute, important insights to fundamental scientific knowledge and understanding of cognition and learning and have influenced educational practices significantly. While each perspective is valuable, they frame theoretical and practical issues in distinctive and complementary ways ….

Tables 3.1 and 3.2 also illustrate some of the potential ambiguities associated with the notion of participation as a concept-in-use in education, alongside differences in their epistemological groundings, and the diversity of views available on knowing, knowledge and learning, motivation and engagement, purposes of learning and

Table 3.2 Learning perspectives and participation, as applied to examples of environment-related learning. (Adapted from Greeno *et al.* 1996.)

Perspective	Behaviourist	Cognitive	Situative
Focus on…	participation in socially acceptable behaviours e.g. *can recycle*	participation in one's own learning process: Concept of metacognition as 'capacity to reflect upon one's own thinking, and thereby to monitor and manage it'… 'self- conscious management of one's own learning and thinking processes' (p. 19) 'beliefs and understanding of themselves as *knowing agents*' (p. 19) e.g. *understand recycling processes and imperatives and links to other environmental issues and themes*	participation in practices of communities Collective knowing (groups are composed of individuals and considering knowing as abilities of groups in their practice) *individual knowing* (considering knowing of individuals as their ability to participate in those practices) …participation in social practices is needed for learning and knowing (apprenticeship learning) participation and identity linked e.g. *action competence displayed through socially critical actions related to recycling* (e.g. *investigating the benefits and drawbacks of reducing and reduced consumption*)
Motivation for active participation in the learning examples	'Engagement in activities can also be considered as a decision based on expected utilities of outcomes of the engagement, which depend on the individual's subjective probabilities and utilities regarding outcomes of alternative participation in different ways in learning activities.' (p. 24) e.g *via positive and negative reinforcement*	'Engagement is often considered to be a person's intrinsic interest in a domain of cognitive activities….' (p. 25) Elements of intrinsic motivations might be: challenge, fantasy, curiosity e.g. *via problem- solving and inquiry*	'Students can become engaged in learning by participating in communities where learning is valued.' (p. 26) e.g. *learning one's native language, learning to read well in order to access cultural and s ocial capital* Identity is viewed as critical to engagement in learning activities. In other words, 'the motivation to learn the values and practices of the community of learners is tied up with establishing their identities as community members.' (p. 26) e.g. *informal as well as formal learning experiences*

Note: This table is intended to be illustrative rather than exhaustive.

assessment, and so forth. Thus, while comparison can help distinguish one particular perspective on participation from another (in terms of participating in one's own learning and that of others, for example), a sole focus on one approach or perspective can occlude the possibility or desirability of pursuing others. Consequently, while some perspectives clearly complement others at some points, other aspects of participation and learning remain distinctive and conspicuous by their stark contrast and relative incompatibility with other understandings (see Chapter 19, by Shallcross and Robinson, as an example). This situation lends support to the notion that different modes and approaches to participation in environmental learning are required, i.e. there is no single best approach for all situations.

The risk in all this is similar to that in the field of development: participation in education becomes a widely used and fashionable term, degraded in its meanings and uses. For example, in England, participation and citizenship in schools are often equated, but the curriculum and the structures of schooling tend to limit conceptions of participation towards the formal mechanisms of democracy (an education in civics and schools councils, for example), rather than engage with broader or alternative notions, like participatory democracy and ecological citizenship (see later, and Chapter 20 by Carlsson and Sanders, this volume). Taking account of diverse perspectives on learning is clearly one possibility for helping evaluate instances of participation and participatory education in the fields of environment, health, and sustainability. But we argue that this situation demands that other social, political, ideological, and instrumental functions underlying their presence in environment-related learning should also be considered and explored. We illustrate these possibilities in the next section, laying out the groundwork for developing an analytical framework for evaluating participation and participatory approaches in environment-related learning.

3.5 Why do Participation?

> Even if we know someone else, and know ourselves, we still have to grasp the "truth of our interrelationship, the truth of the unitary and unique event which links us and in which we are participants" (Bakhtin 1993:17). Accordingly, to understand an object means that we have to understand our "ought" in this relationship, the attitude or position that we ought to take with respect to it and other individuals. For our participation in interaction, we are responsible; each act "presupposes my answerable participation, and not an abstracting from myself. It is only from within my participation that Being can be understood as an event, but this moment of once-occurrent participant does not exist inside the content seen in abstraction from the act qua answerable deed" (Bakhtin 1993:18). We are answerable for each act, every moment of our lives, every act is an answerable act: life itself "can be consciously comprehended only in concrete answerability" (Bakhtin 1993:56).
> (Roth 2003, para. 40)

In the preceding part of this chapter, we outlined some of the conceptual distinctions that might encourage a comparative evaluation of participation as a concept-in-use in both theory and practice in environment-related learning, and that might also assist in the evaluation of various instances of participatory approaches in this field. To reiterate our main purposes in this chapter and to pull together some of the threads of the preceding sections, we note that Rapoport (1985:256) observes:

> Conceptual frameworks are neither models nor theories. Models describe how things work, whereas theories explain phenomena. Conceptual frameworks do neither; rather they help to think about phenomena, to order material, revealing patterns – and pattern recognition typically leads to models and theories.

As suggested earlier, the sites and routes for how one might evaluate participation-related phenomena are potentially wide ranging, given that participation exists in a variety of forms in the 'dreams, mouths, and lives' of its advocates and critics, as well as those of the participants (or in some cases, 'recipients') of these processes. Furthermore, the political 'gravity' that may accrue to such an evaluative task becomes apparent in that, if achieved successfully, it may present serious challenges to a range of interests (including the vested or entrenched), of those for or against a particular practice, model, theory, or ideal, in participatory environment-related learning.

For us, the conceptually, ethically, and philosophically loaded quotation that opens this section illustrates one particular way in which those interested in participation can be invited to inquire and reflect deeply on the ethical purposes of education and the qualities of participation and participatory approaches in teaching and learning. Whether one understands or agrees with Roth (2003) or not, an understanding of teaching and learning that is informed by such considerations reveals that teaching and learning convey a serious ethical enterprise, exhibiting substantial ontological, epistemological, and relational dimensions, i.e. in terms of what it means or might mean to be, know and interact in a participatory learning situation. As Simovska (2000, and Chapter 4 by Simovska, this volume) argues, this is where we might begin to distinguish the authentic and inauthentic, as well as the genuine and tokenistic, in participatory activities and discourses of participation.

Given this, developing a rigorous yet open evaluative framework for participation can require practitioners, advocates, and researchers to think *with* participatory discourse, but also to think *beyond* it and *against* it. Thus, in relation to any particular 'utterance' or 'articulation' of a discourse of participation, we would do well to consider asking, how fixed or culturally accepted is that pattern of meaning for participation (in theory and practice), and what are the expectations and justifications for what counts as participation? In addressing such questions, we use this section of the chapter to illustrate how a set of 'typological heuristics', in the form of key questions, might be developed to promote a critical appreciation of the strengths and weaknesses of different expressions of participation in environmental learning. We then illustrate its application, to draw attention to some of the conceptual, theoretical, and political commitments and contexts, as well as the ideological and pedagogic groundings of participatory approaches, to open up considerations for analysis of the potential for 'tyranny' and 'transformation' in participation in the area of environment-related learning (see also Cooke and Kothari 2001; Bühler 2002; Hickey and Mohan 2004).

3.6 Developing an Evaluative Framework for Complex and Contested Concepts

Our approach in working towards an evaluative framework for understanding diverse expressions of participation has been to develop a set of analytical questions. We have drawn on the methodological work of Andrew Dobson (1998, 2001) who uses this approach for analysing environmental politics in relation to concepts of distributive justice. It requires identifying and delineating a set of 'Principal Organising Questions' (POQs); these are the kinds of questions that a comprehensive delineation of a concept should be able to answer or address (see Table 3.3). In Dobson's case, the POQs are developed in relation to two key concepts: *environmental sustainability* and *social justice*. Dobson organises the possible responses to the POQs that constitute both key concepts in relation to three conceptions of environmental sustainability and dimensions of social justice, and then uses this to work towards a typology, combining and discussing the compatibility of the conceptions with different responses to the dimensions of social justice. Thus Dobson's aim is to demonstrate how grouping various responses to questions through a combination of questions and 'family-related answers' will lead to a limited number of viable and comprehensive conceptions for his typology.

In clarifying the aims of this typological process, Dobson argues that developing an analytical framework for dealing with concepts that are vague, complex, and contested is preferable to simply cataloguing them. In this case, he focuses on the diverse conceptions of sustainable development, inspecting the literature on the concept and the discursive differentiations made vis-à-vis *social justice* and *environmental sustainability*. He claims that this analytical method, which he applied originally in the context of environmental sustainability and sustainable development, is 'in practice applicable to any political-theoretical concept', (Dobson 2001:62) including, by his own extension, participation and environment-related learning, e.g. in the area of ecological citizenship.

Dobson's intention for such a framework, *sensu* Rapoport, is to enable the development of 'a typology of theories' about the concept under investigation. The

Table 3.3 Asking questions about environmental sustainability and social justice. (Adapted from Dobson 1998:39, 63.)

Questions addressed in a conceptualisation of environmental sustainability	Dimensions of a conceptualisation of social justice
What to sustain?	What is the community of justice? (dispensers, recipients)
Why?	What is the basic structure (the options)?
How?	What is distributed?
Objects of concern (primary/secondary)?	What is the principle of distribution?
Substitutability between human-made and natural capital?	

development of such a typology entails that the components of the concept under study are made explicit. Dobson claims that the key advantages of such an analytical approach are: first, typologies do not go out of date immediately; and second, typologies provide plural answers to contested concepts, which by their very nature, guarantee and promote plural understandings. Of course, typologies may serve programmatic and normative purposes, but Dobson's overarching purpose is to sidestep this as via a focus on heuristics, he is able to provide a form of orientation for finding one's way around the territory of a multiply interpretable concept, rather than propose fixed meanings, or police them. This is achieved by developing a typological heuristic that should provide comprehensive coverage of its multiple and possible meanings and interpretations. Thus, in this case, a suitably and successfully developed typology could be regarded as the basis for a mapping tool that can then be used to explicate and orientate assumptions about a concept. Thus whilst remembering that the map is not the territory, the ability to make and understand maps is more the order of the day with a POQ approach.

Thus to develop a POQ-based typological framework, Dobson (1998:37) explains that in his reading of the literature on environmental policy he was guided by the question, 'What are the implicit or explicit questions being asked in these texts?' This led to a list of questions to which any 'theory' (e.g. as one 'utterance' from an array of actual and possible utterances) about a certain concept can be subjected. Differences in interpretation and understandings of the key concepts are explained as shifting and alternative answers to the questions. This, he argues, brings more clarification to the debate over vague concepts (such as sustainable development) as controversial components can be more easily identified. The questions are then used to compose a framework for the grouping of various answers, where 'a combination of questions and "family-related answers" will lead to a limited number of conceptions' (ibid). (To see this in practice, see Dobson's work in Tables 3.3 and 3.4.)

We initially followed Dobson's methodological approach in terms of developing a set of POQs for the concept of 'participation' in a learning context, primarily in light of the literature and material represented in Tables 3.1 and 3.2, but also in relation to the literature discussed elsewhere throughout this book. However, instead of following Dobson's approach to the letter in attempting to identify a limited number of conceptions,[1] our focus has been on clarifying POQs for participation.

As a result, we have categorised a preliminary list of POQs into three broad levels of typological 'interrogation': a level that questions practicalities, a level for explicating underlying theoretical perspectives, and a level of questioning that

[1] This would have required, for example, (a) developing POQs for a second key concept – environment-related learning – which is beyond the scope of our purposes in this chapter, although see Scott and Gough 2003 and Nikel 2005 for some discussion here, as well as (b) connecting the various responses together, few of which seemed to be able to withstand critical scrutiny because they were neither extensive nor robust enough.

Table 3.4 Principal Organising Questions and 'answers' for the sustainable development community. (Based on material from Dobson 1998.)

POQs for environmental sustainability	'Answer' from the sustainable development community	POQs for social justice	'Answer' from the sustainable development community
What to sustain?	*Critical Natural Capital*	What is the community of justice?	*All human beings*
		Dispensers?	*Present and future generation of human beings*
		Recipients?	
Why?	*Human welfare*	What is the basic structure (options)?	*International and intergenerational justice, predicated on impartial, consequentialist and universal theories of justice*
How?	*Renewing/substituting/ protecting* *An instrumental attitude to the value of the non-human natural world*	What is distributed?	*Environmental goods and bads*
Objects of concern (primary/secondary)	1. *Privileges present and future generation human needs over human wants,* 2. *Present generation non-human needs, future generation non-human needs*	What is the principle of distribution?	*Needs*
Substitutability between human-made and natural capital	*Not always possible between human-made capital and critical natural capital*		

looks towards meta-theoretical horizons. We will return to this later, but for now we note that evaluating participatory education discourses and practices across all three dimensions rather than one or another also encouraged our critical consideration of (a) their grounding in and appeals to epistemological and ideological perspectives about learning (e.g. Greeno *et al.* 1996), (b) how they are exposed to 'power relations and structures' (Hickey and Mohan 2004), and (c) how they mobilise different conceptions of environment and sustainable development (Dobson 1998). This can be considered as both an extension to and rupturing of Dobson's methodology, while attempts to delineate more specific conceptions of participation in relation to environmental, health, and sustainability education must remain a goal for further work, for example, when the literature affords such a comparative approach (see footnote 1).

3.7 Developing an Heuristic for Investigating Participation in Environment-Related Learning

With Dobson's procedures in mind then, we developed a set of possible POQs for addressing the various components of participation in theorising and practising environment-related learning. We reviewed a range of literature associated with the RIPEN initiative (Research in Participatory Education Network, described in Chapter 1, this volume) regarding participation in environmental and health education, and used database, library, and citation-index searches on key terms. The outcomes of this process are presented in the left hand column in Table 3.5. The questions address the following components towards identifying a comprehensive conception of participation:

- Defining and delimiting participation
- The nature of participation and non-participation
- Participation criteria
- Selection of participants
- Nature of activity to participate in
- Participants' individual contributions
- Justification for participation
- Expressions of hegemony in participatory settings
- Decision-making in participatory settings
- Participants' views on participation

Analysing a participatory programme or activity through these POQs should aid the identification of key components of a conceptualisation of participation at a number of levels. However, in operating as a heuristic device, it is not meant to imply that the various elements in the table are discrete; indeed, the interaction and interrelationship of responses across questions can be key to understanding patterns in conceptions of participatory education theory and practice, including the initiation, conceptualisation, and outcomes of such activities, within a case study, and across case studies.

The right hand column in Table 3.5 illustrates the range of responses we found in our literature set, and owing to their variety, as yet we have not been able to identify specific, coherent, and separate conceptions (suggesting perhaps that further conceptual analysis and open debate are required in this field). More importantly, the range gives a vivid impression of the diverse settings and activities that are possible under the broad umbrella of participation and participatory learning in environment-related learning, and the components that go to

Table 3.5 POQs and participation in environment-related learning

POQs for participation	Exemplar ranges of response
Who defines what we call participation? *cf. How would you recognise a group of people participating?*	organiser, participants, sponsor, state, industry, political activist… *language, education, dominant discourse…*
What/who is implicated in participating? *How is participation happening already?* *How will/might participation happen?*	people/resources/materials… *as learners as individuals (age, gender, role…)* *as individuals as members of group institutions* *as communities (how defined/identified?)* *as present generation (PG) human beings (HBs)* *as future generation (FG) HBs* *PG sentients and/or FG sentients* *PG non-sentients and/or FG non-sentients…*
What is the degree of freedom that the participant has to participate? *Who is not participating and why not?*	compulsory or voluntary participation…, *not participating as a positive choice or option* *not participating as a negative choice or option…*
What are the criteria for being a participant?	as an individual, as a representative of a sub group, as an official representative (formal role)…
What is the basic structure of the conceptualisation of the participation?	procedural – consequentialist substantive – impartial particular – universal…
Participation in what?	in teaching/learning process (classroom activity, projects) as own learning process, as small group learning (family, peer group)…, in community development in society development in global development in political processes, in preservation, conservation, restoration…
How important is the participants' participation within the complete process?	at the level of consultation, taking part, being involved in decision-making, having a say, synergy…
How is the participation justified, if at all?	efficiency/effectiveness…
(criteria for legitimacy)	morality: *autonomy, community, solidarity, well-being/health of individual and/or society…,*

(continued)

Table 3.5 (continued)

POQs for participation	Exemplar ranges of response
	intertextuality/intersubjectivity *(influence of past forces or others on present)* … spirituality (*sensitivity, sacredness, communion, stewardship, suffering, compassion, goodness, love, hope, existence…*)
Whose (rather than what) reality counts in the process of participation? (*people, knowledges and powers*)	…dominant discourse and legitimated voices… …*role of mediators, facilitators and interpreters* …*structure of the process of decision-making* …*sources of information and access to it…*
How is the process of decision-making organised? *What construction of democracy is underlying the process?*	…priority given to whom or what? … …*an individual has priority* (e.g. *teacher*) …*majority decides* …*discussion until common agreement* …*everybody counts the same* …*a quota/critical mass* …*majority decision takes account of minority rights…*
What is the participants' view on the role of their participation within the process of environment-related learning?	degree of personal/moral/social/ecological/etc. impact, meaningfulness and relevance… e.g. *ownership, economic incentive, nature, conservation, social change, learning, compulsion….*

constitute the elements of a typology in this field. For example, environment-related learning in the past has been differentiated as either education *about* the environment, education *in* the environment, or education *for* the environment; it will be interesting to see whether learning and teaching as part of the UN Decade of ESD (2005–2014) will provide ample data and grounds for delineating the constitutive components of ESD approaches in similar ways in relation to participation.

To test out the POQs in the framework, we conclude this chapter with an illustration of their application to an 'Expertise' report ordered by the *Bund-Länder-Kommision* (BLK) *für Bildungsplanung und Forschungsförderung* (State–Federal States-Commission for educational planning and research promotion) in Germany. Our choice here depends more on our perception of its interest value and potential to illuminate the issues of conceptualising and evaluating participation in environment-related learning, than on its representativeness for the field in general (Stake 1995).

3.8 Applying the Framework to ESD Policy and Frameworks in Germany

The term, *Expertise*, refers to a report written by 'experts' (such as academics) with professional experience and expertise in an area, and usually involves advice and opinion based on evaluation and analysis of prior work and research to inform future projects and innovations in that field. In this example, the *Expertise* document is positioned as central to the upcoming preparation of a *Förderprogramm* (supporting strategy) for implementing and disseminating ESD as an 'innovative task' in schools across Germany (de Haan and Harenberg 1999). For further details, see de Haan (2006) and Bolscho and Hauenschild (2006).

The *Expertise* documentation contains evaluation, discussion, and recommendations on policy, planning, and implementation, and is addressed to people with responsibility for ESD (e.g. federal and state officials, researchers, and environmental educators), and members of the public with an interest in the field, including teachers (*fachinteressierte Öffentlichkeit*).

Significantly for this document, sustainable development is viewed as a 'concept of modernisation' (*Modernisierungskonzept)*. The terminology signals a set of technical, economical, political, and social tasks in creating/modifying/modelling/forming/designing the future of society (*Gestaltungsauftrag*) (de Haan and Harenberg 1999:62). In other words, experts argue that sustainable development is a complex task for individuals and institutions, amongst others, of (re)creating society that combines the global and local dimensions of *Zukunftsgestaltung* (the creation/modification of the future). Citizens' competence is required in communication and decision-making processes (p. 62), while more specifically, *Gestaltungskompetenz für nachhaltige Entwicklung* relates to the necessity of developing a citizen's 'modelling competence' for sustainable development. This *Gestaltungskompetenz* involves citizens working as participants who look ahead, planning the way forward for a society. It articulates the belief that people can and must create and plan for the future in an optimistic way, rather than looking backwards or viewing the future or people's capacity pessimistically, in the face of the ongoing, and in many cases, deepening socioecological crisis.

Three major 'lesson planning and organisational principles' (*Unterrichts und Organisationsprinzipien*) are presented in the *Expertise* to achieve the overall goal of *Gestaltungskompetenz*. These are:

1. Interdisciplinary knowledge (*interdisziplinäres Wissen*)
2. Participatory learning (*Partizipatives Lernen*), and
3. Innovative structures (*Innovative Strukturen*) with particular emphasis on school profile (corporate identity), learning organisations, and cooperation with the outside/wider community

Participation as a term is most explicitly mentioned in relation to the second principle, 'participatory learning'. The *Expertise* calls for further development and evaluation of various methods and forms of participatory learning, arguing, for instance, that in the best of circumstances, it is not experienced as an occasional event but as an integrated part of daily school practice. A variety of learning methods are recommended, such as interdisciplinary learning arrangements, projects with practical application, self-initiated and self-directed learning, learning in different groupings and teams, and learning situations such as Agenda 21-type 'futures workshops' and planning projects – each of which may use a variety of participatory tools, such as interactive presentations, games, simulations, and group work (pp. 64–65). 'Participatory learning' is further operationalised in relation to four aspects (pp. 77–82) which suggest a focus on modelling the 'sustainable city'; investigating rural areas and regions as a key stimulus for learning about sustainable development in the round; participating in Agenda 21-related activities; and participating in a locally based process of identifying, developing, and using sustainability indicators.

In order to test out the POQs in terms of their scope and value for analysis and evaluation, the *Expertise* has been reviewed in light of the typology's questions and components. We set out our analysis in Table 3.6, right hand column.

In accordance with the explicit purpose of an analytical framework, we asked ourselves whether and in what ways it had fulfilled our expectations, as a mapping tool for explicating and orientating assumptions about conceptions of participation in environment-related learning in this particular document.

Working through the document guided by the framework's questions suggested a number of issues that we may not have otherwise recognised. We highlight two here for the purposes of our discussion. On page 20 of the *Expertise*, one reads that citizens' commitment to sustainable development as a modernisation concept is vital. Therefore, to implement and fulfil Agenda 21 it is important to increase the 'participation of groups in society who have not participated as much so far or who have not yet been considered (e.g. children, youth, women)' [our translation and paraphrasing throughout this section]. In the first instance one might question the contents of the list in parenthesis, namely 'women', 'youth', and 'children'. Each group is involved with daily decisions about, for example, consumption, transportation, work…de facto, they are 'participating' in key aspects of sustainable development, so why draw attention to these categories? On page 80, the authors of the *Expertise* offer a resolution when they are more explicit about 'participation in what?' It becomes clear that the authors use the term 'participation' and 'to participate' only in a specific context, namely, in relation to Agenda 21 and Local Agenda 21. In more detail, it is about processes of consultation in setting up a Local Agenda project (including communication processes, decision-making processes, and evaluation) and in the support for this at the level of small communities in local communities, rather than in education and/or daily life *per se*. Arguably, children, youth, and women appear to be less represented or involved in local, collaborative planning, and in the implementation work of Agenda 21 (via Local Agenda 21). Hence they are targeted for inclusion in the programme (cf. Chapter 9 by Læssøe, on sustainable development and participation, and Chapter 20 by Carlsson and Sanders, this volume, on the notion of the 'everyday maker').

Table 3.6 ESD and participation

POQs for participation	Responses in BLK programme 'Expertise'*
Who defines what we call participation? *cf. How would you recognise a group of people participating?*	*Agenda 21* (*and expert interpretation of it*): 'Participation is a central idea of the Agenda processes. Without involvement in decision-making processes, without changing lifestyles and without interest in global justice, sustainable development is not realised.... It has a second meaning which involves the ability of having a sense of community (*Fähigkeit zur Gemeinschaftlichkeit*), a sense of helping and supporting communities on a local and global level...participation is also almost impossible without the ability to solve conflicts' (pp. 62–63)[2] 'The Principle of Participation' (*der Grundsatz der Partizipation*): '...all people "having a share"/"involved"/ "affected" (*alle Beteiligten*) have to be included according to legal possibilities and according to their ability in an equal way. (p. 93)[3] Sub-question: Who is deciding that participation is needed? ...according to the common opinion of the expert world and political agents....[4]
What/who is implicated in participating?	Politicians Pedagogical experts
How is participation happening already?	Citizens (having competences)
How will/might participation happen?	Groups in society which have not been participating as much so far or have not been considered (e.g. children, youths, women)[5]
What is the degree of freedom that the participant has to participate?	Not addressed (more: access to participation has to be ensured, no consideration about not participating) – sustainable development is seen as a society-wide task to create/to model/to modify society/ future (*Gestaltungsauftrag*)[6]
Who is not participating and why not?	– 'All people having a share have to be involved...' According to 'the principle of participation', all people having a share have to be included according to legal possibilities and according to their ability in an equal way. (p. 93)
What are the criteria for being a participant?	Being citizen; everybody; 'all people having a share' (*alle Beteiligten*)
What is the basic structure of the conceptualisation of the participation?	'Substantive' in terms of Agenda 21; 'procedural' in terms of creativity and learning activities, and development of local indicators, but also 'consequentialist' in terms of application of pre-specified or general indicators;

(continued)

Table 3.6 (continued)

POQs for participation	Responses in BLK programme 'Expertise'*
	'universalist' view of all needing to participate though developing 'particularist' competences within a general drive for *Gestaltungskompetenz* across different members of society (e.g. youth will contribute differently to adults)
Participation in what?	The term 'participation' as a term is specifically named in relation to participating in… 1. 'Processes of consultancy aiming to setting up a local agenda project'[7]…including communication processes, decision-making processes, evaluation 2. 'Support at the level of small communities in local community' (solidarity and activity)[8] (see Agenda 21) Implicit emphasis in participation in relation to:…search for innovative solutions, change in consumption, process of reflection
How important is the participants' participation within the complete process?	To model and modify (*gestalten*), to implement, to evaluate[9]
How is the participation justified, if at all?	Necessity for success as modernisation concept (*Modernisierungskonzept*) of society, 'inevitable'
(criteria for legitimacy)	(Education – not seen as additive and therefore maybe it can be left out)
Whose (rather than what) reality counts in the process of participation? (*people, knowledges and powers*)	Agenda 21
	The expert world
	Principle of Participation
	Competent citizens
How is the process of decision-making organised?	Not stated but two other processes in addition to decision-making are emphasised:
What construction of democracy is underlying the process?	• Process of communication and processes of decision-making
	• Implementation and evaluation of sustainable development
	…competences required to participate
What is the participants' view on the role of their participation within the process of environment-related learning?	Not expressed (as it is a policy document) It will be meaningful for them through 'effect on daily life' and 'relevance for future' (*alltagsorientiert, zukunftsrelevant*)

* Expert work on the BLK Programme 'Education for Sustainable Development' (de Haan and Harenberg 1999)

[2] *Partizipation ist ein zentrales Leitbild des Agenda-Prozesses. Ohne Teilhabe an Entscheidungsprozessen für eine nachhaltige Entwicklung, ohne veränderte Lebensstile und das Interesse an globaler Gerechtigkeit, so wird immer wieder betont, sei die nachhaltige Entwicklung nicht zu realisieren….*

(continued)

Table 3.6 (continued)

Der Begriff verfügt über eine zweite Bedeutung. Diese wird sichtbar, wenn man sich auf die Gerechtigkeitsthematik in der Expertise Förderprogramm Agenda 21 konzentriert: Es ist dieses die Fähigkeit zur Gemeinschaftlichkeit, zur Hilfe und Unterstützung im Nahbereich wie das Verstehen, die Verständigung mit und Unterstützung von fremden Kulturen..... Partizipation ist zudem ohne die Fähigkeit zur Konfliktlösung in einer pluralen Gesellschaft kaum möglich. (pp. 62–63).

[3] *Soll der Grundsatz der Partizipation verwirklicht werden, müssen alle Beteiligten im Einklang mit rechtlichen Möglichkeiten und entsprechend ihrer Fähigkeiten gleichberechtigt einbezogen werden: neben den interessierten Lehrern ebenso Schüler und Eltern* (p. 93) – the principle was mentioned in an explanation of the organisational structure of the overall BLK programme.

[4] *Im Zuge der Entwicklung zur Nachhaltigkeit als Modernisierungskonzept ist – nach einhelliger Meinung der Fachwelt und auch der politischen Akteure – ein intensiviertes Engagement der Bürger unverzichtbar.*

[5] *Bisher wenig beteiligter oder berücksichtigter Bevölkerungsgruppen (z.B. Kinder, Jugendliche, Frauen)* (p. 20).

[6] *Als gesellschaftlichen Gestaltungsauftrag den Bürgern erhebliche Fähigkeiten...bei der Beteiligung an Verständigungs- und Entscheidungsprozessen abverlangt* (p. 62).

[7] *Konsultationsprozesse mit dem Ziel einzuleiten sind, eine lokale Agenda zu erstellen* (p. 80).

[8] *Unterstützung auf der Ebene der kleinen Gemeinschaften, Gemeinsinn and Teilhabe* (p. 77).

[9] *Partizipationsfähigkeit: Bereitschaft, sich an Planungen, Projekten und Programmen mitgestaltend zu beteiligen* (p. 59).

The document also draws on 'the principle of participation' (*der Grundsatz der Partizipation*): that '...all people "having a share"/"involved"/"affected" (*alle Beteiligten*) have to be included according to legal possibilities and according to their ability in an equal way' (p. 92). The principle is mentioned within an explanation of how the overall BLK programme might be structured to ensure broad involvement at the school level. Perhaps not unsurprisingly for a policy document, no reference is provided as to the derivation of 'the principle'. But we have to ask, who defined it? Who designated it a principle? And, what meanings are intended?

The principle is discussed in the text in ways that suggest there are legal considerations, ability considerations, and issues of equality in deciding whom to include or exclude in the participation. Here too the choice of words raises interesting questions and tensions. The 'population' from which a selective decision has to be made is described as *alle Beteiligte*. This is a highly plastic phrase and requires qualification as the term can refer to 'persons concerned', 'persons involved', 'persons having a share in', 'persons having a part', 'persons having an interest', 'persons who contribute to', and 'persons to help in'. Consider a motor vehicle accident. The term *alle Beteiligte* refers to all those involved, including any eyewitnesses. It does not differentiate or delimit the status or role of the participants. As such, it raises questions as to the grounds on which participation is to be understood to require people in bringing their own 'interest/s' or 'concern/s' with them, rather than rely solely on those of, for example, the policymaker, educator, or convenor of the participatory event. Related questions include whether it might mean that participation requires explicitly defined 'shares' in the process, event, or outcome, and 'parts' for people to play within these; and whether participation necessarily involves activities such as helping out, or if it can be considered equally legitimate to be a bystander, onlooker, or observer as a 'participant' in, by extension, sustainable development.

In our concluding comments we return to the relationships and interconnections between the POQs, and the implications these have for using such evaluative frameworks in relation to conceptions and examples of participation.

3.9 Identifying the Levels and Depth of POQs

The list of POQs in Table 3.5 represents a collection of possible analytical questions. In our attempt to further clarify similarities and differences between the questions, and consequently to group the questions in light of their use with the BLK example, we can consider who might ask a particular question, and what distinctive assumptions underlie each one?

To reframe the POQ framework along these lines leads to a differentiation that distinguishes between questions at a more descriptive level, questions driven by certain theoretical assumptions, and finally questions that look towards a meta-theoretical perspective.

Table 3.7 groups the questions and suggests that one can analyse and evaluate a conceptualisation of participation, participatory programme, or activity, at at least

Table 3.7 Three levels of questions for analysing participation

Level 1: Practice level–- delineating the practicalities of engagement	
Related POQs	Participation in what? How is participation happening already? How important is the participants' participation within the complete process? What are the criteria for being a participant? How is the process of decision-making organised? What is the participants' view on the role of their participation within the process of environment-related learning?
Level 2: Theory Level – delineating the participation by engaging in theoretical, epistemological, and ideological theories	
Related POQs	What construction or understanding of democracy underlies the decision-making process? What is the basic structure of the conceptualisation of the participation? How will/might participation happen? How is the participation justified, if at all? (criteria for legitimacy)
Level 3: Meta-theoretical Level – delineating the involvement, outcomes, and impacts	
Related POQs	Who defines what we call participation? Who/what is implicated in participating? What is the degree of freedom the participant has to participate? Who is not participating and why not? How would you recognise a group of people participating here? Whose (rather than what) reality counts in the process of participation?

three levels. Questions at the first level attempt to address a variety of issues related to the practicalities of participation. They are akin to a description of a research study where research instruments, sampling, and data collection procedures are described comprehensively but without reference to guiding theories or research paradigms.

Questions contributing to the second level inquire into such theories and assumptions, in terms of what may have guided or influenced decisions and work at the level of practice. It may be that the conceptualisation or practical approach is informed by more than one theory or model, and these may not always be in harmony. For example, theoretical assumptions about what constitutes 'democratic' in a decision-making process, or what constitutes learning (see Table 3.1), can vary immensely, and a simple iteration of the possibilities suggests that some will fit well with the other, while others will be in tension.

Finally, questions at the third-level prompt consideration of the degree to which the example of participation illustrates an attentiveness towards acknowledging, challenging, or transcending hegemonic structures, discourses, and practices. Adding this meta-level signals that evaluation might make a direct response to the kinds of critique of participatory work from the development field in which, for example, Rahnema (1992) pointed to the risk of participation and declaimed participatory approaches acting solely as a deceptive myth or a dangerous tool for manipulation.

Returning to our earlier comments then, we can suggest that in light of this, current conceptualisations and practices of participatory environment-related learning do not necessarily address, focus on, or take into account these issues as they relate to all three levels. Thus, we would also suggest that responding to each group of questions and examining the interplay of the levels may go some way towards helping analyse and evaluate that which determines the outcomes of participatory planning, implementation, and evaluation in education, particularly as it relates to education, the environment, health, and sustainability.

The third group of questions in particular has become an increasingly important matter for shared awareness within development studies, and tackling the issues they raise is often perceived as a way forward for participation and participatory processes – out of tyranny towards transformation (Hickey and Mohan 2004:4). However, as yet, such an awareness is not as recognisable in environment-related learning discourses that invoke conceptualisations and practices of participation, and thus in the attempt to plug this gap and stimulate further research and debate, we hope the POQs outlined here, and the process by which they were developed, may be of value to all those interested in practising and developing participatory environment-related learning.

Acknowledgments We would like to express our thanks to Kelly Teamey, Bjarne Bruun Jensen, Venka Simovska, Mark Rickinson, and Bob Jickling for their helpful comments and suggestions on earlier drafts of this chapter.

References

Arnot, M. and Reay, D. (2001) *Participation and control in learning: a pedagogic democratic right?* Unpublished paper, Department of Education, University of Bath, 6 February.

Arnstein, S.R. (1979) Eight rungs on the ladder of citizen participation. *Journal of the American Institute of Planners*, 35, 4: 216–232.

Bakhtin, M. (1993). *Toward a Philosophy of the Act*. Austin, TX: University of Texas Press.

Bolscho, D. and Hauenschild, K. (2006) From environmental education to Education for Sustainable Development in Germany. *Environmental Education Research*, 12, 1: 7–18.

Brierley, G., Hillman, M., and Devonshire, L. (2002) Learning to participate: Responding to changes in Australian land and water management policy and practice. *Australian Journal of Environmental Education*, 18: 7–13.

Bühler, U. (2002) Participation 'with Justice and Dignity': beyond the 'New Tyranny'. *Conflict and Development*, 1: 1–16.

Chambers, R. (2004) *Notes for Participants in PRA-PLA Familiarisation Workshops in 2004*. Brighton: Institute of Development Studies.

Cooke, B. and Kothari, U. (eds) (2001) *Participation: The New Tyranny?* London: Zed Books.

Cornwall, A. (2002) Participation in development: tracks and traces. *Currents*, 28: 4–8.

Cornwall, A. and Brock, K. (2005) *Beyond Buzzwords: 'Poverty Reduction', 'Participation' and 'Empowerment' in Development Policy*. Overarching Concerns: Programme Paper No. 10 – November 2005, Geneva: United Nations Research Institute for Social Development.

Danby, S. and Farrell, A. (2004) Accounting for young children's competence in education research: new perspectives on research ethics. *Australian Educational Researcher*, 31, 3: 35–49.

Dobson, A. (1998) *Justice and the Environment: Conceptions of Environmental Sustainability and Theories of Distributive Justice*. Oxford: Oxford University Press.

Dobson, A. (2001) *Green Political Thought*, 2nd edn. London: Routledge.

Escobar, A. (1995) *Encountering Development: The Making and Unmaking of the Third World*. Princeton, NJ: Princeton University Press.

Esteva, G. (1985) Beware of participation. *Development: Seeds of Change*, 3: 77–79.

Farrell, A. (ed) (2005) *Ethical Research with Children*. Milton Keynes: Open University Press.

Fielding, M. (2001) Students as radical agents of change. *Journal of Educational Change*, 2, 2: 123–141.

Gramsci, A. (1971) *Selections from the Prison Notebooks*. London: Lawrence & Wishart.

Greeno, J.G., Collins, A.M., and Resnick, L.B. (1996) Cognition and learning. In: Berliner, D.C. and Calfee, R.C. (eds), *Handbook of Educational Psychology*. London: Prentice-Hall, pp. 15–46.

Haan G. de, and Harenberg, D. (1999) *Gutachten zum BLK Programm: Bildung fuer eine nachhaltige Entwicklung* [Translated: Expert work on the BLK Programme 'Education for sustainable development'] Bonn: BLK. Available online in German (http://www.blk-bonn.de/materialien.htm, select Heft 72, pdf file)

Haan, G. de (2006) The BLK '21' programme in Germany: a 'Gestaltungskompetenz'-based model for Education for Sustainable Development. *Environmental Education Research*, 12, 1: 19–32.

Hart, P. (2000) Searching for meaning in children's participation in environmental education. In: Jensen, B.B., Schnack, K., and Simovska, V. (eds), *Critical Environmental and Health Education*. Copenhagen: Danish University of Education, pp. 7–28.

Hart, R. (1992) *Children's Participation: From Tokenism to Citizenship*. Florence: International Child Development Centre/UNICEF.

Hart, R. (1997) *Children's Participation: The theory and Practice of Involving Young Citizens in Community Development and Environmental Care*. London: Earthscan.

Hickey, S. and Mohan, G. (eds) (2004) *Participation: From Tyranny to Transformation? Exploring New Approaches to Participation in Development*. London: Zed Books.

IDRC (1998) Citizen participation and environmental management. *Reports Magazine*, 23 December 1998 (Lois Sweet). (http://reseau.crdi.ca/en/ev-5093–201-1-DO_TOPIC.html)

Iozzi, L., Laveault, D., and Marcinkowski, T. (1990) *Assessment of Learning Outcomes in Environmental Education*. Paris: UNESCO.

James, A., Jenks, C., and Prout, A. (1998) *Theorising Childhood*. Cambridge: Polity.

Jensen B.B. (2000) Participation, commitment and knowledge as components of pupil's action competence. In: Jensen, B.B., Schnack, K., and Simovska V. (eds), *Critical Environmental and Health Education – Research Issues and Challenges*. Copenhagen: Danish University of Education, pp. 219–238.

Kirby, P. (2001) Participatory research in schools. *FORUM*, 43, 2: 74–77.

Laclau, E. (1990) *New Reflections on the Revolution of Our Time*. New York: Verso.

Lardner, C. (2001) *Youth Participation — A New Model*, (www.lardner.demon.co.uk)

Mannion, G. (2003) Children's participation in school grounds developments: creating a place for education that promotes children's social inclusion. *International Journal of Inclusive Education*, 7, 2: 175–192.

Matthews, H., Limb, M., and Taylor, M. (1999) Young people's participation and representation in society. *Geoforum*, 30: 135–144.

Mordock, K. and Krasny, M. (2001) Participatory action research: a theoretical and practical framework for environmental education. *Journal of Environmental Education*, 32, 3: 15–20.

NAAEE (2004) *Excellence in Environmental Education — Guidelines for Learning*. Second edition. (Pre K-12). Rock Spring, GA: North American Association for Environmental Education.

Nederveen, P.J. (2001) *Development Theory: Deconstructions/Reconstructions*. London: Sage.

Nikel, J. (2005) Ascribing Responsibility: A Three Country Study of Student Teachers' Understanding(s) of Education, Sustainable Development, and ESD. Unpublished doctoral thesis, University of Bath.

Rahnema, M. (1992) Participation. In: Sachs, W. (ed), *The Development Dictionary*. London: Zed Books, pp. 116–131.

Rapoport, A. (1985) Thinking about home environments: a conceptual framework. In: Altman, I. and Werner, C.M. (eds), *Home Environments*. New York: Plenum Press, pp. 255–286.

Roberts, M. (2001) Childcare policy. In: Foley, P., Roche, J., and Tucker, S. (eds), *Children in Society: Contemporary Theory, Policy and Practice*. Basingstoke: Palgrave, pp. 52–64.

Roth, W.-M. (2003). Autobiography as Scientific Text: A Dialectical Approach to the Role of Experience. Review Essay: Harry F. Wolcott (2002). Sneaky Kid and Its Aftermath [42 paragraphs]. *Forum Qualitative Sozialforschung/Forum: Qualitative Social Research* [On-line Journal], 5(1). (http://www.qualitative-research.net/fqs-texte/1-04/1-04review-roth-e.htm)

Scott, W.A.H. and Gough, S.R. (2003) *Sustainable Development and Learning: Framing the Issues*. London/New York: RoutledgeFalmer.

Shier, H. (2001) Pathways to participation: openings, opportunities and obligations. *Children and Society*, 15: 107–111.

Simovska, V. (2000) Exploring student participation within health education and health promoting schools. In: Jensen, B.B., Schnack, K., and Simovska, V. (eds), *Critical Environmental and Health Education*. Copenhagen: Danish University of Education, pp. 29–45.

Stake, R.E. (1995) *The Art of Case Study Research*. Thousand Oaks, CA: Sage.

Tobin, J. (1995) Post-structural research in early childhood education. In: J. Hatch (ed), *Qualitative Research in Early Childhood*. Westport, CT: Praeger, pp. 223–243.

Treseder, P. (1997) *Empowering Children and Young People, Training Manual*. London: Save the Children Fund.

UNESCO-UNEP (1976) The Belgrade Charter: a global framework for environmental education. *Connect: UNESCO-UNEP Environmental Education Newsletter*, 1, 1: 1–2.

UNESCO-UNEP (1978) *The Tbilisi Declaration: Final Report Intergovernmental Conference on Environmental Education*. Organized by UNESCO in cooperation with UNEP, Tbilisi, USSR, 14–26 October 1977, Paris, France: UNESCO ED/MD/49.

Williams, R. (1983) *Keywords: A Vocabulary of Culture and Society*. Revised and expanded edition. London: Fontana.

Chapter 4
Learning *in* and *as* Participation: A Case Study from Health-Promoting Schools

Venka Simovska

Keywords learning, student participation, health education, health-promoting schools, action competence

4.1 Introduction

Drawing on theoretical discussion and the vitality of an empirically-based case study, this chapter documents, explores, and reflects on processes of learning about health through participation and action. The study is positioned within the democratic health-promoting schools tradition which emphasises a critical approach to the issue of student participation and the importance of taking action as part of learning about health. The chapter begins with discussion of the health-promoting schools initiative in Europe as exemplified by the European Network of the Health Promoting Schools, the position of the concept of participation within the frames of the health-promoting schools approach, and its implications for the ways we look at learning. Then, a model distinguishing two different qualities of participation, (token and genuine), is considered. The model builds on two complex sets of theoretical concepts – the democratic approach to health-promoting schools on the one hand, and the sociocultural perspective on learning on the other. The model is used as the main analytical framework in the case study. The findings from the case study are discussed in several sections, shedding light on the different processes of knowing in which students were involved. This includes illuminating the forms of peer collaboration and mutual interactions as well as the activity structures and forms of participation in which students were engaged, e.g. investigations, identifying problems, solution ideas, and taking action to bring about changes with respect to two overall health topics. At the end of the chapter, a few dilemmas and challenges for future research arising from the study are outlined.

DPU, Research Programme for Environmental and Health Education, Danish School of Education, University of Aarhus, Tuborgvej 164, DK-2400 Copenhagen NV, Denmark, vs@dpu.dk

A. Reid et al. (eds.), *Participation and Learning*, 61–80.

4.2 Health-Promoting Schools, Participation and Learning

The core principles underlying the health-promoting schools initiative as discussed and adopted at the first two conferences of the European Network of Health Promoting Schools (ENHPS) include: *democratic practices, participation and partnership; equity and access; empowerment and action competence; safe and supportive school environments; curriculum* (health knowledge and understanding as well as health-promoting teaching and learning methodologies); *teacher training; evaluation for building on successes; collaboration with stakeholders, communities and parents*; and *sustainability* (WHO 1997, 2002). These principles clearly indicate a move away from the traditional, disease-focused approach to health education and health promotion, towards an empowering, social model. The health-promoting schools approach brings together the strategic guidelines outlined in the Ottawa Charter (WHO 1986) and the principles stated in more recent WHO documents, such as Health 21 – the Health for All policy for the WHO European Region, which sets out targets for the 21st century. Health 21 draws on the values of health for all, including, health as a fundamental human right, equity in health, and participation of individuals, groups, institutions, and organisations in health promotion. One of the key strategies that this policy document emphasises is a participatory health development process that involves relevant partners for health, at all levels – home, school, and workplace, local community and country – and that promotes joint decision-making, implementation, and accountability (WHO 1999).

Accordingly, health promotion in schools is construed as a social process of individual and collective empowerment. A health-promoting school is defined as an educational setting that attempts to constantly develop its capacity for healthy learning, working, and living (WHO 1993). Health is interpreted positively and holistically, encompassing the living conditions related to health as well as dimensions of physical, social, emotional, spiritual, and mental well-being. The development of an individual's skills, self-determination and agency with regard to health matters is always considered within a given context in connection to the surrounding living conditions. The whole-school environment is viewed as an important arena for promotion of health and for learning about health.

Interpreted in this way, the health-promoting schools approach inevitably brings the issue of meaningful student involvement in teaching and learning processes to the fore. Moreover, 'student participation' has become one of the trendy, captivating terms within the ENHPS, holding the central position in portraying the health-promoting schools initiative. In reality, however, the ideology underpinning the health-promoting schools initiative is to a large degree influenced by elements of professional power and the need for public accountability (Denman *et al.* 2002). The concept of health-promoting schools has been interpreted differently in different cultural, geographical, and educational contexts and thus obtained a wide range of, sometimes contradictory, meanings (Simovska 2000). A number of models of health-promoting schools have emerged over recent years reflecting different educational priorities, ideologies, needs, and systems of meaning within the national networks (Jensen and Simovska 2002). Often, in spite of or parallel to the rhetoric emphasising participatory and empowering nature of the health-promoting schools approach, the practice remains dominated by a behaviouristic paradigm focusing primarily on individual students and modification of their lifestyles.

One of the significant challenges to the behaviouristic perspective is characterised by the distinction between 'moralistic' and 'democratic' health education and health promotion conceptualised within the Danish Network of Health Promoting Schools (Jensen 1997). The democratic perspective suggests that it is important that a health-promoting school accepts the challenge to revisit its structures and environment and improve its potential to enhance students' capacities for visionary thinking and social responsibility, and their competence to tackle health-related problems. This is instead of endorsing empty participationism while aiming solely at knowledge transmission and behaviour change. Thus, the main aim of democratic health-promoting schools is construed as the development of students' action competence, that is, the ability to act and bring about positive change with regard to health. Action competence is operationalised through integration of cognitive and affective components such as knowledge, commitment, visions, and action experiences (Jensen 2000, 2004). Participation is interpreted as a transformative process focused on making a difference, as opposed to conforming to the status quo. It is viewed in connection to the characteristics of the school environment, e.g. in terms of appropriate democratic and inclusive structures, supportive relationships, positive social norms and values, opportunities for achieving success and developing skills and competences, and so on. Accordingly, one of the key tasks of a democratic health-promoting school is providing an appropriate space for students to participate actively in relevant rather than trivial aspects of decision-making processes at school. Moreover, it is considered essential that a health-promoting school should ensure resources and opportunities for students to develop, enhance, exercise, and exert their competences to act as qualified agents in democratic environments. This presupposes fostering students' self-awareness, critical thinking, decision-making, and collaboration skills, connecting students among themselves and with the school, and empowering both students and school communities to deal with health determinants and other health matters that concern them (Simovska 2000).

Thus, the democratic approach to health-promoting schools can stimulate the introduction of fundamental changes to school approaches to teaching and learning as well as school management, which move away from top-down hierarchical school structures towards more participatory and empowering systems on all levels. Consequently, as will be discussed in what follows, this perspective points to controversial processes of challenging traditional power imbalances in schools and also implies a different view of the nature of learning. Both taking into account the whole-school environment along with the classroom as an arena for learning, and highlighting the close links between the school, the family, the local community, and society at large, emphasises a view of learning as situated in a sociocultural context and located in processes of participation or co-participation rather than solely with the individual.

When we think about participation from a variety of perspectives in learning theory, the meaning of it varies substantially (see Table 3.1, and Reid and Nikel's discussion of learning theory and participation in Chapter 3). Conventional learning theories typically attempt to explain the ways individuals learn and to discuss the implications of these explanations by considering teaching strategies that would foster an isolated individual's learning. In contrast, the sociocultural theory of learning and development inspired by the ideas of Vygotsky, among others, interprets learning as a profoundly social process, linked closely to the processes of psychological development. The central educational concept in Vygotsky's theory (1978) is the concept of the zone of proximal development (ZPD), i.e. the distance between the actual and the potential developmental level. While the actual developmental level is determined by independent learning, the potential level is determined by the amount of guidance, from adults or more experienced peers, needed in problem solving. The ZPD concept points to a change of focus in learning theories, suggesting deeper consideration of the interaction between cognition, context, and practice. The change in focus also includes that the unit of analysis is not the individual but the dynamic integration of the individual and the social environment; this change radically reorients learning theory from an individualistic to a relational and sociocultural perspective. The following oft-cited words of Vygotsky (1978:57) highlight his view on the essentially social nature of psychological development, which has profoundly influenced theories of learning:

> Every function in the child's cultural development appears twice: first, on the social level, and later on the individual level; first, *between* people (*interpsychological*) and then *inside* the child (*intrapsychological*). All higher functions originate as actual relations between human individuals.

The developmental processes become part of the individual's independent development through the processes of internalisation, or as suggested in latter interpretations of Vygotsky's theory (e.g. Rogoff 1995; Rogoff *et al.* 2001), through the processes of appropriation. Appropriation refers to a personally active – and at the same time – multidirectional process; it indicates that new knowledge and competence are actively transformed rather than simply interiorised by the learner. The process of guided participation provides link between previous experience and competences and the skills and information needed to solve new problems (Rogoff 1993; Rogoff *et al.* 2003). Intersubjectivity and participation-in-meaning are therefore considered to be core elements of participatory learning. These two concepts serve to emphasise that creation of meaning and understanding is relational, that is, it happens between people (see, for example, Chapter 8 by Vare, this volume). Both the concepts of intersubjectivity and participation-in-meaning refer to a process in which participants reach an agreement and common, dialogical understanding of actions with which they are faced. In this perspective, knowledge is interpreted as a social process of knowledge construction rather than an object for students to internalise. Meaning and knowing are negotiated and dynamically created and re-created through participation in socially organised activities. Accordingly, both authentic student participation in teaching and learning processes and social guidance that builds on students' perspectives, are considered essential dimensions of personally meaningful learning.

Thus, in the context of health-promoting schools, one can argue that participation in dialogue, changes of perspective, reflecting on, and co-constructing shared meanings about health problems, their determinants and strategies for solutions, are equally important in the development of action competence as undertaking specific actions.

4.3 Token and Genuine Student Participation

Inspired, on the one hand, by Hart's categorisation of participation into different levels illustrated by the metaphor of the ladder (Hart 1992, 1997), and on the other, by the sociocultural perspective on learning as an underlying theoretical framework, two distinctive qualities of student participation are identified in this chapter by drawing on the experience from the Macedonian Network of Health Promoting Schools and its collaboration with other networks of the ENHPS, namely: *token* and *genuine* student participation. Unlike Hart's ladder which sets up more procedural democratic criteria for involving children and distinguishing between different degrees of participation, this distinction focuses on the quality of participation apart from its presumed position on the ladder (the participation part). It deals with values which are often implicitly embedded in socially organised participatory activities involving students at school but repeatedly neglected when researching the processes of teaching and learning. The underpinning values or principles that this distinction endorses as essential to participatory health education and health promotion in schools include self-determination, democracy, and diversity (Simovska 2000, 2004). As presented in Figure 4.1, three main points serve to differentiate between token and genuine student participation: *focus, outcomes*, and *target of change*.

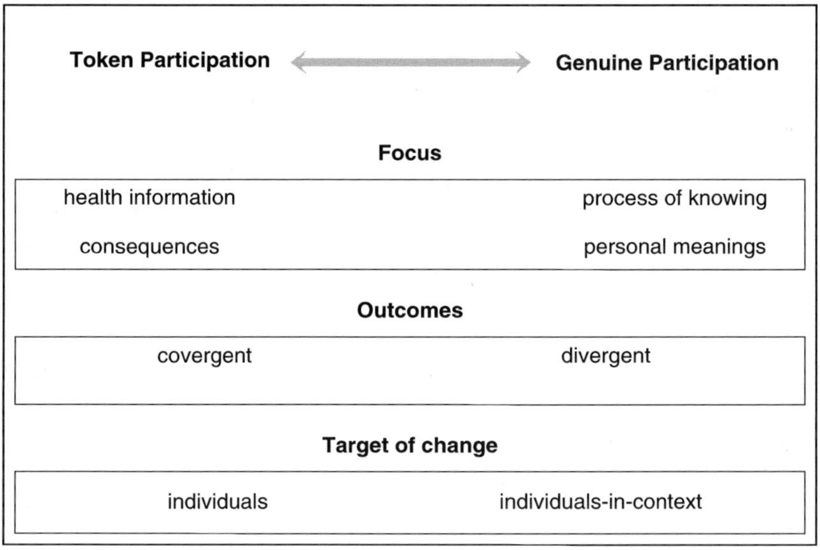

Figure 4.1 Three points of differentiation between token and genuine student participation

The first point of differentiation is the focus of the health-promoting and learning activities in which students participate. Two metaphors can be used to illuminate differences in the understanding of the process of learning in terms of the focus of participation: (a) individual acquisition of knowledge, and (b) participatory knowledge building. Token participation would have its focus on acquisition of curriculum content that has to be learned, accepted, and utilised. In the context of the health-promoting schools, such content involves the traditional factual knowledge relating to health and the hazardous effects of different behaviour styles. Students do not have much influence on the knowledge with which they are supposed to work. However, they participate in an interactive methodology that helps them acquire that knowledge.

Genuine participation, on the other hand, focuses on knowledge building through reflection on meanings and on different ways of constructing knowledge within the health domain. Factual information is addressed too, but it is the processes that lead to legitimation of information and its integration in a system of economic, historical, and ideological aspects that are considered essential. Students are involved in processes of knowing which are social and relational in their essence. These processes take place in communities of learners, consisting of both asymmetric relationships of students with teachers as more experienced dialogue partners and symmetric relationships with more or less equally skilled peers.

In contrast to the views of participation as merely a motivational tool, the experience from the health-promoting schools that rely on genuine student participation shows that it is possible – and in the long run more conducive to health – to build on the view of learning as a process primarily seeking and constructing meaning, as seeing something from different perspectives (Marton and Booth 1997) and changing as individuals, while initiating changes in the surrounding environment. The development of competence to act intentionally requires not only knowledge but also the ability to regulate one's own cognition and action in a way that identifies, makes use of, and improves the potentials and possibilities of the environment. Thus, the challenge is to look at learning as a 'way of being in the social world, not coming to know about it' (Hanks, in Lave and Wenger 1991:24).

The second point of differentiation between token and genuine student participation is in the expected outcomes of the health-promoting school activities in which students are engaged. The outcomes of token participation could be defined as acceptance of pre-existing healthy lifestyles that correlate with facts describing what is healthy and what is not. The learning outcomes are closed or convergent: rules and facts regarding health are fixed, prescribed by experts on the basis of scientific evidence, without much room for personal choice and determination. Student participation within these frames means active practice in making 'healthy' decisions and developing assertive and other personal and social skills, in order to avoid health 'risks' and possible negative pressure by classmates, peers, or the media.

In terms of genuine participation, again the aims would be to encourage students' autonomy, their critical consciousness with regard to health matters, and their potential to deal with the complexities of their own lives and the world in active, creative, and socially responsible ways. Consequently, the expected outcomes would be open and divergent, depending on the ideas and interests of individuals or groups of students, as well as on the constellation of power relations, needs, and possibilities existing in a particular school environment at a given moment. In other words, the expected outcomes of genuine participation would be the students' lived identities as active agents in health domains, based on negotiated, social, and imaginative learning experiences. The motivation and competence to engage in further learning also represents an important dimension of the expected outcomes.

The third point of differentiation between the two forms of participation is the target of change of the participatory activities. Token participation tends to target individuals with a view to changing their lifestyles, while within genuine participation the target would be individuals-in-context. In the latter, individual behaviour is closely intertwined with interpersonal involvements and organisational structures. In the words of Rogoff (1990:193):

> To act and communicate, individuals are constantly involved in exchanges that blend 'internal' and 'external' exchanges characterized by the sharing of meaning by individuals. The boundaries between people who are in communication are already permeated; it is impossible to say 'whose' an object of joint focus is, or 'whose' a collaborative idea is. An individual participating in shared problem solving or in communication is already involved in a process beyond the individual level.

As discussed earlier, the point of departure is that students' competences are not only their own property. The development of skills and competencies includes processes that occur at three levels – personal, interpersonal, and cultural. Students are as competent as their context (schools for instance) affords them the opportunity to be (Pianta 1999) and, at the same time, they are able to influence these circumstances and to initiate positive change. Therefore, it could be argued that if students have opportunities to participate actively in improving their surroundings as part of their education and thus be agents of their own learning, they are enabled to assume responsibilities for their own lives, to deal with change, and also to participate competently in the social web.

Arguably, health-promoting schools that are based on genuine participation hold the potential to achieve a better balance between the long-debated individualistic and structural (social) approaches to health promotion in schools (Simovska 2000). Health and health promotion are seen holistically without neglecting either the environment and health conditions or the individual and the importance of personal meanings. In the spirit of Vygotsky (Holzman 1997), a student participating genuinely in school health-promoting processes is looked upon not as an individual but rather as a 'person-and-environment', where the school and the environment are not abstractions but real entities with real people. Consequently, indicators for successful learning about health would not be only what a student knows, but rather what she or he wants to and can do alone or in collaboration with others.

Inherent to the conceptualisation of teaching and learning through genuine participation are issues of power and ownership. Genuine student participation allows for more room for student ownership of the learning process. Ownership presupposes that the potential for effective individual and group action is embedded in the knowledge that is acquired. In contrast to traditional school knowledge, 'owned knowledge' positions its possessor as an acting subject, able to employ his or her knowledge in a dynamic way (Paechter 2001) by visualising different alternatives and dealing with complexities of change.

4.4 Case Study: 'Young Minds' Learning Through Participation and Action

The case study draws on the educational development project 'Young Minds – exploring links between youth, culture and health'. Young Minds is an international web-based project in which students from a number of schools in different European countries collaborate on issues related to health. The project as a whole has been organised in different rounds or phases, with students from different countries and schools taking part in each phase. Even though each project phase has a different content focus, they all follow the same overall educational design (for more about the project, see Simovska and Jensen 2003; Simovska 2005; Jensen *et al.* 2005).

The overall stated purpose of the project is to generate new, action–research-based knowledge on effective methods for engaging primary and early secondary school students in learning about health in an action- and collaboration-focused way. Democratic teaching and learning processes allowing for an adequate and flexible level of student participation shape the educational framework of the project. Further, the educational framework is characterised by action-focused teaching and use of information and communication technology (ICT) as an interactive platform for cross-cultural communication and collaboration. The web site, www.young-minds.net, created and administrated jointly by the students in all the participating classes, provided the main mediational tool defining the project's shared context.

An additional important feature of the project is the collective 'real life' action outside of the school frames, at international conferences with a high political and professional profile. This action was planned as part of the project from the outset; it was construed as a special kind of student action contributing to the project's main aims. In accordance with the conceptualisation of action suggested by Jensen and Schnack (1994), the action at the conferences as well as the actions taken as part of the classroom work were characterised by: (a) intentional mutual efforts of the participants, and (b) directedness towards initiating positive changes or making a difference with regard to the health problems in question.

The present case study is limited to: (a) the first project phase as a whole (YM1), and (b) the project work of a few selected classes from the second project phase (YM2). Table 4.1 summarises the main aspects of the two project phases constituting the case, that is, the duration, the overall topic, the participants, and the related conference.

Data were generated through document and web content analysis, observation and interviews with the participating teachers and students. The material used in this chapter forms part of a larger body of data collected for a doctoral research

Table 4.1 The boundaries of the case: duration, focus, participants, and related conference

	Duration	Overall topic	Participants (students, teachers, facilitators)	Related conference ('real life' action)
Young Minds 1	June 2000–January 2001	Youth, culture, and alcohol consumption	Approximately 100 students in four classes from schools in Denmark, the Czech Republic, Macedonia, and Sweden; their respective teachers and a facilitator in each of the countries	WHO ministerial conference *Young People and Alcohol* 19–21 February 2001, Stockholm, Sweden
Young Minds 2	February–September 2002	Well-being and the school environment	Approximately 100 students in four classes from schools in Iceland, Macedonia, Portugal, and Slovenia; their respective teachers and one facilitator for the whole group	ENHPS conference *Education and Health in Partnership*, 25–27 September 2003, Egmond, the Netherlands

project (Simovska 2005). In what follows I discuss the findings from the web content analysis concerning the participation structures that the students were engaged in over the course of the project, and the focus of their delineations of the contents, particularly in terms of health problems that student identified in their work, ideas for solutions of these problems, and actions taken with an aim to initiate positive change.

Peer collaboration and forms of interaction

The analysis of the web site content showed that in their work with the health topics, students were engaged in diverse types of inquiry activities, gathering information about the health issues at hand from a number of sources, including: surveys and questionnaires; experts', teachers', and other adults' opinions; literature and the Internet; and peer-generated information. The open-ended inquiry activities in which students were engaged assumed different participant structures and a plethora of forms of peer collaboration that were non-hierarchical, that is, mutual. As shown in Table 4.2, the range of classroom as well as cross-class activities was broad, resulting in diverse structures of interaction. The interaction structures were common for the two Young Minds phases and involved the full range of possibilities: (a) small group work, (b) working in pairs, (c) whole-class discussions, and (d) individual work. The analysis showed that in both the project phases most of the

Table 4.2 Enquiry methods, participation structures, and forms of collaboration

Inquiry methods	Participant structures	Forms of peer collaboration
Cross-cultural surveys and questionnaires	Small groups, whole class	Negotiating and formulating areas of enquiry and questions to be used; administrating questionnaires; summing up findings; negotiating modes of graphical presentations of the findings on the web site; sharing reflections and comments on the findings; formulating conclusions and recommendations
School-based surveys and questionnaires	Small groups, whole class	
Surveys and questionnaires in the local community	Small groups, pairs, whole class	
Interviews with peers and teachers at school	Individual, small groups	Negotiating content and focus; formulating questions; conducting interviews; transcribing; formulating comments and reflections
Interviews with key people (politicians, policymakers, health professionals) in the local community	Individual, pairs	Negotiating content and focus, developing strategies to approach the informants; getting help from teachers, parents, and other adults; conducting the interviews, presenting and commenting joint comments
Photo narratives	Individual, small groups	Selecting places and objects, taking photos, selecting and putting photos on the web site, formulating the narrative
Mapping out the school and local environment	Whole class	Brainstorming ideas, suggestions, division of the work in small teams, negotiating teams and subtopics; selecting methods
Essays	Individual	Getting feedback from others
Web and literature search and review	Individual, whole class	Debate in the class, feedback, negotiating how to present the contents on the web site
Creative workshops involving drawing, modelling	Whole class	Modelling, drawing together, providing feedback mutually
Brainstorms and focused class debates	Whole class, individual	Mutual feedback, support, and criticism, complementing and confronting each other's ideas
Cross-cultural debates in the Forum and over email	Individual, pairs, groups	Exchanging ideas, comments, providing feedback

investigative work was done in small groups and pairs, and substantial time was devoted to whole-class discussions.

As shown in the table, there were two major forms of peer collaboration in these activity structures, focused on: (a) creating common frames of reference and shared focus, and (b) acting together to generate data for investigations or to prepare content for the web site.

This in fact meant that a variety of learning situations were created for the students to be engaged in joint productive activities[1]: *joint* for the reason that almost all the tasks that students had – in conducting inquiries and presenting them in a way that they could be communicated with the other students in the project – required goal-oriented student collaboration; *productive* because the investigative activities were aimed at producing specific joint products, i.e. material representations of their work with the project topic to be presented on the web site and discussed across classes.

Obviously, the individual inquiries were also embedded within these joint productive activities. Moreover, the teacher guidance and assistance were invaluable if the mutual interactions were to create shared discourse among the students, conducive to intersubjectivity. The common goals that students had in these dynamic forms of interaction helped create learning situations in which all of the participating students were in a position to both receive and provide assistance to others in certain aspects of the task, depending on their interests, skills, and preferences.

Furthermore, through cross-class collaboration on the Internet, the communication and mutual feedback students provided for one another cross-culturally widened the amount of interaction and assisted available performance in the classroom exponentially. Peers, both within the class and across classes, were seen as significant resources for learning, in addition to teachers. The inquiries made over the course of the project in all the classes, for example, the cross-cultural surveys and questionnaires, required mutual coordination and joint work. As students communicated about their activities and re-presented their findings and reflections for the benefit of their peers in the other classes, there were more opportunities for self-discovery of the tasks in which they were involved, and appropriation of their actions.

Health problems, solution ideas, and actions to bring about change

Given the fact that the action orientation was an integral part of the educational approach employed in the project, in the course of ongoing project activities the students identified a number of determinants of health problems and suggested various ideas for their solution. Table 4.3 summarises the causes of health problems

[1] The use of the term 'joint productive activity' is inspired by Dalton and Tharp (2002:183).

Table 4.3 Problems and possible solutions described by students

	Health problems and their causes	Ideas of solutions
Young Minds 1	Social pressure, norms of youth culture, traditions related to drinking;	Raising awareness among young people;
Alcohol consumption and young people	Family relationships;	Creating new, innovative structures for having fun without alcohol;
	Social conditions (opportunities, the 'feeling of society' legislative, traditions, advertising, access);	Improving family relationships; better understanding between parents and students; greater dialogue;
	Individual conditions (self-confidence, self-esteem, identity, coping skills, need to be independent, different, to belong, to connect, to prove oneself)	Involving governments, local authorities and school management to listen more to students' voices, involving students in decision-making process;
		Improving the psychosocial and physical school environment to foster personal development;
		Improving individual awareness, behaviour and lifestyle
Young Minds 2	Relationships at the school (both between students and teachers and among students);	More dialogue to foster trust, connectedness, belonging;
Well-being and the school environment	School architecture (uncomfortable classrooms, lack of space, lack of places for socialising and creative activities, inappropriate temperature, etc.);	Improving teachers' listening and communication skills;
	Stress related to schoolwork (assessment, exams, relationships);	Reducing learning-related tensions in classroom by using interactive teaching, teamwork, project work, open discussions;
	Inappropriate decision-making mechanisms at school (lack of student participation, inappropriate punishment strategies)	Improving the school building to address identified problems (involving other institutions such as city authorities, department of education, etc.;
		Reducing examination stress, improving relationships between students and teachers;
		Enhancing student participation in decision-making processes at school and in general in everyday school life;
		Organising more social events in the school;
		Motivating innovation

linked to the two overall project topics and ideas for solving these problems, which the students in YM1 and YM2 articulated and discussed on the web site.

As shown in Table 4.3, in relation to the issue of alcohol consumption and young people, the students participating in YM1 discussed related living conditions as well as lifestyles and personal determinants. The students reflected on, negotiated, and mapped out the variety and complexity of alcohol-related problems as well as their root causes. The scope of issues addressed by the students in their discussions included: (a) personal factors such as self-esteem, self-confidence, and identity issues, (b) concerns linked with the 'youth culture' – a sense of belonging, connectedness, and peer pressure, and (c) family circumstances, structural determinants, and the overall quality of life in society.

The students working with the issue of well-being and the school environment (YM2) almost exclusively discussed the social determinants of health-related problems. They pointed to four – in their view – very important categories of root causes relating to well-being: (a) the social relationships in school, (b) the physical (built) school environment, (c) the 'culture' of the schoolwork (i.e. examinations, assessment, etc.), and (d) democracy (i.e. student involvement) in everyday school life.

Furthermore, as shown in Table 4.3, in both the phases of the project the students did not simply identify problems or discuss their determinants in the manner of a scholastic, intellectual exercise. On the contrary, the participation and action orientation of the project framework, emphasising the four-dimensional knowledge model suggested within the democratic health education paradigm (Jensen 2000, 2004), provided stimulating space for the students to envisage different alternatives, including solution-focused ideas and areas of the students' potential impact. Consequently, the ideas that students developed about alternatives addressed root causes rather than merely symptoms of the problems: the solutions were seen in relation to determinants of the problems and both direct and indirect improvements were suggested in these areas. In both the Young Minds phases, improvements considered beneficial embraced psychosocial as well as physical living conditions and emphasised the value of active participation of young people in decision-making in this regard. In relation to the issue of alcohol, the students suggested a few additional strategies to approach some of the alcohol-related problems, including awareness raising and individual empowerment.

Table 4.4 summarises the variety of actions documented on the web site, which the students in the different classes planned jointly and carried out 'locally' in their schools or local communities as a part of the project. The table also demonstrates the specific changes they expected as outcomes of these actions.

Examples of actions, as shown in Table 4.4, include a change in school policy (a ban on school-based alcohol advertising), establishing new spaces in the school for students to socialise, and new, more inclusive mechanisms for decision-making at the school, amongst other things.

Evidently, in their work with health-related problems, students in both phases of the project were encouraged to consider the links between lifestyles, living conditions, culture, and context. More importantly, they were guided by their teachers to consider these in meaningful and purposeful ways, by reflecting on their own possible roles

Table 4.4 Examples of actions and expected changes

	Actions	Expected changes
Young Minds 1	Organising alcohol-free party at school	Raised awareness of young people; school-based parties as a good example
	Debate on alcohol between parents and students organised at school	Improved mutual understanding between parents and students; young people's voices heard; new structures at school for sharing ideas between students, teachers, and parents
	Suggesting (to the school management) changes to school policy on advertisements in school	A ban on using free drinks in school-based advertising of junior parties; alcohol-ad-free school
	Conducting interviews with the local mayor and the minister of health suggesting ideas for new policies and laws	Improved dialogue between the school and local community, voices of the young people heard by local authorities
Young Minds 2	Student led 'communication workshops' for all students in the school, teachers participate too	Improved communication among students and between students and teachers; better conflict management; improved social climate at school
	Suggesting (to the school management) specific changes in the school architecture, collaboration with an architect	Improved school building; more places for socialising; more flexible and student-friendly school environment; improved general feeling of the school, the school physical environment
	Suggesting (to the school management) improvements in the decision-making mechanisms in the school, e.g. establishment of a student council	Improved student participation in everyday school life; improved school ethos and democracy
	Presentation of Young Minds in the school	Raised awareness in the school community about the benefits of student active involvement and international collaboration; students and teachers encouraged to take similar initiatives; dissemination of the project principles and outcomes
	Presentation of Young Minds in the local media	Raised awareness; dissemination and lobbying; inspiration for other teachers and schools

and areas of influence, and by taking concrete action to bring about health-promoting changes. The forms of representation of the content used over the course of the project served not only to articulate information, ideas and concepts in the health domain with an aim to publish them on the web site, but also to help students learn how to use these representational systems in meaning-making, communicating, sharing, and discussing their understandings of the issues and arguments in the health domain.

Summing up

The analysis of the case study illuminates the trajectories of participation in which students learned about health in intentional, relational, and purposeful ways. These participation trajectories are viewed as situated in activity structures consisting of a variety of mutual interactions and different forms of participation, emphasising:

- Dialogue, i.e. suggesting, exploring alternative ideas, explanations, and problem solutions
- Action, i.e. envisioning and producing the most satisfactory outcomes possible at the given time in a given community of learners

In other words, students were engaged in a variety of processes of knowing, that included exploring, and envisaging solutions to the problems, and acting to bring about positive changes with regard to health. The classroom discourse and cross-class communication consisted of what Lave and Wenger (1991) call 'situated negotiation and re-negotiation' of ideas, concepts, meanings, and solution strategies as a basis for creating a common focus and frame of reference.

The students were involved in decision-making processes relating to both the process and the content of learning about health. The main areas of decision-making in which the students took active part included:

- Selecting relevant aspects/issues relating to the overall topic
- Deciding on the variety of questions to be researched
- Negotiating the methods of investigation and different modes of representing the findings
- Creating peer teams in which to work and organising the work within the teams
- Deciding about the representation of the content on the web site
- Selecting and planning specific actions to be taken to bring about positive change

Table 4.5 gives a summary using the participation model discussed earlier. It shows how the aims and expected outcomes of the student participation in teaching and learning activities over the course of the project were open and divergent, and that they depend on the choices that students made, together with their teachers, during the teaching and learning process.

The case study also shows that the participatory and action-oriented teaching approach, as employed in the project, can extend beyond the traditional focus on the subject matter prescribed by the curriculum. There was no pre-formulated, fixed content, or body of knowledge in the health domain that the students had to learn, memorise, recall, and employ. Even though the overall project topics were decided outside the project's frames and were assigned to students, the students investigated the area in their own ways, guided by their teachers and using the broad possibilities of ICT and cross-cultural collaboration.

The analysis also reflects the fact that the focus of the participation was on processes of critical reflection, goal-oriented dialogue, and negotiation of meanings related to health matters, rather than on moulding students' health-related behaviour and lifestyles.

Table 4.5 Characteristics of student participation in Young Minds

Student participation was *focused* on	Investigation in the broad area of the project's overall health topics, creating shared frames of reference, developing common understandings, exploring alternative ideas, explanations and problem solutions, and creating visions across classes (i.e. cultures) to construct problem solutions
The expected *outcomes* concerned	Planning and taking action together with others, bringing about changes as a part of learning, students' enhanced awareness about local and global aspects of health problems, critical thinking, creative articulation of ideas, responsible collaboration, sense of the other
Students' actions *targeted*	Everyday school life, policies and decision-making mechanisms at a whole-school level, policies in the local community, links between school and the local community, awareness of teachers, parents and policymakers about young people's voices concerning project's topics

Furthermore, the fact that the students shared the responsibility for selecting those aspects of the topics to be investigated and the methods they would use to do so resulted in an increased sense of ownership of their learning activities. This led further to increased student intent and responsibility and to the development of new strategies for mutual collaboration, which contributed to successful completion of the learning tasks at hand, and, arguably, to building knowledge as well as competence to take action.

The collaborative knowledge-building activities in which students were engaged in Young Minds were action focused. This involved working with a more comprehensive and complex landscape of knowledge encompassing insights into causes rather than only consequences of health problems, as well as visions about the future, and knowledge about solution strategies (Jensen 2000, 2004). This knowledge was of interdisciplinary character and built in a shared process of critical dialogue, reflection, development of shared visions, and taking joint actions. Consequently, the health issues that students explored, articulated, and represented on the web site evidently belonged to the democratic rather than to the moralistic health education/promotion discourse. Evidently, the students worked with an open health concept, addressing the social determinants of health and suggesting structural as well as individual solutions for selected health problems.

All these point to a genuine participation discourse whereby the participatory teaching and learning (as opposed to a transmission teaching model) was clearly directed towards facilitating, encouraging, and extending the educational dialogue about health issues that were of relevance to the community of learners. Learning was situated in students' everyday lives and experiences. Moreover, it made use of a variety of cultural resources, local community knowledge, and more global cross-cultural norms, differences, similarities, and traditions.

The student involvement aimed primarily at their socialisation to the democratic processes of making decisions, creating meanings and visions together with others, and acting to reach shared goals, but also at knowledge building as well as development of social, emotional, and personal competences with respect to health. In this sense, the content of the curriculum served the role of a 'mediating resource' (Wells 2002; Wells and Claxton 2002) for shaping the processes of learning by participating in 'communities of practice' (Lave and Wenger 1991), rather than being an end in itself. As a result, opportunities were provided for creating dynamic and overlapping collective 'Zones of Proximal Development', where students moved in and out as they appropriated – that is, transformed as well as internalised – health-related cultural resources – knowledge, competences, skills, and strategies for change – and practices.

Within these learning zones the students' individual choices were interdependent: they constituted one another and also depended on the possibilities that existed at the level of the group or the community of learners. The community of learners was heterogeneous with regard to experience, competence, skills, and knowledge, which created a specific dynamic structure of the learning zones consisting of more as well as less experienced participants, complementing one another's learning. In other words, within these collective zones of proximal development, meaning and knowledge were co-constructed within a cooperative environment that included various forms of social interaction and interpersonal (both asymmetric and symmetric) relationships. The processes of collaborative production allowed for the processes of collective learning to take place, leading gradually to the establishment of common frames of reference and a common foundation for knowledge building. One of the crucial aspects in this regard was externalisation or objectification of jointly created ideas and meanings about health into products or 'works' (Bruner 1996). Representation of one's thoughts, understandings and still-to-be-formulated ideas as part of teaching and learning process, as well as their communication with others through discussion, sharing, and receiving reactions from others in a critical but collaborative spirit, fosters learning at both individual and group level.

In these ways, it is argued, teaching and learning about health by participating in democratic learning communities can serve as a primary means of initiating students into an appropriation of the values, beliefs, ways of knowing, and rituals of the health education/promotion discipline, which, ultimately, can also be conducive to the development of their action competence or their potential to participate in creative, critical, and responsible ways in health matters that concern them.

4.5 Future Challenges

Although 'Young Minds' could be seen as an exemplary case of involving young people in learning about health within the health-promoting schools initiative, there are a number of challenges to be addressed if the principles of genuine participation and the action-focused teaching and learning strategies are to become embedded in the everyday praxis of the health-promoting schools, rather than representing an isolated example of 'good practice'. Given the various theoretical considerations, insights and reflections, empirical findings, and strengths and limitations of the study, the dilemmas and challenges for further research arising from the study include:

- What parallels can be drawn between the project-based teaching and learning about health as documented in the present case study and regular health education and health promotion in schools? What, for example, are the possibilities for and barriers to creating diverse classroom structures and mutual interactions that encourage intersubjectivity and participation in meaning? In other words, how can transferring the project-based principles of classroom organisation and cross-cultural collaboration into regular health education and the health-promoting schools curriculum be supported, which would allow for genuine student participation and which would seriously take into account students' as well as their community's concepts, ideas, concerns, and everyday experience in relation to health?
- What is the adequate and efficient balance between different participation structures in classroom teaching and learning processes (in terms of teamwork, individual work, work in pairs, whole class discussions, direct instruction, etc.) if the aim is to utilise the benefits of peer collaboration and design teaching and learning situations that are in advance of students' current developmental level? In this respect, what is the role and impact of voluntary non-participation on students' learning and competence development?
- In the context of school health education, what is the realistic and beneficial interplay between dialogue and taking 'real life' action to initiate positive change with regard to health, given the typical curriculum workload, the number and diversity of students in a class, and the existing tensions between standardised learning outcome requirements on the one hand, and participatory teaching approaches on the other?
- What constitute adequate teacher competences for guiding teaching and learning processes and fostering learning and the development of action competence within democratic communities of learners composed of dynamic zone(s) of proximal development? What forms and strategies of professional development and teacher support can more efficiently help teachers shape their professional identities as facilitators, consultants, and moderators of the processes of knowing, that is, as knowledge makers rather than transmitters? With this regard, what is the role and value of supporting mutual collaboration, relationships, and social networks among teachers, on different levels – school, national, international?

References

Bruner, J. (1996) *The Culture of Education*. Cambridge, MA: Harvard University Press.

Dalton, S.S. and Tharp, R.G. (2002) Standards for pedagogy: research, theory and practice. In: Wells, G. and Claxton, G. (eds), *Learning for Life in the 21st Century*. Oxford: Blackwell, pp. 181–195.

Denman, S., Moon, A., Parsons, C., and Stears, D. (2002) *The Health Promoting School: Policy, Research and Practice*. London and New York: RoutledgeFalmer.

Hart, R. (1992) *Children's Participation: From Tokenism to Citizenship*. Florence: UNICEF International Child Development Centre.

Hart, R. (1997) *Children's Participation: The Theory and Practice of Involving Young Citizens in Community Development and Environmental Care*. London: Earthscan.

Holzman, L. (1997) *Schools for Growth: Radical Alternatives to Current Educational Model*. Mahwah, NJ: Lawrence Erlbaum.

Jensen, B.B. and Schnack, K. (eds) (1994) *Action and Action Competence as Key Concepts in Critical Pedagogy*. Didaktiske studier, Studies in Educational Theory and Curriculum, vol.12. Copenhagen: Royal Danish School of Educational Studies.

Jensen, B.B. (1997) A case of two paradigms within health education. *Health Education Research*, 12, 4: 419–428.

Jensen, B.B. (2000) Participation, commitment and knowledge as components of pupil's action competence. In: Jensen, B.B., Schnack, K., and Simovska, V. (eds), *Critical Environmental and Health Education: Research Issues and Challenges*. Copenhagen: Research Centre for Environmental and Health Education, the Danish University of Education, pp. 219–239.

Jensen, B.B. and Simovska, V. (eds) (2002) *Models of Health Promoting Schools in Europe*. Copenhagen: WHO Regional Office for Europe.

Jensen, B.B. (2004) Environmental and health education viewed from an action-oriented perspective: a case from Denmark. *Journal of Curriculum Studies*, 36, 4: 405–425.

Jensen, B.B., Simovska, V., Larsen, N., and Glud-Holm, L. (2005) *Young People Want to Be Part of the Answer: Young Minds As an Educational Approach to Involve Schools and Students in National Environment and Health Action Plans*. Copenhagen: ENHPS International Planning Committee, WHO Regional Office for Europe.

Lave, J. and Wenger, E. (1991) *Situated Learning: Legitimate Peripheral Participation*. Cambridge: Cambridge University Press.

Marton, F. and Booth, S. (1997) *Learning and Awareness*. Mahwah, NJ: Lawrence Erlbaum.

Paechter, C. (2001) Schooling and the ownership of knowledge. In: Paechter, C., Preedy, M., Scott, D., and Soler, J. (eds), *Knowledge, Power and Learning*. London: Paul Chapman, pp. 167–181.

Pianta, R. (1999) *Enhancing Relationships Between Children and Teachers*. Washington, DC: American Psychological Association.

Rogoff, B. (1990) *Apprenticeship in Thinking: Cognitive Development in Social Context*. Oxford: Oxford University Press.

Rogoff, B. (1993) Children's guided participation and participatory appropriation in sociocultural activity. In: Wozniak, R.H. and Fisher, K.W. (eds), *Development in Context: Acting and Thinking in Specific Environments*. Mahwah, NJ: Lawrence Erlbaum, pp. 121–155.

Rogoff, B. (1995) Observing sociocultural activity on three planes: participatory appropriation, guided participation, and apprenticeship. In: Wertsch, J., Del Rio, P., and Alvarez, A. (eds), *Sociocultural Studies of Mind*. Cambridge: Cambridge University Press, pp. 139–165.

Rogoff, B., Turkanis, C.G., and Bartlett, L. (eds) (2001) *Learning Together: Children and Adults in a School Community*. Oxford: Oxford University Press.

Rogoff, B., Paradise, R., Arauz, R.M., Correa-Chavez, M., and Angelillo, C. (2003) Firsthand learning through intent participation. *Annual Review of Psychology*, 54: 175–203.

Simovska, V. (2000) Exploring student participation within health education and health promoting school. In: Jensen, B.B., Schnack, K., and Simovska, V. (eds), *Critical Environmental and Health Education: Research Issues and Challenges*. Copenhagen: Research Centre for Environmental and Health Education, the Danish University of Education, pp. 29–45.

Simovska, V. and Jensen, B.B. (2003) *Young-minds.net/lessons learnt: Student Participation, Action and Cross-cultural Collaboration in a Virtual Classroom*. Copenhagen: Danish University of Education Press.

Simovska, V. (2004) Student participation: a democratic education perspective – experience from the health-promoting schools in Macedonia. *Health Education Research*, 19, 2: 198–207.

Simovska, V. (2005) Learning by InterAction: Learning About Health Through Participation and Action – the Health Promoting Schools Perspective. Doctoral dissertation. Copenhagen: The Danish University of Education.

Vygotsky, L.S. (1978) *Mind in Society*. Cambridge, MA: Harvard University Press.

Wells, G. (2002) Learning and teaching for understanding: the key role of collaborative knowledge building. In: Brophy, J. (ed), *Social Constructivist Teaching: Affordances and Constraints*. Oxford: Elsevier Science, pp. 1–43.

Wells, G. and Claxton, G. (2002) Introduction: sociocultural perspectives on the future of education. In: Wells, G. and Claxton, G. (eds), *Learning for Life in the 21st Century*. Oxford: Blackwell, pp. 1–19.

World Health Organization (WHO) (1986) *The Ottawa Charter for Health Promotion*. Health Promotion, 1, iii–v. Copenhagen: WHO Regional Office for Europe.

World Health Organization (WHO) (1993) *The European Network of Health Promoting Schools, Resource Manual*. World Health Organization, Council of Europe and Commission of the European Communities. Copenhagen: WHO Regional Office for Europe.

World Health Organization (WHO) (1997) *Conference Resolution, 1st Conference of the European Network of Health Promoting Schools*. Copenhagen: WHO Regional Office for Europe.

World Health Organization (WHO) (1999) *Health 21 – Health for all in the 21st century, the Health for All Policy Framework for the WHO European Region*. European Health for All Series, No. 6Copenhagen: WHO Regional Office for Europe.

World Health Organisation (WHO) (2002) *The Egmond Agenda: A Tool to Help Establish and Develop Health Promotion in Schools and Related Sectors Across Europe*. Egmond aan Zee, the Netherlands, 25–27 September 2002/ENHPS Technical Secretariat. Copenhagen: WHO Regional Office for Europe.

Chapter 5
Environmental Learning and Categories of Interest: Exploring Modes of Participation and Learning in a Conservation NGO

William Scott[1] **and Stephen Gough**[2]

Keywords environmental learning, categories of interest, participation, conservation NGOs

5.1 Introduction

This chapter explores ideas of participation by considering some of the issues surrounding frameworks for environmental learning. We begin by reviewing a recent analysis of a range of categories of interest in environmental learning. This indicates that the people and groups who promote or encourage environmental learning can have widely differing assumptions about both its purposes and processes, and about participation by learners in both their learning and thence in any social action they might take. We develop this analysis by examining how that which might be expected of the learner and the teacher/instructor in such environmental learning processes, and in ensuant participatory social and/or environmental change, can vary markedly across categories.

We then present an explorative case study of a membership non-governmental organisation (NGO) with a remit to enhance biodiversity and a mission to draw the public into actively participating, not only in the organisation itself, but also in society, through changing how they live. In this case, we examine the tensions between the interests of the organisation, with its need for people to participate in its work and for it to achieve its goals, and the interests of the participants themselves with their own values, goals, and imperatives. In particular, we examine the tension experienced by educators in the organisation between their values as educators and their work to support organisational goals. Finally, the work of the organisation is examined in relation to recent work on the relations between different types of learning and sustainable development.

[1]Centre for Research in Education and the Environment (CREE), Department of Education, University of Bath, Bath, BA2 7AY, UK, w.a.h.scott@bath.ac.uk

[2]Centre for Research in Education and the Environment (CREE), Department of Education, University of Bath, Bath, BA2 7AY, UK, s.r.gough@bath.ac.uk

A. Reid et al. (eds.), *Participation and Learning*, 81–97.

5.2 Environmental Learning and Categories of Interest

As we have explored in some detail elsewhere (Scott and Gough 2003a, b; Gough and Scott 2005), environmental education can usefully be thought of as a broad church whose congregation and ministers represent the many different facets of what has come to be called *environmental learning*. This is the learning which accrues or is derived from a study of the environment or environmental ideas, as the extensive environmental education literature confirms. Such learning can be the outcome of formal or non-formal educational programmes in schools and/or communities, or of designated environmental education interventions and/or personal or incidental learning where no teacher or instructor was involved. All these involve participation by learners in one form or another.

Table 5.1 sets out nine categories of interest which show the range of possible foci and objectives of those teachers, lecturers, non-formal educators, trainers, interpreters, field studies officers, conservation scientists, environmental activists, environmental philosophers, and researchers who value, espouse, and promote such environmental learning. It will be noted that this group varies considerably, not only in obvious ways relating to the kinds of professional responsibilities they have towards learners (and learning), but also in how they and learners interact. What all constituents represented here have in common is that, one way or another, they use the environment as a stimulus to learning. It will be observed that this common denominator – an interest in environmental learning – may link people whose principal concerns and interests, and perhaps whose underlying intentions (Lundholm 2004) in terms of desired social or environmental goals, may actually prove to be very different. It is certainly worth noting that this categorisation is merely a snapshot (at an early point in the 21st century) and that, had a similar analysis been attempted 30 years earlier (and were one to be attempted in 30 years time), the results would certainly have been (and would perhaps be) quite different. Indeed, carrying out the former at intervals would be an interesting retrospective way of mapping the development of the field, and the various pressures and influences on it.

Table 5.1 shows how emphasis varies across different interests. It sheds light upon the relationship between development education, and environmental education, and the interconnection of each of these with sustainable development and learning. All these contribute something to learning about the human condition, and about our co-evolving (Norgaard 1984, 1994) relationship with nature.

Of course, any such categorisation has to be a simplification, but this heuristic does allow us to consider how those interested in environmental learning can have widely differing assumptions about both purpose and process and, as we shall see later, about participation. For example:

- From #1 to #8, interest in nature *per se* decreases markedly along with a shift from a realist view of nature to a metaphorical one. There is also a shift from an interest in the individual learner to the social context.
- From #3 to #7, the environment (natural or otherwise) is viewed mainly heuristically, i.e. as a means of exploring issues and achieving particular goals.

Table 5.1 Categories of interest in environmental learning

Categories of Interest	Focus and outcomes	Exemplified by …
1. Those *interested in sharing the joy and fulfilment derived from nature*, in order to bring about significant life-enhancing and life-changing experience for learners e.g. Van Matre and Weiler (1983)	*Nature values and feelings*	*Non-formal educators* and *interpreters* seeking attitudinal and/or value change; possibly seeking to introduce and extend particular philosophies of living
2. Those *interested in the study of the processes of nature* in order to understand, or to teach about them e.g. Research sponsored by the National Science Foundation – www.nsf.gov, or by the UK's Engineering and Physical Sciences Research Council – www.epsrc.ac.uk	*Nature understanding*	*Teachers* of ecology, (physical) geography, the earth sciences, and rural studies; *researchers* in these (and other) areas
3. Those *using nature as an heuristic* to foster the development of knowledge, understanding, skills, and character which, although situated, are transferable to other contexts and through time e.g. Higgins (2002)	*Nature skills*	*Teachers, environmental interpreters, and field studies officers* seeking to develop students' cognitive/conative/affective/ psychomotor skills related to environmental work
4. Those *using the natural and/or built environments as heuristics* to achieve conservation and/or sustainability goals e.g. Foster *et al.* (2004), The *National Association for Interpretation* – www.interpnet.com	*Conservation understanding*	*Conservation/heritage scientists* (and others) working for government or NGOs bringing communication and education strategies to bear on conservation and sustainability issues
5. Those *advocating/promoting individual behaviour changes* in order to achieve conservation/ sustainability goals e.g. McKenzie-Mohr and Smith (1999)	*Conservation behaviours*	*Environmental activists* and *teachers* who have clear views on what the problems are and on their solutions

(continued)

Table 5.1 (continued)

Categories of Interest	Focus and outcomes	Exemplified by …
6. Those *advocating/promoting particular modes of social change* in order to achieve environmental/conservation/ sustainability goals e.g. Fien (1993), Huckle (1993)	*Social change Social justice*	*Environmental activists* and *teachers* with clear views on the form of social organisation needed to achieve sustainability
7. Those *using environmental, conservation, and/or sustainability issues as contexts* for the development of skills and knowledge related to the exercise of democratic social change e.g. Hungerford and Volk (1990)	*Social change democratic citizenship skills*	*Teachers* and others interested in helping (young) people acquire democratic and citizenship skills which will enable them to participate in open-ended social change relating to human–environment relationships
8. Those *promoting nature as a metaphor for a preferred social order* – which may be 'cooperative' or 'competitive', according to world view e.g. Devall and Sessions (1985), Lovelock (1979), Wilson (1975)	*Social change values*	*Sociobiologists, deep ecologists, social Darwinists, Gaianists* who engage in communication and informal education in relation to the relationship between humans and nature
9. Those *interested in the study of environmental learning* (and environmental education) itself e.g. Researchers such as ourselves – www.bath.ac.uk/cree	*Learning, learning about learning*	*Educational researchers* interested in various aspects of learning and teaching related to environmental and sustainability issues

- From #4 to #8, interest in social change increases strongly.
- #4, #5, and #6 are marked by an emphasis on activism which can complement (and sometimes supplant) educational goals.
- #8 sees the natural world as providing the foundations of a coherent and liveable philosophy that explains our social and ecological obligations. Some of those we have placed in this category may have very little else in common.
- #9 is a mix of those interested primarily in social/environmental issues, and those whose focus is on educational issues. #9 can usefully be further subdivided along methodological lines.

What Table 5.1 does not show, however, is the way that each of these categories has implications for participation, that is, for the engagement of people with these ideas and in learning. Yet all demand participation because all involve learning – and all learning involves participation, one way or another, whatever theory or model is invoked to explain this. Those teachers whose belief in technique is matched by a faith that the learner's mind is simply waiting to be filled have to believe that the empty vessel is waiting, open – even if the learner's only participative act is to aid this or not resist it.

In Table 5.2, a range of modes of participation in learning is explored in relation to the nine categories of interest set out in Table 5.1. These variously show what might be expected of the learner (and the teacher/instructor) in the learning process and, in some cases, what might then be expected in the use of any such learning in processes of social and/or environmental change. It will be seen that what is expected can vary from being better prepared to think about and act in novel contexts (e.g. where complexity, risk, and necessity are dominant features), to putting into action what one has newly learned to do (e.g. changing in probably fairly narrow ways one's use of energy), to thinking differently in respect of certain issues (e.g. about social justice). Such a learning–action menu clearly ranges from education to training – with a little bit of conditioning on the side.

5.3 Case Study

What follows now is a brief exploratory case study inspired by a local membership Non-Governmental Organisation (NGO) (hereafter called 'the Organisation') whose remit is to enhance biodiversity in its locality, and with a mission to draw the public into actively participating not only in the work of the Organisation (e.g. through using facilities and volunteering time), but also through changing how they live to ensure a sustainable future for wildlife and people.

In the Organisation there are many different professional roles and kinds of expertise, for example, conservation scientists, land managers, nature reserve wardens, communicators, fund-raisers, educators, trainers, and community outreach workers (not forgetting managers, accountants, secretaries, trustees, etc.). Whilst there is a tendency to see such roles as separate, all are actually part of an integrated whole, working towards the vision of the Organisation and its goals.

Table 5.2 Categories of interest and modes of participant in learning

Categories of interest	Modes of participation in learning
1. Those *interested in sharing the joy and fulfilment derived from nature*, in order to bring about significant life-enhancing and life-changing experience for learners	Here, participation involves being immersed and confronted with that which can provoke, inspire and affect; there is something of the waiting empty vessel idea here: nature as instructor, as expressed by Romantic poets such as Wordsworth
2. Those *interested in the study of the processes of nature* in order to understand, or to teach about them	The learning here is largely cognitive (and then may be pedagogical) and may well involve the sort of contact with nature outlined in [1]; it may involve social or individual engagement with nature, but will be focused on acquiring certifiable knowledge, understanding, and skills that will enable participation in society in appropriate ways
3. Those *using nature as an heuristic* to foster the development of knowledge, understanding, skills, and character, which, although situated, are transferable to other contexts and through time	Such heuristical devices demand involvement – a willing participation in order to acquire knowledge, etc.; transferrence to novel contexts is then possible through further participation
4. Those *using the natural and/or built environments as heuristics* to achieve conservation and/or sustainability goals	Although more narrowly focused than [3] and not necessarily involving novel contexts, learning is unlikely to be achieved without active participation by those involved; further, such learning is typically expected to enable participation in conservation/sustainability projects
5. Those *advocating/promoting individual behaviour changes* in order to achieve conservation/ sustainability goals	Where specific behaviours are sought as, for example, in social marketing, the learner's role is to take part and (in effect) do as instructed; such limited scope for involvement suggests that extended participation will be unlikely
6. Those *advocating/promoting particular modes of social change* in order to achieve environmental/ conservation/sustainability goals	Here, participation involves a degree of induction into the insights of the teacher. Where techniques to remove 'false consciousness' (Braybrooke 1987) are in use, participation is not even-handed
7. Those *using environmental, conservation and/or sustainability issues as contexts* for the development of skills and knowledge related to the exercise of democratic social change	Given that such endeavours might reasonably involve the practice of such skills by learners, it is here that levels of participation might be expected to be very high
8. Those *promoting nature as a metaphor for a preferred social order* – which may be 'cooperative' or 'competitive', according to world view	Though superficially similar to [1] in the sense of nature as teacher, here nature cannot be expected to do the job unaided; nature needs to be interpreted and its ideas persuasively presented; the participation involved is close to that required in [6]; note how very different messages may be presented; anything from 'learn to compete at all times' to 'learn to cooperate at all times' – in each case because it is *natural*

(continued)

Table 5.2 (continued)

Categories of interest	Modes of participation in learning
9. Those *interested in the study of environmental learning* (and environmental education) itself	Such people are mainly researchers and participation here will have various meanings – depending on the nature of the research being carried out; action research, ethnography, critical enquiry, and even surveys all involve participation – though very different kinds

Box 5.1 Possible benefits of participating in the organisation

Psychologically

For example, enjoyment, stimulus, succour, contemplation, release, escape, appreciation of nature, shared purposeful activity, mental well-being [these span a number of categories in Table 5.1.]

Physically

For example, fresh air, exercise, fitness, de-stressing, physical well-being

Ecologically

For example, maintaining the Web of Life, regulating the homeostasis between the quality of life and the quality of the environment, ecological well-being

For example, *conservation scientists* might be thought to have (or might see themselves as having) a fairly unambiguous focus on wildlife and biodiversity, maintaining/extending habitat, protecting species, constructing wildlife corridors, increasing acreage under conservation, etc., and this fits very well with the core aims of this, and similar organisations. Biodiversity is an essential aspect of the biosphere that supports human life and civilisation. Thus, the work of scientists and land managers in the Organisation enables participation by a wide swathe of people who benefit from the work of this and other organisations (see Wildlife Organisations UK Office 2001) at several levels, as shown in Box 5.1.

Thus, what conservation scientists do is crucial for people's lives, both now and in the future and here and elsewhere, as biodiversity enhanced locally is biodiversity enhanced more widely. This anthropocentric view has the merit of ensuring that human welfare (here and now, elsewhere and in the future) remains in view when policy is made and enacted. However, the point, ultimately, of advancing these arguments about participation is tightly focused for the conservation scientist – it is to conserve nature, and not, for example, to make the world more just, or the human species better educated in the hope that conservation interests might somehow be served. Not surprisingly, the Organisation spends a great deal of effort and resource working with landowners and managers on the ground, encouraging, persuading, and helping such individuals and groups to adopt pro-conservation/

biodiversity practices. In this, conservation scientists and land managers are to the fore encouraging landowner participation in pro-conservation schemes.

Educators might be thought to play a different, but still vital role. Box 5.2 shows how they might contribute to the goals of the Organisation (see CEE 1997). Each role is qualitatively different. They get more complex from 1 to 7, offer greater/deeper scope for participation in social/environmental decision-making, and hence become more valuable. Through this transition they also form an increasingly good fit with what people interested in sustainable development do. These seven levels of purpose suggest different learning outcomes and different kinds of educational approaches/methodologies/modes of participation. Of course, it may not be sensible to think of seven separate levels (and/or there may be more than seven). Table 5.3 explores these ideas by examining possible goals for the Organisation and outcomes in relation to those goals, in terms of the indicators, measures, and proxies that it might use to examine the degree of participation that ensues. This is, essentially, a means of evaluating outcomes against goals, and it needs to be stressed that not everything in reality will be as neat and sequential as implied in this table.

All this, however, serves to highlight three major difficulties. The first is that the chances of success (whether in terms of conservation, biodiversity, or sustainable development) are limited because educational interventions rarely seem to be directed at the main issues, rather they address proxies – and sometimes poor ones at that. Thus, educators within the Organisation are much more likely to engage

Box 5.2 Possible contributions to the goals of the organisation

Helping people to:

1. Raise their knowledge and awareness of what the Organisation does, how it does it, and why.
2. Have first-hand experience and engagement: viz., visiting nature reserves, working in local conservation groups, contributing to practical conservation, including developing social/practical/cognitive skills for use in their own lives.
3. Realise what biodiversity is, and how valuable and important it is to all life on Earth, and in particular to human quality of life and well-being.
4. Think how their own lives affect (positively and negatively), and are affected (positively and negatively) by biodiversity issues (historically, culturally, spiritually, psychologically, socially, environmentally, economically, etc.)
5. Consider how they might change the life they lead, helping them understand options, benefits, and drawbacks.
6. Work through such changes, to enhance their awareness and understanding of how their lives are different and the impacts that this has on them, other people, and nature.
7. Work with others to have an effect on how wider social groups and institutions (including government) view such issues.

Table 5.3 Goals, outcomes and degrees of participation

Possible goals	Possible outcomes	Degrees of participation: indicators, measures, and proxies
1. Raising people's knowledge and awareness of the work of the Organisation	*Knowledge* of what the Organisation does, *awareness/ understanding* of why this is important, leading to *involvement* and enhanced *empathy* towards conservation as an idea	Membership and funding from individuals/ families increases; people think about the issues and what it means to their lives
2. Encouraging/enabling people to have first-hand experience of the work that the Organisation does and the facilities and experiences it offers	*Involvement*/activity: visiting nature reserves, contributing to *practical* conservation and other Organisation activities, being part of the Organisation's work	People get out more; they visit the Organisation's nature reserves and they begin to value what there is and want to conserve it
3. Helping people to realise how valuable and important the Organisation's focus on biodiversity and conservation issues are	*Understand* that having healthy (and dynamic) ecological systems is important for all species and *especially* for human well-being, i.e. for *sustainable development*	People volunteer and get involved, they begin to want to make a difference (to their locality) and improve matters
4. Helping people to think how their own lives affect, and are affected by, biodiversity issues	People *understand* that what they do affects all other aspects of the biosphere and can see how they benefit from healthy ecosystems, and how their (and other's) descendents will as well	People realise that what they do can make a difference to ecosystem health, and that ecosystem health (and dynamism) makes a difference to human lives
5. Helping people to consider ways in which they might change the life they lead	People *think* carefully and critically about biodiversity/ conservation issues at home, at work and at leisure, and *seek to change* how they live accordingly	People begin to think that they should do something and have ideas about what they might do in how they live their lives; they begin to evolve strategies for change
6. Helping people support themselves and others as they work through changes in lifestyle and philosophy	*Making changes* to the way lives are led (from purchasing, to political engagement, transport use, investment, and the like) and shared *understandings* of what can be achieved (and *how*); increased *skills* of analysis, planning, evaluation, and reflection	People can demonstrate progression in how they live towards modes more 'in tune with nature', they show increasing abilities to analyse and plan, and to work collaboratively
7. Helping people to work with others to have an effect on how wider social groups and institutions (including government) view such issues	Local campaigns are begun or strengthened to *influence* the thinking and practice of institutions	People see that, whilst what they do and how they live matters, what others do matters more, and take steps to influence institutions and other organisations

people (and to be encouraged to do this by the Organisation) on specific though relatively marginal issues, such as composting and recycling, each of which can readily be accommodated within a normal, business-as-usual lifestyle, rather than attempting to engage people with how they live, and to think about their lives, in a more comprehensive and radical way. They do this for a number of reasons, prominent amongst which is that funding is more readily available for such marginal activities than for more ambitious and admittedly less well-defined (and definable) goals. Another powerful set of arguments which results in a focus by educators on marginal issues (and which substitutes one sort of participation for a more limited kind), rests on a long-standing conviction amongst conservation organisations that funding educators is a relatively poor use of resources (Fien *et al.* 2001). In part, at least, this is because, though results from educational work may be enduring, they are uncertain and long-term. Other uses of time and money, such as lobbying or campaigning on specific issues or information dissemination about practical tasks, can offer quicker and much more measurable returns.

The second difficulty is that it is quite clear that the correspondence between more education on the one hand and more conservation on the other, is at best imperfect, even when expressly targeted in the way outlined above. This is partly because an education which places any value at all on conceptions such as autonomy or independent thinking must allow for the possibility that educated individuals will elect to take risks, value short-term over long-term considerations, deploy environmental assets for the purpose of securing competitive advantage for themselves or their families/communities (perhaps in the interests of justice), and/or prefer human-made to 'natural' surroundings. It is also because many other factors, apart from what particular groups of learners think and/or believe, can determine actual outcomes in the environment. In consequence, it seems increasingly and properly accepted that attempts to obtain predefined conservation (or sustainability) outcomes from particular educational interventions are unlikely to be successful, except through good luck – no matter how participative an education is on offer. The issue here, perhaps, is to be alert to possibilities and to intervene to take opportunities as they occur, thus maximising the chances of success. As Foster (2005:13) notes:

> [I]t must surely by now be obvious to anyone confronting the issues without illusions: that a sustainable human future, if it comes about at all, will come about essentially *by chance* – or, at best, through the quality of our responses to the chances which present themselves.

And, as Foster goes on to argue (p. 133), although we may be:

> At the mercy of such happenstance, we can nevertheless strive to make our own luck: not just by continuing the vital work of building sustainability understanding and practice where we can, but also – and crucially – by ensuring that we build the optionality, social intelligence and heuristic learning capacity to apply our knowledge adaptively and creatively in situations of perhaps extreme turbulence, and at comparatively calmer junctures to seize the unattended favourable opportunity, the suddenly available option.

Here, of course, the need for participation is writ large.

The third difficulty is the considerable tension contained within Box 5.2 and Table 5.3. Here, the degree to which the learner's participation is devoted to the Organisation's prime (and local) goals (which may include social ones such as more recycling/composting and lower energy/water use) changes rapidly from 1 to 7, to a focus on social/global concerns. The irony here is that the more the Organisation allows and enables this sort of personal learning, and the more successful it is, the less likely it is that it will directly benefit itself. We return to this dilemma later.

To illustrate these issues, we now look at the Organisation's reported activities to members (and the nature of the participation these involved) over one recent three-month period (Table 5.4).

It will be seen from this (opportunistic) analysis that, setting aside participation through financial donation (which is, of course, an important activity), the majority of ways in which people are engaged are found in roles 1 to 4, as set out in Box 5.2 (shown in the left-hand column in Table 5.4). There is considerable challenge for

Table 5.4 Activity, the organisation, and participation

Box 5.2 roles	Activity reported	Nature of participation by members and others in the activity
1	Report on the AGM	• Members attended the meeting
2	Practical conservation work	• Volunteers for manual work
2	Wildlife surveys and observations	• Members take part
2	Guided walks	• Members take part
2	Encouragement of getting out in Winter to see wildlife on an Organisation reserve	• Individuals and families encouraged to use the Organisation's facilities for enjoyment and edification
2	Practical skills classes; e.g. composting, growing plants	• Members take part
2	Award of lottery grants to the Organisation to fund land acquisitions and to manage conservation activities	• Purchase of lottery tickets by people (many of whom participated unwittingly) • Donation of time by volunteers to provide 'in-kind' contributions • Financial donations by individuals and companies to match lottery funding
2	Recording wildlife in the region	• Volunteers to record wildlife (especially indicator species) in specific areas
2	Restoring industrial land to a meadow	• Practical clearing activities by volunteers • Financial donating by individuals and companies
2	Encouragement to organise participatory projects locally, to get involved with Organisation activities, and to volunteer	• Organisation staff provide information and support • Everyone is encouraged to get involved under a 'saving the planet' slogan
3	Illustrated talks	• Members take part
4	Encouragement to take more exercise, be more energy efficient, buy green electricity and local food/milk, reuse paper, make compost (and encourage others to do these), in order to reduce climate change, and its adverse effects on conservation	• Organisation staff provide information and support • Everyone is encouraged to get involved under a 'saving the planet' slogan

(continued)

Table 5.4 (continued)

Box 5.2 roles	Activity reported	Nature of participation by members and others in the activity
4	Defeating a proposal to build on a conservation site	• Lobbying by the Organisation • Lobbying by individuals
-	Recruiting volunteers	• The Organisation employed someone to increase the number of volunteers
-	Opportunities provided by the reform of the EU's Common Agricultural Policy	• Organisation managers work directly with landowners • Landowners can take part
-	Fund-raising walk/run for the Organisation	• Taking part with sponsorship • Volunteering to support walkers/runners
-	Assisting the production of a television programme to encourage people to get outdoors and look at wildlife	• This was in support of a communications exercise (television programme)
-	Books for sale with a percentage of income going to the Organisation for dedicated conservation work	• Financial donation through purchase

such organisations to shift the emphasis away from awareness-raising, practical skill development, and marginal (if fashionable) activity to the deeper kinds of considerations that many commentators now suggest are necessary (see Foster 2005; Sterling 2001). Sterling sees the current situation as a crisis in which our habitual ways of thinking limit our ability to cope with problems positively. He argues for radically different thinking within an 'integrative, holistic, systemic, connective and ecological' mode (Sterling 2001:61) and advocates a change from a mechanistic educational (and learning) paradigm to a 'more humanistic, democratic and ecological (holistic)' one which focuses on human–ecological values. For Sterling, the way forward is to focus on 'third-order learning' (transformative learning) which is 'creative and involves a deep awareness of alternatives worldviews and ways of doing things' (Sterling 2001:15). It is this sort of approach which is increasingly represented as we shift from Roles 1/2/3 to 4/5/6 in Box 5.2. Sustainable development and sustainable living are seen essentially as a process of learning with sustainability as the context for the learning. A similar notion can be found in Scott and Reid (2001:24) who propose that significant personal transformation happens when individuals start 'to think about their lives in relation to sustainable development, and thus think about sustainable development itself, not in the abstract, but in the crucible of everyday decision-making'.

Nikel (2005) summarises these issues in her adaptation of the ideas of Scott and Gough (2003a). This is shown in Table 5.5 which, in conjunction with Table 5.4 and Box 5.2, illustrates that the bulk of participatory learning activities actually undertaken by the Organisation are Types 1 and 2, and not the Type 3 approaches advocated by Sterling and explicated by Scott and Gough. The notes in Box 5.3 set out the essential differences between these three types.

As noted above, not only is there considerable challenge for such organisations to shift the emphasis away from awareness-raising, practical skill development and marginal (if fashionable) activity, but there is also something of a disincentive, as such shifts detract from core Organisation goals which tend to be both conservation focused and locally directed.

Table 5.5 Sustainable development, learning, and learning design

Type	Sustainable development problem definition	Role of learning	Role of learning design
1 and 2	Pro-environmental or sustainable behaviour can be specified – either based on perception of sustainable development as caused by environmental conflicts [Type 1], or by social conflicts [Type 2]	Bringing about appropriate (pre-described) knowledge, skills, action	Sustainable development problem definitions can be selected supporting the development of pro-environmental behaviour (adjudicative decision is made beforehand)
3	Pro-environmental or sustainable behaviour can NOT be specified (emphasis on complexity and uncertainty); people's opinions, actions, and feelings are often confused or contradictory	Bringing individuals to begin to reflect on their perceptions of 'sustainable behaviour' in the context of their own and other's institutional affiliations	Learners have to make adjudication for themselves and there is therefore a need to confront learners with competing problem definitions within changing context and changing affiliation assumptions

Box 5.3 Notes on Types 1/2/3 approaches

Type 1 approaches assume that the problems humanity faces are essentially environmental, can be understood through science, and resolved by appropriate environmental and/or social actions and technologies. It is usually the 'understanding through science' that is seen as the clever bit of this, and it is often assumed that learning will simply lead to change once facts have been established and people told what they are. Type 1 approaches see learning as a tool for the achievement of environmental maintenance where people turn objective knowledge into social action. This can be an effective approach when the scientific facts *can* be clearly established, and when there is wide agreement about the desirability or otherwise of the consequences of action or inaction. However, instances of failure are far more common (Kollmuss and Aygeman 2002).

Type 2 approaches assume that our fundamental problems are social and/ or political, and that these problems produce environmental *symptoms*. Such

(continued)

Box 5.3 (continued)

fundamental problems can be understood by means of anything from social–scientific analysis (at one extreme) to an appeal to indigenous knowledge. The solution in each case is to bring about social change, where learning is a means *inter alia* to:

- Create an environmentally responsible citizenry
- Disseminate the ideas of global (e.g. UN) governance
- Enable bottom-up social change through emancipation of the poor and weak

In Type 2 approaches, the proposed role of learning is to facilitate choice between alternative future 'end-states' which can be specified on the basis of what is known in the present. Such approaches may be useful, for example, where there are clear opportunities for citizenship action, or uncontroversial evidence of environmental damage resulting from systematic social oppression. Once again, however, they more usually fail.

Type 3 approaches to learning and the environment are grounded in the notion of the co-evolution of society and its environment (Norgaard 1984, 1994). They assume that the exact nature of many problems may be incapable of precise specification for the foreseeable future. This is to accept that what is (and can) be known in the present is not adequate, and that desired 'end-states' cannot be specified. This means that any learning must be open-ended. In Type 3 approaches, therefore, the role of learning is to enable learners to develop their ability to make sound choices in the face of the inherent complexity and uncertainty, and in key respects the indeterminacy, of environmental futures, and so acquire, as Sen (1999:74) puts it: 'the substantive freedoms – the capabilities – to choose a life [they have] reason to value'.

In relation to participation, although Type 2 approaches may seem more participative that Type 1 ones, and are often hailed as such by practitioners who favour them, we think that Type 3 approaches have to be inherently more participative as these give the learner a central role in setting agenda for learning and action, and value the contribution that differing perspectives bring to this. Thus, Type 3 approaches are also inherently more educative.

Further, Type 3 approaches seem to be essential if the uncertainties and complexities inherent in how we live now are to lead to useful learning about how we might live in the future.

Finally, if we return to Table 5.1, we can say that, while a range of positions is possible, in both cases, the conservation biologist's focus can only ultimately be on nature or conservation, and the educator's on learners and what they do. Each will evaluate the results of any participation differently. Thus, an underlying tension is likely to exist and, from time to time, emerge between them.

5.4 Concluding Comments

We have argued here that there is a continuum of approaches to thinking about how participation in conservation and sustainability initiatives might be brought about through education and learning. At one end lies the view that the educators (or designers of learning opportunities) know best about what to do – and how to do it. Thus, participation is on their terms and in relation to their (existing) values, not the learners' – and this low-trust approach might be seen both to characterise the Organisation's work and represent its interests.

At the other end would lie an approach consistent with Sen's (1999, 2002) view of both rationality and freedom. Here, a major purpose of education is to facilitate people's development of preferences over what preferences to have. Sen calls these 'metapreferences'. This is necessarily much more of a high-trust approach where participation is on learners' terms and in relation to their (emerging or developing) values. Looking again at Table 5.2, it will be clear that some categories of interest are much more likely to promote high-trust participation (e.g. #1/3/7), than are others (e.g. #5/6/8).

This does not always create a dilemma, however. As we have seen in Table 5.3, at particular times and places it may be that the perceptions of learners and educators about what needs to happen coincide. A good example of this within the Organisation is its programme to promote home composting through leaflets, its web site and a newsletter. The Organisation knows about composting best practice and is effective at disseminating this – at least to those who want to know. The learner (the householder in this context) is an expert in the practicalities and limitations of their own context and is able to interpret and implement the advice provided. This is a good example of Type 1 learning approaches whereby simple information, skilfully provided, enables practical conservation/sustainability practice. Similarly, the creation and growth of a car-share scheme (e.g. www.citycarclub.co.uk) for people with the occasional need for a car for journeys where public transport, walking, or cycling just are not suitable, exemplifies a successful Type 2 approach.

And it seems quite understandable in economic terms if a conservation scientist or an NGO Chief Executive is not prepared to commit resources to helping people participate on their own terms, that is, to trust people to use the outcomes of the educational process in ways that make sense to them as social beings – whether as a direct result of participating in the Organisation's work, or as a result of something more complex – rather than as some kind of agent *for* the Organisation. After all, the scientist or Chief Executive probably have the same bottom line to contribute to: one that sees conservation outcomes and impacts as *the* priority. If you can do this *and* enable the public to learn (especially if it contributes to Organisation goals on your terms/values), all well and good (Figure 5.1).

However, this calculus is more difficult for educators inside such an NGO. Not to enable learners to learn on their own terms and in relation to what they value, might be seen as a mark of impoverishment and failure in an educator – whatever your employer tells you is for the best.

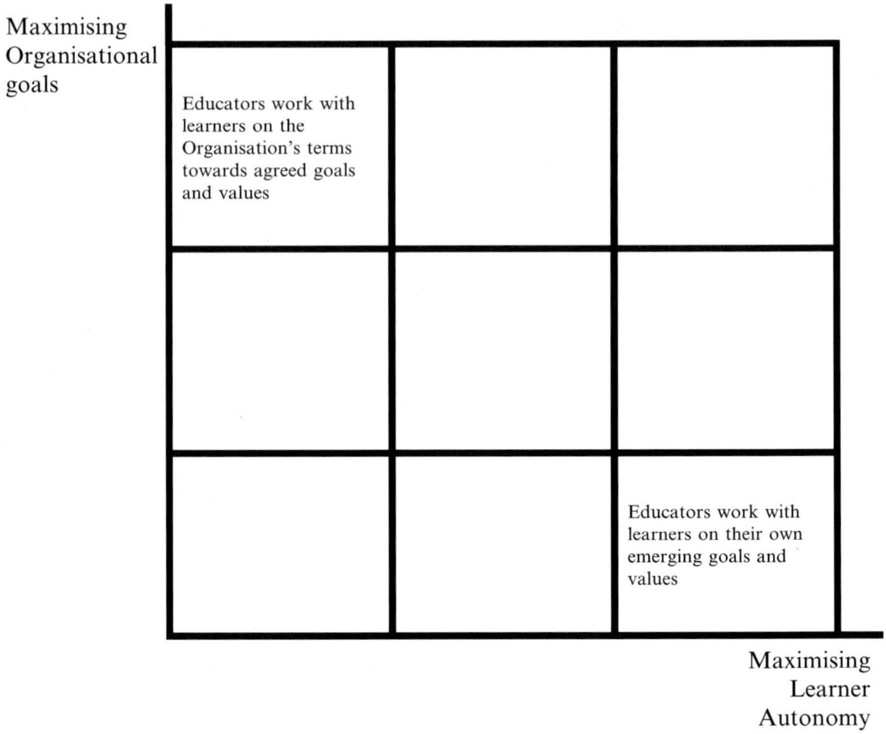

Figure 5.1 Exploring tensions between NGOs and their educators

References

Braybrooke, D. (1987) *Philosophy of Social Science*. Englewood Cliffs, NJ: Prentice-Hall.

CEE (1997) *The Aims of Biodiversity Education*. Reading, MA: Council for Environmental Education.

Devall, B. and Sessions, G. (1985) *Deep Ecology: Living As If Nature Mattered*. Salt Lake City, UT: Peregrine Smith.

Fien, J. (1993) *Education for the Environment: Critical Curriculum Theorising and Environmental Education*. Geelong: Deakin University Press.

Fien, J., Scott, W.A.H., and Tilbury, D. (2001) Education and conservation: lessons from an evaluation. *Environmental Education Research*, 7, 4: 379–395.

Foster, J., Grove-White, R., Scott, W.A.H., and Gough, S.R. (2004) *Aiming for Sustainability – Can We Keep on Track? A Note on Indicators, Uncertainty and the 'capital' Model in Sustainable Development*. Lancaster University: Institute for Environment, Philosophy and Public Policy [produced as part of the ESRC Environment and Human Behaviour Programme, www.psi.org.uk/ehb].

Foster, J. (2005) Options, sustainability policy and the spontaneous order. *Environmental Education Research*, 11, 1: 135–155.

Gough, S.R. and Scott, W.A.H. (2005) Promoting environmental citizenship through learning: towards a theory of change. In: Dobson, A. and Bell, D. (eds), *Environmental Citizenship: Getting from Here to There*. Cambridge, MA: MIT Press, pp. 263–285.

Higgins, P. (2002) Country focus: outdoor education in Scotland. *Journal of Adventure Education and Outdoor Learning*, 2, 2: 149–168.

Huckle, J. (1993) Environmental education and sustainability: a view from critical theory. In: Fien, J. (ed), *Environmental Education: A Pathway to Sustainability*, Geelong, Australia: Deakin University Press, pp. 43–68.

Hungerford, H. and Volk, T. (1990) Changing learner behaviour through environmental education. *Journal of Environmental Education*, 21, 3: 8–21.

Kollmus, A. and Agyeman, J. (2002) Mind the gap: why do people act environmentally and what are the barriers to pro-environmental behaviour? *Environmental Education Research*, 8, 3: 239–260.

Lovelock, J. (1979) *Gaia: A New Look at Life on Earth*. Oxford: Oxford University Press.

Lundholm, C. (2004) Case studies – exploring students' meanings and elaborating learning theories. *Environmental Education Research*, 10, 1: 115–124.

McKenzie-Mohr, D. and Smith, W. (1999) *Fostering Sustainable Behaviour: An Introduction to Community-based Social Marketing*. Gabrioloa Island, BC: New Society.

Nikel, J. (2005) Ascribing Responsibility: A Three-country Study of Student–Teachers' Understanding(s) of Education, Sustainable Development, and ESD. Unpublished doctoral thesis. University of Bath.

Norgaard, R.B. (1984) Coevolutionary agricultural development. *Economic Development and Cultural Change*, 32, 2: 525–546.

Norgaard, R.B. (1994) *Development Betrayed: The End of Progress and a Coevolutionary Revisioning of the Future*. London: Routledge.

Sen, A. (1999) *Development as Freedom*. Oxford: Oxford University Press.

Sen, A. (2002) *Rationality and Freedom*. Cambridge, MA: Harvard University Press.

Scott, W.A.H. and Reid, A.D. (2001) Exploring our responsibilities: a critical commentary on education, sustainability and learning. *Environmental Education*, 66: 23–24.

Scott, W.A.H. and Gough, S.R. (2003a) *Sustainable Development and Learning: Framing the Issues*. London/New York: RoutledgeFalmer.

Scott, W.A.H. and Gough, S.R. (2003b) Categorizing environmental learning. *NAAEE Communicator*, 33, 1: 8.

Sterling, S. (2001) *Sustainable Education: Re-visioning Learning and Change*. Totness: Green Books (on behalf of the Schumacher Society).

Van Matre, S. and Weiler, B. (1983) *The Earth Speaks a Powerful Tribute to Our Home in Space*. Greenville, West Virginia: The Institute for Earth Education.

Wildlife Organisations UK Office (2001) *Wildlife Organisations' People and Wildlife Policy Statement*. Newark: UK Wildlife Organisations.

Wilson, E.O. (1975) *Sociobiology: The New Synthesis*. Cambridge, MA: Harvard University Press.

Chapter 6
Participation and the Ecology of Environmental Awareness and Action

Louise Chawla

Keywords community development, children's agency, ecological psychology, environmental learning, environmental citizenship

6.1 Introduction

This chapter takes up two questions essential to participatory environmental education: *What experiences prepare children to be aware of their environment and to take action on its behalf? And, how can communities support children's environmental learning and action?* I suggest answers to these questions based on an ecological approach to psychology and show how research on the significant life experiences of people committed to environmental education and action can be understood within this framework. I also argue that environmental education can most productively encourage children to know, value, and protect the diversity of life on this planet if it builds on a theoretical foundation that embeds human development in an ecological context.

6.2 Learning to see and Learning to Take Action

Growing up to know and value the diversity of life is partly a matter of *learning to see*: learning to see communities of plants and animals, details of their individual existence and interactions, and patterns of their ever-changing habitats. In cities and towns it includes learning to see the diversity of human communities and the ways in which people interact with their place. As the wildlife biologist Aldo Leopold observed many years ago, environmental education faces a great challenge, 'how to bring about a striving for harmony with land among people, many of whom have forgotten there is any such thing as land, among whom education and culture have

---

College of Architecutre & Planning, University of Colorado, Campus Box 314, Boulder, Co 80309-0314, USA, louise.chawla@colorado.edu

A. Reid et al. (eds.), *Participation and Learning*, 98–110.

become almost synonymous with landlessness' (Leopold 1949, 1966:210). Therefore this chapter begins with a discussion of theories of perception.

Yet, learning to see brings risks as well as fascination and delight. What if a person sees painful things? As Leopold (1949, 1966:197) also observed, 'one of the penalties of an ecological education is that one lives alone in a world of wounds.' He compared attentiveness to the land, in the midst of a culture that is largely oblivious to the effects of its actions on other forms of life, to the condition of 'the doctor who sees the marks of death in a community that believes itself well and does not want to be told otherwise' (ibid.). Therefore, learning to see needs to be balanced by *learning to take action* on the environment's behalf – learning that one is not, in fact, alone, learning how to organise and collaborate with others in order to investigate the sources of problems, and learning how to build a common vision for a better world and take steps to bring it to reality. This goal is comparable to the concept of *action competence* that has been defined by Jensen and Schnack (1997) and Breiting and Mogensen (1999). It is the theme of the second half of this chapter.

For both learning to see and learning to take action, participation in the life of one's community is essential – where 'community' is meant in an expansive sense of the plants and animals as well as the people and cultures of one's locality. To maintain this broad meaning, this chapter is equally concerned with formal and informal participation. Informal participation involves freedom to move about and explore natural and built environments, to gather with others and to observe and try out roles in public places. Children cannot see the diversity of life unless they have a chance to venture out into it. By getting out, they come to find out about issues taken up in formal channels for participation, like the loss of habitat or homelessness. Thus formal and informal participation are two interdependent sides of the inclusion of children in their communities. Through freedom to move about in their localities and become part of diverse settings, children gain authority to speak about the conditions that they find. They also gain reasons to become involved. If they have benefited from open and accessible communities with a variety of resources, they are likely to defend these advantages. If they live in places that are boring or dangerous, but they have had opportunities to see alternatives, they are likely to advocate a more generous plan (see, for example, Chapter 18 by Barratt and Barratt Hacking, this volume).

One advantage of ecological psychology for environmental education is that it directs attention to the importance of informal as well as formal learning. In doing so it accommodates the learning of very young children, from infancy up, who primarily know their communities through the informal contexts of everyday life. Typically, when formal mechanisms for participatory planning involve young people they focus on pre-adolescents and adolescents. Younger children, however, often have thoughtful views as well. Even before they learn to talk, their disposition to engage with the world and to make their mark – a disposition essential for community participation – is rooted in infancy. Ecological psychology provides a foundation for a conception of environmental education that extends beyond the school to all areas of a child's life and furthermore, to all periods of the life span. It lays responsibility for the quality of environmental learning not only on the shoulders of school administrators and teachers, but also on parents, city planners, public officials, and other people whose decisions shape the places where children live and the opportunities that these places afford.

6.3 An Ecological Theory of Perception and Action

Ecological psychology offers a strong foundation for conceptualising children's environmental learning and participation, in both the formal and informal sense, for at least three reasons. It focuses on children's agency, it describes the environmental context of action and development, and not least, it places children and the environment together in a common realm.

Although people caught up in the natural attitude of everyday life generally assume that they perceive their surroundings directly, this is not the dominant epistemology of psychology and the social sciences. The view that has gained wide currency in these fields is that people never know the world as such but only mental representations of it. Social constructivist theory – a version of this representationalist view that has gained particular prominence – posits that individuals inhabit separate subjective realms that are socialised by gender, class, ethnicity, and all of the other contingencies of social existence. This view places the world and people's awareness of it in separate physical and mental realms and emphasises divisions in human experience.

Ecological psychology does not deny the variety of human cultures and individual differences, but it does maintain that as humans we share a single and singular planet where we have evolved fundamentally common characteristics. Our co-evolution with other elements of the environment, as well as our cultural evolution, points to a relational theory of perception and action. According to a central principle of ecological psychology developed by James and Eleanor Gibson (Gibson and Pick 2000; Gibson 1966, 1979), to be a living organism – to be animate – means to have agency that functions in the service of maintaining an awareness of the environment. Action is a means of staying in touch with the environment's significant properties: whether this be an infant following its mother's face with its gaze or an adult striding through the landscape. The evolutionary history of each species includes selection for systems of perception and action that detect aspects of the environment with functional significance. Awareness of the world always occurs within this animal–environment relationship, and thus the mind and the world are not two separate realms, mediated somehow by a mental representation, but reciprocal facets of a shared functional system. Rather than being mediated, this relationship is direct.

To take the case of vision, which James Gibson (1979) studied intensively, the properties of objects are carried to the eye as structures in reflected light. These structures serve as potential information about the objects and their qualities, available for all perceivers who are present. Because reflected light carries a multitude of information from all sides and as different perceivers may be attuned to different features of it, they may not see the same thing. What they see, however, *are* features of the world, not mental fabrications. The stimulus information around them constitutes a common ground. Therefore, with experience, people can learn to see the same features of their shared world.

This view of perception leads to another contribution of the Gibsons' ecological approach: the concept of *affordances*. Affordances are relational, functionally significant properties of the environment. By this view the meaning of environmental features is inherent in the relationship between the environment and an individual, rather than being a mental construction imposed on the environment. For example, children can build a tree house only if the specific properties of the construction material and the tree match the competencies of the children and the goals of the task. They can climb into the tree only if its branches begin low enough relative to each child's height and reach. They can build a platform to stand on only if the boards are light enough for these children to lift but sturdy enough to bear the children's weight. From the perspective of *these* children, *these* environmental features are functionally meaningful and their meaning is specified relative to the properties of the objects, the task and the children considered together.

This example also serves to illustrate the relationship between perceiving and acting discussed previously. Perceiving and acting proceed together in a system that is simultaneously defined by the environment's properties and by a creature's body, capabilities, and goals. In addition, this example demonstrates Edward Reed's (1996a) distinction between *exploratory* and *performatory* activities. In exploratory activities, an animal scans and tests the properties of the environment without significantly altering what it finds. In performatory activities, it alters the environment according to its goals. Yet successful performance depends on experience gained from exploration. The children had doubtless tested the properties of trees, boards, and their bodies before they carried out their work. In the process, they discovered values and meanings inherent in the environment that they could put to use. This example also shows that the world of culture and the natural world are inseparable. People typically learn performatory actions in social groups but the cultural world that they construct depends on the properties of the natural world.

This view of agency and perception avoids a matter–mind dichotomy. It places humans in a world in which they have co-evolved with other living things, dependent on the intrinsic qualities of the world, its resources and its limits, but equipped to discover what these qualities and limits are. It recognises that humans, like other animals, alter their environment, but in the process they need to preserve the resources that are the conditions for their well-being. Because ecological psychology emphasises these common dimensions of human life it forms a basis for collective action for the environment.

This emphasis on agency is in harmony with the goals of participatory environmental citizenship, which seeks to engage people in actively learning about, monitoring and managing their surroundings. According to both ecological psychology and this view of participation, people flourish more fully when they have a rich range of opportunities to realise their capabilities, and their capabilities include seeing the environment accurately and knowing how to take effective action in response. For all of these reasons, ecological psychology forms a sound theoretical foundation for environmental education.

Although it gives due attention to the intrinsic qualities of the physical world, ecological psychology does not deny the social and cultural dimensions of action. As Reed (1996a) has noted, humans have evolved very special modes of action and awareness that are highly interactive and imbued with symbolism. At around six months of age, children begin to pay attention to features of the environment to which another person is attending. This achievement of joint attention forms a basis for all subsequent instruction and learning. From this time on, whether others are teaching deliberately or by chance example, most of children's discoveries build on involvement with others who direct their attention to particular features of their surroundings. At first, these features are primarily concrete, but as children grow, they increasingly include abstract relationships and ideas. Barbara Rogoff (1990) describes this way of learning through joint attention as a series of apprenticeships with people who are more experienced in different domains. This point, too, is vital for environmental education.

These principles of ecological psychology suggest that the following conditions support the development of children's environmental awareness and competence:

- Affordances that promote discovery and responsive person/environment relationships
- Access and mobility to engage with affordances
- Perceptual learning to notice and value the environment
- Opportunities to take responsible roles in community settings

The following sections of this chapter will take up each of these conditions in turn and suggest their implications for environmental education.

6.4 Affordances that Promote Discovery

People are most likely to continue to engage with the environment when they receive immediate information in response to their actions. Even infants, who are so limited in their ability to act on the world, seek to catch their mother's eye and if she smiles back, a coordinated exchange begins in which the infant's wiggles and kicks activate its mother's encouragement (Stern 1985). This relationship is equally true in interactions with the physical environment. Rovee-Collier (1986), for example, compared infants as young as two months under two conditions. One set of infants had a cord from a mobile that hung over their crib tied to their ankle, so that they could make the mobile move by kicking. For another set of infants, the mobile was moved by the experimenter. When the infants caused the mobile's movement, their rate of kicking increased. When their kicking had no effect, it rapidly declined. The infants were motivated to continue to engage with the mobile when they could see themselves producing effects by their own actions.

This reciprocity between self-produced action and environmental events lies at the heart of the development of a sense of competence (Bandura 1997). For this reason, Fuglesang and Chandler (1997) argue that responsive early childhood programmes and parent training to increase interactive experiences of this kind are important precursors to children's readiness for more formal types of participation. In research with adolescents and adults, this experience of self-efficacy promotes learned optimism (Seligman and Csikszentmihalyi 2000) and life satisfaction (Ryan and Deci 2001). As Bandura (1997) has noted, to inspire people to continue working towards goals, nothing is as motivating as 'mastery experiences' of their own capability.

Environmental features vary, however, in the degree to which they provide these effects (Heft 1988; Kytta 2002). For ball play, rollerblading, or cycling, nothing is better than hard, flat surfaces. For most other actions, children need a more malleable world. Natural elements are particularly rich in the effects they offer, and what is more, although they are predictable to a point, they also offer intriguing surprises. Water flows and splashes in fluid forms. Soft earth can be dug and moulded in infinite iterations. Every rock has a different heft and shape for building. Every rotting log reveals a different universe of insect life inside. Convivial city streets and public spaces offer similarly responsive settings for social interactions, which – like natural elements – never do exactly the same thing twice.

In environmental autobiographies, childhood places that afford experiences of this kind resonate in memory. Adults' accounts of special childhood places show that they tend to be on the margins of the adult world, where children can find freedom to take risks and prove themselves, get dirty, make and unmake play worlds with abandon and interact intensively both with the physical world and with friends (Cooper Marcus 1978; Goodenough 2003). These sites include tree houses, play forts, creek banks, and overgrown lots, colonised by children on the edge of the adult world in both cities and rural areas. They also include public spaces where children feel welcomed by adults. Ethnographic work with contemporary children shows that these opportunities for interactive engagement still figure prominently in children's choices of favourite places (Chawla 1992, 2002).

When environmental activists and environmental educators are asked about their sources of commitment to their work, one of the most frequent reasons that they give is the experience of natural areas of this kind in childhood or adolescence (Tanner 1998). Not only is the natural world responsive, but it yields alluring sensory information. This quality is evoked by the words of an Australian teacher in a study that asked environmental educators in nine countries to write about significant experiences that led to their environmental concern:

> Sensory flashes of childhood: being swooped by magpies; listening to the rain on our tin roof; smelling pine needles under shadowy trees; building cubbies from bracken fern; caring for and sharing life with pets; awe-inspiring thunderstorms; the taste, smell and chill of the sea. (Quoted in Palmer *et al.* 1999:184.)

Similarly, a study of citizens in England revealed that memories of nature play in childhood emerged as a major reason for people's concern to preserve parkland and wild areas (Harrison *et al.* 1987).

When environmental educators and activists describe their formative experiences, they mention natural areas outside school more often than schools. Yet schools can provide these experiences too if they turn their yards into natural habitats. An evaluation of these changes at a California elementary school showed that children value what they gain. The following statements are representative of 50 students who were interviewed about their reactions:

> When I see another school I think, too bad, they've just got a cement yard and we've got trees and a river and ponds with fishes, frogs, tadpoles, snakes and a turtle.
> You would never say, "Let's go outside and learn about a cement yard." Now there is always something new to find out.
> I know how it *feels* to have ponds. I know what lives there. I've seen the way ponds change. (Quoted in Moore 1989:203, 205.)

This emphasis on affordances that promote discovery owing to their responsive nature also applies to the design of participatory projects with children. The wheels of community change often grind slowly and the realisation of children's recommendations often depends on many other stakeholders. It is therefore important to build in a sequence of goals, from those that are securely within a group's own power to more ambitious distant goals. For example, while children are working on the clean-up of a local stream, they can restore a wetland in a corner of their school grounds.

6.5 Access and Mobility

No matter how rich the array of affordances may be in children's surroundings, they are of little consequence unless children can reach them. As Kytta (2004) has observed in her study of affordances for children in a variety of communities in Finland and Belarus, the best places for children provide 'positive interactive cycles'. In this case, children enjoy independent mobility to explore their surroundings and when they get out, they discover responsive affordances. As a consequence, they feel motivated to explore further and in doing so, they discover more. In the process, they build a growing repertoire of environmental competencies.

Children themselves consider freedom to move about safely an important measure of a good place in which to grow up (Chawla 2002). In contrast, studies from industrialised nations around the world indicate that children's independent mobility and access to natural areas is eroding. In a world of rapid urbanisation and rising populations, a number of barriers impede access – the hazards of automobile traffic, crime and pollution, parents' fears of strangers and children's fears of bullies (see, for example, Chawla 2002; Rissotto and Tonucci 2002; Kytta 2004). In one of the most systematic studies, Gaster (1991) examined children's free play in northern Manhattan from 1910 to 1980. He found vanishing natural play areas and, since the 1940s, a decrease in the number of places that children visited, combined with a steady increase in the age at which they were first allowed out alone and in the amount of adult-supervised play. Recent studies of children's free range show that it is not unusual for urban children to stay within their block or immediate housing site, even in adolescence (Chawla 2002).

Considering how often people cite childhood play in nature as a reason for environmental concern and commitment, provision for environmental learning needs to include a broad vision of urban planning. In addition to naturalising schoolyards, much can be done to increase children's free access to nature even in densely built cities. Nature can be brought near through landscaping, site design, and affordable and secure public transportation to resources like parks and ecological reserves (Chawla and Salvadori 2003; Louv 2005).

6.6 Perceptual Learning

Children are rarely alone as they detect information present in their surroundings. Even if no other person is nearby, children exist in worlds that are structured by others. Often it is a structure that is deliberately designed to guide children's behaviour such as a toy placed near at hand or the bars of a crib to contain movement. In other words, even when children enjoy *fields of free action* where they explore autonomously, they usually move within *fields of promoted action* or *constrained action*, where a distinction between the physical and social world is artificial (Reed 1996a; Kytta 2004).

When adults and other children are nearby, they often collaborate to help a novice child learn a new activity or joint task. By placing resources within a child's reach, by direct aid or by example, they 'scaffold' learning (Vygotsky 1978). In a variety of ways, the principle function played by more experienced people is to direct the learner's attention to relevant features of the task. They may do this by verbal instruction, but what the guide knows is often tacit and not easily articulated in words. In this case she may point, position an object so that the critical feature will be noticed, or simply say 'look!' or 'listen!'. These are the conditions of apprenticeship in the broadest sense and the contexts for social learning that is so characteristic of human beings (Rogoff 1990, 2003). They are also the conditions for perceptual learning, which involves becoming selectively attuned to particular information out of a larger field of potential information.

Representations of the environment in books and on television and computer screens can contribute to this process of learning to see, but Reed (1996b) argues that they can never replace the role of direct experience. *Primary* or first-hand experience of the world exposes a person to inexhaustible possibilities for learning, and the richer the environment, the richer the possibilities. A person outdoors encounters a dynamic, dense, multisensory flow of diversely structured information. In *secondary* experience, when people learn about the world second-hand through texts, images or someone else's story, this information is radically reduced. The realm of full-bodied primary experience is also where people form personal relationships and place attachments – where they find people and places to care for and others to join with them in action. Despite the growing importance of the internet for political and environmental organising, people who work to defend wild places report that they draw strength from their bonds with real places and from face-to-face networks of supportive people (Zavestoski 2003).

The power of simply drawing attention to elements of the natural world in an appreciative way is suggested by the fact that another reason given by environmental educators and activists for their concern and commitment, often mentioned as frequently as a special childhood place, is an influential role model such as a parent or other family member (Tanner 1998). What people recall are rarely didactic messages such as 'you ought to protect wild areas'. Instead, they remember people who drew their attention to the natural world as something worth valuing. As a Norwegian biologist who fought against the damming of wild rivers recalled, everyone in Norway in the 1950s went hiking, berry picking, and fishing. What distinguished her family was that, 'My mother knew the names of the plants more than other mothers did. So we talked more deeply about things. We didn't only fetch berries and fish, but talked about it' (quoted in Chawla 1999:20). Sometimes this instruction was wordless. Another activist, who fought against the damming of a wild river in Kentucky, tried to understand why he was different than others of his age who also grew up hunting and fishing. He described a father who, 'could teach you how to make a willow whistle or a pop gun out of certain things or how to find the fishing bait under the rocks and appreciate what's there. Or who takes you out on the porch when a thunderstorm comes in so you could enjoy it' (quoted in Chawla 1999:20). This combination of special childhood places and people is exactly what the naturalist Rachel Carson advised for an undying sense of wonder. In addition to abundant time in the natural world, she wrote that each child, 'needs the companionship of at least one adult who can share it, rediscovering with him the joy, excitement and mystery of the world we live in' (Carson 1956:45).

Certainly, teachers too can direct children's attention to the natural world. Although time in nature in childhood and influential family members or friends are the most frequent reasons that environmental activists and educators give for their concern and commitment; education is another reason mentioned, often third or fourth in frequency (Tanner 1998). At their best, schools use the processes of scaffolding, directed attention, and apprenticeship to encourage children to notice the environment in new ways and value what they see. However these concepts point to the importance of fieldwork beyond the classroom, where children can be surrounded by the world that they are learning to observe with closer attention. Because teachers can rarely equal the influence of parents or other family members in a child's life, schools should also reach out to families, communicating the importance of sharing appreciation for the natural world and seeking to include family members in environmental education activities as often as possible.

6.7 Participation in Community Settings

Aldo Leopold (1949/1966) remarked that as people learn to notice places around them with a keener eye, they begin to see wounds on the landscape. As young people discover problems, if they are to believe that they can help solve them, they need to feel like a part of their community and find channels open to them to participate in environmental decisions. Another tool from ecological psychology – the concept of *behaviour settings* developed by Roger Barker – provides a means to assess opportunities for individuals to participate in the social life of their community and to play a role in charting its direction (Barker 1968; Schoggen 1989).

Behaviour settings are places where individuals gather to engage in coordinated activities. They are constituted by individuals' actions together, as well as the affordances of the place that make these actions possible (Heft 1989, 2001). Like affordances, behaviour settings are real physical entities with perceptually meaningful features, for example, a gathering of an environmental organisation at a particular time and place. They are also similar to affordances because some behaviour settings are more responsive than others to an individual's actions and, in this way, offer better support for the development of competence.

Barker (1968) distinguished different 'zones of penetration' that an individual can occupy in a setting, which correspond to varying levels of involvement and responsibility in contributing to a group's goals:

1. *Onlooker*, who takes no active role in the setting
2. *Audience*, who has a recognised role, but very little power
3. *Member*, who has potential rather than immediate power, such as an organisation member who is eligible to vote for officers
4. *Active functionary*, with power over part of the setting, such as someone engaged in group activities like conducting a bird count or monitoring a stream
5. *Joint leader*, with immediate but shared power over the entire setting
6. *Single leader*, with immediate and sole authority over the operations of the setting

Because the concentrated power of a single leader violates the spirit of participatory decision-making, its presence in a programme is a sign of dysfunction. Otherwise, the deeper that young people penetrate into a setting, from peripheral levels that primarily involve observation to central levels with considerable influence over activities, the more engagement and responsibility they take on and the correspondingly greater opportunities they enjoy to develop a range of competencies.

Research in large and small schools (Barker and Gump 1964) and different communities (Barker and Schoggen 1973) has shown that the availability of responsible positions affects the likelihood that children will take on multiple roles and experience leadership. This research provides a model for how the idea of zones of penetration could be applied in planning and evaluating participatory programmes with children. It shows how to identify existing behaviour settings where children may already be playing active roles in their community and where action for the environment might be introduced. It also indicates the importance of planning programmes that give positions of influence to as many children as possible. For example, a weekly child-run radio programme on community issues requires a large number of ongoing roles such as researchers, reporters, informants, editors, and technicians. Child-to-child teaching, such as high school students preparing and presenting environmental lessons to elementary school students or older children leading younger children on field trips, provides other examples. In addition to creating many responsible roles for young people to fill, activities of this kind create many occasions for apprenticeships, not only between novice children and more experienced children, but also between child leaders and adult facilitators. Thus programmes of this kind provide key ingredients for the development of competence: role models who include peers and encouraging adults, guided participation in fields of promoted action, opportunities to influence the setting, and experiences of achievement.

In two comprehensive reviews of what motivates people to take action for the environment, some similar variables emerge. In the analysis of Hungerford and Volk (1990), 'environmental sensitivity' is the major entry-level variable that predicts responsible environmental behaviour and it is derived from significant life experiences such as positive experiences of the natural world in childhood, and early role models who communicate nature's value. Other critical variables include a personal investment in issues, knowledge and skill in environmental action strategies, and a belief that one's actions can have an effect. In a review by Stern (2000), important variables include biospheric values (care for plants, animals, and communities of living things) and the self-perceived ability to reduce threats to the environment. In both analyses, valuing the natural world and confidence in one's own ability to make a difference on its behalf are critical predictors of responsible action.

These conclusions correspond to the ecological view of human development presented in this chapter. Ecological psychology presents a theoretical rationale for the importance of enabling children to know the natural world through their own exploration of it and for the company of people who direct their attention to other living things in a way respectful of their inherent value. It also indicates the importance of providing children with opportunities to take increasing levels of responsibility for environmental decision-making and action, and thus develop their competences. In this way, children find occasions to develop confidence in their capabilities, including their ability to join with others for collective action, as they follow issues of their own interest and concern.

Acknowledgment This chapter owes a great debt to collaboration with the ecological psychologist Harry Heft. It draws on ideas that we have developed together in Chawla and Heft (2001) and Heft and Chawla (2005), and on Heft's clear articulation of the work of James and Eleanor Gibson and Roger Barker. It has also benefited from his careful reading and suggestions.

References

Bandura, A. (1997) *Self-efficacy*. New York: W. H. Freeman.

Barker, R. (1968) *Ecological Psychology*. Stanford, CA: Stanford University Press.

Barker, R. and Gump, P. (1964) *Big School, Small School*. Stanford, CA: Stanford University Press.

Barker, R. and Schoggen, P. (1973) *Qualities of Community Life*. San Francisco, CA: Jossey-Bass.

Breiting, S. and Mogensen, F. (1999) Action competence and environmental education. *Cambridge Journal of Education*, 29, 3: 349–353.

Carson, R. (1956) *The Sense of Wonder*. New York: Harper & Row.

Chawla, L. (1992) Childhood place attachments. In: Altman, I. and Low, S. (eds), *Place Attachment*. New York: Plenum, pp. 63–86.

Chawla, L. (1999) Life paths into effective environmental action. *Journal of Environmental Education*, 31, 1: 15–26.

Chawla, L. (ed) (2002) *Growing Up in an Urbanising World*. London/Paris: Earthscan Publications/UNESCO.

Chawla, L. and Heft, H. (2001) Children's competence and the ecology of communities: a functional approach to the evaluation of participation. *Journal of Environmental Psychology*, 22: 201–216.

Chawla, L. and Salvadori, I. (2003) Children for cities and cities for children: learning to know and care about urban ecosystems. In: Berkowitz, A., Nilon, C., and Hollweg, K. (eds), *Understanding Urban Ecosystems: A New Frontier for Science and Education*. New York: Springer, pp. 294–314.

Cooper Marcus, C. (1978) Remembrance of landscapes past. *Landscape*, 22, 3: 34–43.

Fuglesang, A. and Chandler, D. (1997) *Children's Participation: A Case for a Strategy of Empowerment in Early Childhood*. Oslo: Save the Children Norway.

Gaster, S. (1991) Urban children's access to their neighbourhoods. *Environment and Behavior*, 23, 1: 70–85.

Gibson, E. and Pick, A.D. (2000) *An Ecological Approach to Perceptual Leaning and Development*. Oxford: Oxford University Press.

Gibson, J.J. (1966) *The Senses Considered as Perceptual Systems*. Boston, MA: Houghton-Mifflin.

Gibson, J. J (1979) *The Ecological Approach to Visual Perception*. Boston: Houghton-Mifflin.

Goodenough, E. (ed) (2003) *Secret Spaces of Childhood*. Ann Arbor, MI: University of Michigan Press.

Harrison, C., Limb, M., and Burgess, J. (1987) Nature in the city: popular values for a living world. *Journal of Environmental Management*, 25: 347–362.

Heft, H. (1988) Affordances of children's environments: a functional approach to environmental description, *Children's Environments Quarterly*, 5: 29–37.

Heft, H. (1989) Affordances and the body: an intentional analysis of Gibson's ecological approach to visual perception. *Journal for the Theory of Social Behavior*, *19*: 1–30.

Heft, H. (2001) *Ecological Psychology in Context*. Mahwah, NJ: Lawrence Erlbaum.

Heft, H. and Chawla, L. (2005) Children as agents in sustainable development: the ecology of competence. In: Spencer, C. and Blades, M. (eds), *Children and Their Environments*. Cambridge: Cambridge University Press, pp. 199–216.

Hungerford, H. and Volk, T. (1990) Changing learner behavior through environmental education. *Journal of Environmental Education*, 21, 3: 8–21.

Jensen, B.B. and Schnack, K. (1997) The action competence approach in environmental education. *Environmental Education Research*, 3, 2: 163–178.

Kytta, M. (2002) Affordances of children's environments in the context of cities, small towns, suburbs and rural villages in Finland and Belarus. *Journal of Environmental Psychology*, 22: 109–123.

Kytta, M. (2004) The extent of children's independent mobility and the number of actualized affordances as criteria for child-friendly environments. *Journal of Environmental Psychology*, 24: 179–198.

Leopold, A. (1949/1966) *A Sand County Almanac*. New York: Oxford University Press. (Originally published in 1949.)

Louv, R. (2005) *Last Child in the Woods*. Chapel Hill, NC: Algonquin Books.

Moore, R. (1989) Before and after asphalt: diversity as an ecological measure of quality in children's outdoor environments. In: Bloch, M. and Pelligrini, T. (eds), *The Ecological Context of Children's Play*. Norwood, NJ: Ablex, pp. 191–213.

Palmer, J., Suggate, J., Robottom, I., and Hart, P. (1999) Significant life experiences and formative influences on the development of adults' environmental awareness in the UK, Australia and Canada. *Environmental Education Research*, 5, 2: 181–200.

Reed, E.S. (1996a) *Encountering the World*. Oxford: Oxford University Press.

Reed, E.S. (1996b) *The Necessity of Experience*. New Haven, CT: Yale University Press.

Rissotto, A. and Tonucci, F. (2002) Freedom of movement and environmental knowledge in elementary school children. *Journal of Environmental Psychology*, 22: 65–77.

Rovee-Collier, C.K. (1986) The rise and fall of infant classical conditioning research. In: Lipsett, L.P. and Rovee-Collier, C.K. (eds), *Advances in Infancy Research*, 4: 139–159. Norwood, NJ: Ablex.

Rogoff, B. (1990) *Apprenticeship in Thinking*. Oxford: Oxford University Press.

Rogoff, B. (2003) *The Cultural Nature of Human Development*. Oxford: Oxford University Press.

Ryan, R.M. and Deci, E.L. (2001) On happiness and human potentials. In: Fiske, S.T., Schachter, D.L., and Zahn-Waxler, C. (eds), *Annual Review of Psychology*, 52: 141–166. Palo Alto, CA: Annual Reviews.

Seligman, M. and Csikszentmihalyi, M. (2000) Positive psychology: an introduction. *American Psychologist*, 55: 5–14.

Schoggen, P. (1989) *Behavior Settings*. Stanford, CA: Stanford University Press.

Stern, D. (1985) *The Interpersonal World of the Infant*. New York: Basic Books.

Stern, P.C. (2000) Toward a coherent theory of environmentally significant behavior. *Journal of Social Issues*, 56, 3: 407–424.

Tanner, T. (ed) (1998) Special issue on significant life experiences. *Environmental Education Research*, 4, 4: pp. 365–463.

Vygotsky, L.S. (1978) *Mind in Society*. Cambridge, MA: Harvard University Press.

Zavestoski, S. (2003) Constructing and maintaining ecological identities. In: Clayton, S. and Opotow, S. (eds), *Identity and the Natural Environment*. Cambridge, MA: MIT Press, pp. 297–315.

Chapter 7
Participation, Situated Culture, and Practical Reason

Heila Lotz-Sisitka[1] and Rob O'Donoghue[2]

Keywords environmental education, participation, reflexivity, situated culture, social learning

7.1 Introduction

This chapter examines the emergence of participatory education as both a central feature and a terrain of ambivalence within the developing landscape of environmental education in South Africa. From its roots in nature experience activities through to more socially critical forms of environmental education, participatory imperatives in this area have yet to address sufficiently the conceptual and practical challenges inherent in pedagogies of participation. We argue that more recent developments reveal similar anomalies, such that participatory education in South Africa has now become an idealised and techniqued logic of practice.

7.2 Participation and Environmental Education in South Africa

Today's patterns of participatory educational practice in capacity development processes and curriculum contexts in South Africa are largely shaped by contemporary perspectives on participation in education for social transformation. Our chapter shows how an individualising turn within these processes, inscribed as they are with democratic ideals, constructivism, and education for sustainable development, suggests that participatory processes in capacity development can become increasingly tenuous and self-referential. Recent studies show that environment and

---

[1]Environmental Education and Sustainability Unit, Education Department, Rhodes University, P.O. Box 94, Grahamstown 6139, South Africa, H.Lotz@ru.ac.za

[2]Environmental Education and Sustainability Unit, Education Department, Rhodes University, P.O. Box 94, Grahamstown 6139, South Africa, r.odonoghue@ru.ac.za

A. Reid et al. (eds.), *Participation and Learning*, 111–127.

sustainability activities can become marked by a pre-inscription of an idealised perspective on participatory learning and action in curriculum settings. In particular, some of the social processes of professional development and curriculum planning – cooperative activity, solidarity, and apparent movement – have created an illusion of change, even as participants engage in activities with a heavily mediated script, for participatory forms of engagement, such that participatory education becomes an idealised process that is not open to critical scrutiny.

In asking whether this is this necessary or inevitable, we draw on a range of perspectives on culture,[1] rationality, and change to aid our understandings of some of the anomalies in current idealised perspectives on environment and participation in education, and to suggest alternative configurations and understandings that will move debate and practice forward.

The advent of environmental education in southern Africa in the early 1980s was shaped by and has shaped the developing imperatives for schoolchildren and communities to participate in the resolution of environmental issues. Since then participation in education has become a key concern that has strengthened with the struggle for democracy in a post-apartheid South African state. Within these processes environmental educators and researchers have embraced participatory perspectives amid the debris of colonial intrusion and after an extended period of colonial and apartheid rule.

A concern for participation was already apparent within the focus on providing nature experiences as part of environmental education in the late 1970s (O'Donoghue 1996, 2006), and has more recently been a central concern in education initiatives in response to globalised socioecological risks, poverty, and HIV/AIDS. Of interest here is how legislation and education have co-evolved as the social and self-regulatory constraint[2] to mitigate environmental risk. In South Africa, we see an emerging history that represents the interplay of policy, law, and educational processes around responses to socioecological risk. This has promoted participatory imperatives in fostering change, including wide-scale public participation in environmental education in the southern African region.

As is also the case elsewhere, we must also note here that all education is implicitly participatory, be it in the form of reluctant presence or willing engagement. Even in extreme cases of social alienation and conflict, as typified in the colonial and apartheid struggles that have characterised the social history of the region, there are clear patterns of an intermeshing of a variety of participatory processes, approaches, and outcomes. The shifting politics of participation is thus central in understanding

[1] Culture is a key word in this chapter, as it refers to the world of knowledge, ideas, objects which are the product of human activity. Note that we do not refer to culture in the functionalist tradition in which culture is seen as ideological force or political power for imposing social order (Bourdieu 1968). Grenfell and James (1998) note that this functionalism can take two forms: positivism and critical radical. Bourdieu's reflexive objectivity opposes any research that reduces its object to an undynamic 'thing'. In this context, culture remains a dynamic interplay of knowledge ideas and objects that are in constant review as products of human activity.

[2] This sensitising notion of a balancing interplay between social constraint and self-restraint is derived from the long-term process sociological perspective of Norbert Elias (e.g. 1994). It serves to locate education processes alongside other shaping social processes in a contemporary democratic society.

the contesting solidarities of social struggle and the more recent concerns for the rights and responsibilities of individuals in a democratic society. Thus, for example, we can find in today's environmental education a greater concern for willing, collaborative participation in education activities associated with increasingly diverse environment, and sustainability movements, issues, and contexts.

Yet despite participatory processes being both constitutive of environmental movements and a key process in the education of others (see, for example, Chapter 19 by Shallcross and Robinson, this volume), little research and debate on participatory education was apparent until the early 1990s in South Africa. However, a decentring of education from an early concern about getting across the facts on environmental matters to collaborative meaning-making was already appearing in the 1980s. Earlier perspectives on participation in education and evaluation research were articulated by O'Donoghue (1986) and by O'Donoghue and Taylor (1988), while O'Donoghue and McNaught (1991) reported a process for developing participatory materials and curriculum. It is notable that these early arguments for participation in education emerged alongside a rapid shift towards democratisation during the last years of apartheid in South Africa. During this period, there were calls for more inclusive, participatory approaches to education practice within an expanding liberation of the oppressed and the democratisation of social life and institutions. But with the demise of the apartheid regime, participation was to take on additional meanings and forms, and it is to these that we now turn.

7.3 Participation as a Developing Focus

The 1990s saw the rapid development of participation as a theme in environmental education course design and research in South Africa. Much of this work was influenced by the 'paradigms debate' in environmental education and interpretations and application of critical theory emanating from Australia to environmental education and research (Robottom 1987; Fien 1993; Robottom and Hart 1993), alongside that of Freirian pedagogy (Freire 1972) and liberation politics (Cross 1999). Of note here is the emergence of an open-entry, open-exit participatory certificate course in environmental education that was developed under the auspices of the Environmental Education Association of Southern Africa (EEASA).[3] This course, influenced by critical theory and an understanding that 'we are all learners and educators' when dealing with uncertainty and risk, served to foreground participation in environmental learning amongst adult learners, and later provided the conceptual scaffolding for diverse participatory courses across the region (Lotz 1999).

Earlier work on participatory curriculum and materials development (O'Donoghue 1986; O'Donoghue and Taylor 1988) was taken forward by researchers such as Lotz

[3] The course was later accredited by Rhodes University, and was taken up as a key offering by the Southern African Development Community's Regional Environmental Education Programme. The course has been adapted for use in six southern African countries.

(1996) whose work centred on participatory materials development in a changing curriculum context, and by Wagiet (1996), focusing on the emancipation of teachers through participatory curriculum development. These research initiatives were inevitably undertaken within participatory research designs (most being participatory action research studies) but were also centred on wider issues of participation in education.[4] For example, Lotz (1996) argued that, for participation in education to be meaningful, there was a need to pay attention to the enabling conditions for participation. Her study was one of the earliest to question a somewhat blind conventional wisdom that participation was the salve for all of the ills of the past.

Through working with, in, and beyond a socially critical orientation, Lotz also challenged some of the assumptions underpinning socially critical environmental education practice (largely based on interpreting the critical theories of the Frankfurt School tradition). At the time, socially critical environmental education and research in the critical paradigm (after Fien 1993; Robottom and Hart 1993) was beginning to manifest itself as ideologically loaded in the ways that it sought to structure the participation of 'The Other' for their emancipation. A developing axis of tension around the principles of intentionality and intervention also erupted in an extended international, online deliberation in January 1999, around the notion of Education for Sustainable Development (ESD) (Hesselink *et al.* 2000, see http://www.iucn.org/themes/cec/education/esdebate.htm). This was a somewhat acrimonious debate that was followed with interest by many environmental educators working in southern African contexts, as the concerns debated there on emerging ESD concepts, good practice, work programmes, and evaluation, were central to helping clarify the substance and direction of participatory processes in the region.

The communicative rationality of the Habermasian/Frankfurt School critical theory discourses also influenced environmental education practice, training, and research in southern Africa throughout the 1990s. At policy and curriculum levels, a participatory policy initiative was established called the Environmental Education Policy Initiative (EEPI) which was later followed by a participatory Environmental Education Curriculum Initiative (EECI).[5] In the research arena, a plethora of critical research initiatives emerged that were focused on an interest in furthering emancipation through participation in education (Naidoo 1992); teacher participation in curriculum development (Mbanjwa 2002; Atiti 2003); community participation in development (Lupele 2003); participatory evaluations (e.g. Kristensen 1999; Mukute 2001); and learner participation in action projects (Ashwell 1992; De Lange 1997).

In the broader arena of natural resource management, participatory methodologies also became the norm in community-based natural resource management and

[4] Here we distinguish between participatory research orientations, and research on processes of participation in education. While many projects developed participatory research designs, little was done in the way of research to understand the *processes of participation in education* as a key concern in environmental education.

[5] See Lotz-Sisitka (2002) for a broader review of these processes in the context of curriculum transformation in South Africa.

community education practices (e.g. Pretty 2002; Fabricius *et al.* 2004), and the technologies of participatory approaches characterised much of the adult education offered in the name of transformation. In most of these cases participation was taken at face value, simply falling into place with the democratic imperatives (ideological and political) of social justice and freedom for all.[6]

7.4 Emerging Discourses that Troubled Assumptions About Participation

In 1999, the collaborative research initiative undertaken within the EECI raised the prospect that participation was 'under-theorised' in environmental education research in southern Africa. This led to a review of participation in environmental education and research, which was deliberated at an EECI research colloquium, a research forum at the EEASA Conference, and later published in the *Southern African Journal of Environmental Education* (O'Donoghue 1999). The paper entitled, 'Participation – an under-theorised icon', raised a number of issues associated with the trend towards participatory educational practices in environmental education. These included a rapid popularising and intellectual co-option of action research; a somewhat 'undifferentiated faith in cooperative interaction'; and an observation that the process and its failures shifted the responsibility/blame to the participants. The paper thus opened up concerns about an emerging institutional tyranny in facilitated participatory imperatives (see also Cooke and Kothari 2001). The processes were characterised by participatory learning spaces (normally developed as choreographed educational activities) for 'the other' to participate in learning activities and to clarify 'their' needs and perspectives. An axis of tension between inviting participation in an open-ended sense and specifying or scaffolding activities to enable and shape this, shaped participation as a contested concern that was not well differentiated in any detail. The benign guiding hand of facilitation could thus be seen to retain a somewhat 'hidden' tyranny and a subverting ambivalence that was neither fully understood nor trusted.

O'Donoghue's review stimulated a renewed critical interest in work on participation in environmental education. Many of the underlying theories and assumptions about participatory practice were examined in subsequent debates, as the discourse shifted from participatory imperatives and the use of participatory research methods to a concern for environmental learning and social change through participation in education.

Some of the work in this area included a review of participatory processes in the open-entry, open-exit certificate course (Janse van Rensburg and le Roux 1998) and work by Lotz (1999) on curriculum deliberation in participatory curriculum development

[6] For a critique of this phenomenon, see Babikwa (2004).

for courses in professional development. This work engaged some of the earlier naïve assumptions of participatory curriculum development amongst adult learners and the prevalence of particular technologies of participation in these contexts. Price (2004) illustrated this work further, suggesting that there was a potential for inherent conservatism in participatory deliberations (where everything is accepted at face value, just because it has been said by a participant in the group), and arguing that 'facilitators/educators' have an ethical responsibility for considering the implications of the deliberations of the 'best possible' truth in any given context. Recent research by Lotz-Sisitka and Burt (2004) has also pointed to a number of emerging paradoxes and axes of tension as participation becomes legislated and institutionalised within transformation-oriented policy frameworks,[7] and in this, their research has identified the shifting meaning of participation in institutional narratives. Indigenous knowledge research by O'Donoghue (2003), Masuku (1999), and O'Donoghue and Neluvhalani (2002) has also opened up questions about the relationships between epistemology and participation in an African educational context, noting for example, that forms of participatory and democratic processes were apparent in African societies before colonial intrusions and the decades of marginalising oppression of the indigenous peoples that followed.

Such research initiatives have led to a broadening and strengthening of the contextual focus in research design (Lotz-Sisitka 2003), and researchers wanting to contribute to the democratisation of education and social life have found themselves no longer trapped within the horizons and confines of the 'critical paradigm' and associated participatory research as transformative pedagogy, that had held sway in the field of environmental education in the 1990s. Most of the current research on participation in education is directed at an interest in deepening democracy and enhancing or strengthening meaning-making through participation in education processes to engage environmental issues and risk (see, for example, Lotz-Sisitka and Burt 2004; Babikwa 2004; Price 2004). It is in this socio-historically and contextually informed deliberation that participation remains a central theme in environmental education in southern Africa.

7.5 Some Recent Institutional Developments

Participatory techniques and institutional frameworks

Following the 'ice-breakers' of early participatory training and nature experience fame, the institutionalisation of participation in education in South Africa has more recently been associated with rationales that have developed with the emergence of

[7] The study in question (see Lotz-Sisitka and Burt 2004) is investigating the implementation of participation as an organising structure for integrated water resources management, as legislated in the National Water Act (RSA 1998b). This trend towards institutionalising participation in environmental management practices is also visible in the National Environmental Management Act (RSA 1998a).

a democratic society and an increase in international environment and development aid projects. Within the latter, participation became a key feature through facilitator training in participatory techniques, in particularly, participatory rural appraisal (PRA) and more recently, participatory learning actions (PLA). The two are now seen as synonymous by many and are characterised by a plethora of techniques for fostering participation in education activities (see, for example, Hope and Timmel 1986; Pretty 1992; Sugerman *et al.* 2000).

Within curriculum processes in formal education, participation has become similarly 'techniqued' as teachers are inducted into various strategies for group work and collaborative learning. These techniques used as methods to foster learner participation have become a popular conventional wisdom, driving much professional development for curriculum transformation. As a result, teachers 'equate' the new curriculum with these techniques, and 'success' is equated with group work and collaborative learning (Lotz-Sisitka and Raven 2001; Taylor and Vinjevold 1999). There is also a trend towards inscribing the possibilities for participation and action in policy frameworks. For example, in its assessment standards the new national curriculum suggests that all Grade 3 learners should ideally participate in a recycling project; and all Grade 9 learners should participate in an audit of water use, and contribute to the formulation of a school environmental policy (DoE 2002), while new water legislation requires all stakeholders in a catchment to participate through various catchment management structures and procedures (RSA 1998b, National Water Act).

As noted above, much of the deliberation around participation in education has been closely associated with the features of a developing democracy and the imperatives of post-apartheid transformation. New forms of institutionalised participation in education touched on above are a developing perspective in South Africa that we wish to probe in more depth below. Our particular interest is in examining how these forms of participation in education relate to notions of situated culture and reflexive learning processes in contexts of socioecological risk. In the next section, we explore the trend towards an increased institutionalisation of participation in education through a brief review of emerging pedagogic processes for social transformation and the range of technologies of participation and conventions associated with them.

7.6 The Shaping of Participation in Education for Social Transformation

Popkewitz (2000, drawing on Foucault and Bourdieu) notes that there is a need to consider how institutional narratives in education shape possibilities for participation and action, creating differing patterns of inclusion and exclusion within institutional governmentalities. In South Africa, participation in education developed as a key trend with the emergence of mediated group activities, most notably in training and rural development initiatives. These activities functioned as a scaffolding pedagogy for participants to co-engage with and take up ideas on environment and sustainability before taking these forward in ways that were meaningful to them in their practice. The approach demands that past theory and materials are cleared away and displaced by a series of carefully scaffolded educational activities ideally drawn from local cases and contexts. The process is intended to create a space for revising ideas and developing more relevant materials within an emerging consensus among participants, and as they navigate their way through a series of carefully crafted cooperative learning activities under the mediating hand of the facilitator.

Existing materials and associated dispositions – except those of the participants – are explicitly excluded to create an open-ended agenda for structured deliberations. For example, 'old' or published examples of lesson plans would not be brought into a teacher workshop, as the assumptions guiding this process are that teachers should develop the lesson plans 'on their own' and in social interaction with each other in the workshop, normally using a framework for lesson planning provided by the facilitator. This is designed to allow participants opportunities to construct new possibilities that they define in ways that are personally meaningful. The process includes the use of strategies that ensure a mutual respect for differences of opinion within the group (e.g. setting agreed upon codes of conduct at the start of the meeting). What typically happens is that the established social conventions (of say, lesson planning) are simply held back and mediated into the developing process by some of the participants (as they are asked to come up with new strategies for lesson planning) and with the facilitator playing a subtle but key signifying role (as the facilitator creates the opportunities for the teachers to 'apply' and 'create' knowledge of lesson planning using his/her framework, and as these contributions are affirmed through positive responses from the facilitator and others as the task unfolds).

Within such processes, the post-apartheid period has become characterised by a pedagogical rhetoric of 'out with the old and in with the new'. Of note is an individualising disposition where the new is to be carried into effect by the renewed individual who has cooperatively moved beyond the dispositions of the past. These processes are currently playing out in diverse participatory renewal discourses that are centred on setting aside the past through transformation of the individual[8] so that they give effect to the necessary change in their local context.

These forms of 'personalised, co-constructive reconstruction' through participation in education processes have developed on an institutional terrain of slow and failed change, where policy has not always been successfully brought into effect, and where a lack of capacity has come to explain the problem of failed transformation (see, for example, Lotz-Sisitka and Olivier 2000). Capacity development (also known here as professional development) has thus become the mantra for meaningful transformative change, and participation in education a touchstone for democratic processes of meaningful learning.[9]

---

[8] See also Bauman (2001) and Beck and Beck-Gernsheim (2003) for a broader analysis of this social phenomenon.

[9] The latter, meaningful learning, is often described using the language of social constructivism after Vygotsky (see, for example, Janse van Rensburg and Lotz-Sisitka 2000).

7.7 The Emergence of Participation in Education for Capacity Building

Institutional tensions and an emphasis on participation in education appear to have shaped a participatory turn where environment and sustainability ideals for 'the other' are passed on for re-differentiation in cooperative capacity development activities. This is done within activities that scaffold the interactions under the mediating hand of capacity development trainers (professional developers) with the assumption that the environment and sustainability ideals will then play out in everyday contexts once 'the other' is suitably empowered through the professional development activities. The facilitator, who is committed to a democratic design, is confident in the communicative rationality within these scaffolded learning activities and enthuses participants to make of, and take from, activities, that which is personally meaningful for them to take forward into their practice. These experiences are often positively felt by 'the other', as participants are given the 'freedom' to 'make of' and to 'take from' the activities that which is meaningful to them. The choreography of this curriculum process remains hidden from view, and participants are encouraged to invest in learning interactions within which everyone can achieve success,[10] as clarity emerges.

A key feature of participatory pedagogies for capacity development and social transformation is the explicit exclusion of existing materials and historical conventions. The checks and balances provided by contextual history (involving other experience, opinions, and reflexive social struggle that shapes moral orientation and ethical disposition) are displaced by workshop exhortations to work with what each brings and with what becomes personally meaningful.

7.8 Probing Participation in Education as a Popular Culture of Propositional Play

What we see here is that the ideals of participation in education for democratic, social transformation were simply applied to address the need for capacity development. More of the same thus became a developing game within which failures were ascribed to the challenging task and the demands of transformation. Of note here is

[10] In our experience it is difficult to establish the 'success' of these interactions beyond the 'felt success', given that the activities are often not applied and there is often very little evidence available to construct other indicators of success, beyond the artefacts created through the interactive processes in workshop contexts. The limitations of this 'workshop culture' are being felt in educational circles in South Africa as departments of education seek more sustained approaches to professional development (see, for example, NEEP-GET 2004).

how an enthusiasm for participatory democracy left no space for questioning the culture of the developing participatory ideology and its associated pedagogical assumptions.

What often appears to happen is that participants learn and soon master a culture of participation in education as scaffolded play, often simply playing a democratically mediated rhetorical game of constructing ideas and ideals in an unthreatening propositional arena[11] somewhat removed from the realities of the everyday. The emerging culture of democratic participation in education has thus been shaped as an engaging but somewhat hollow, performative terrain of mediated talk around shared convictions that are not easily carried into practical effect in the everyday. Added to this, because of the deflection of the performative problems of governance to that of a problem of capacity, there are still institutions without the structures to support and give effect to the outcomes of participatory educational activities such as this. For example, Mbambisa's (2005) research has shown that subject advisors (who have attended numerous participatory workshops on lesson planning) are still unable to provide support in lesson planning to educators in schools owing to a range of structural and other constraints that were not considered in the professional development workshops as these focused on the ideals of lesson planning. Similarly Motsa's (2003) research has shown that while meaningful participation has taken place in group-based activities in the context of a course, participants are often unable to apply ideas presented in the professional development course in context, owing to analogous structural and other constraints that were not engaged during the course processes.

To reconcile aspects of these intermeshed axes of tension it is useful to note that in drama education and psychology (the arenas that primarily gave rise to participatory pedagogies of generative role play), there has been a clear differentiation and an interplay between creative play time and reflexive deliberation for signifying meaningful ideas and ideals. Pedagogical dilemmas appear to arise where the processes are conflated within the scaffolded activity as a mediating abstraction (propositional ideals), with the expectation of a personalised, knock-on re-enactment of the process for the democratisation and transformation of work place and community.

The worm in the apple of these emerging pedagogies of participation in education is an equating of a culture of reflexive activity for changed practice with a situated social process of reflexive practice, and a conflating of these within a single process of facilitated participation in scaffolded, choreographed education activity. Despite the enthusiasm of many of those involved, such facilitated participatory workshop processes are proving to have similar limitations to early participatory technologies used in nature experience and critical emancipation processes. First, the withholding and exclusion of the present (including its histories) to create the illusion of reconstructive freedom can often create a hollow and rudderless process. Second,

[11] There is, for example, no imperative to use the lesson plans developed in these processes, as participation in the facilitated process of developing the lesson plan becomes an end in itself.

alongside this, there is the questionable assumption that a facilitatory scaffolding of the arena of 'solicited propositional co-construction' constitutes a constructive engagement with the realities of the everyday (see Moll 2002). Third, there is the questionable assumption that the developing blend of reflexive and collaborative activity will actually roll out into meaningful processes of both work place and community transformation.

Thus, learning in participatory ways within scaffolded activity (a culture of propositional activity for changed practice) cannot be equated with, and assumed to run on into, situated process of struggle for/to change (reflexive change in situated patterns of practice).

7.9 How Participation is Playing Out in Environmental Education

O'Donoghue and Russo (2004) have argued that the participatory induction of individuals into the ideas and ideals of environmental education has come to characterise much of the professional development in environmental education today. These processes often take the form of educators being inducted into a participatory logic for democratic, socioecological change within a participatory process of professional development. The same participatory training process is then carried into continuing environmental education practice to give effect to participatory education, to enable school pupils to foster the desired change in current patterns of practice.

It is only when one looks closely at what is being taught and how this is being done that it is possible to begin to see how participatory pedagogies such as these are prone to all of the dilemmas elucidated above. Indeed, it is notable that in recent professional development activities there are cases where no content is specified as this must be brought into the process by the participants for the process to be meaningful. Similarly, asking what is being done is deferred as a question to a focus on the processes of participation as directed by the developing deliberations, while in cases such as this, there are seldom any published materials to consider after the event.

Such situations have led to a process of participatory induction initiating participants into a participatory culture of propositional ideals. And it is here that the communicative rationality of this movement[12] and its culture of democratically facilitated change (the cultural logic of techniqued participatory practice within the movement) become fused with an intergenerational social habitus and logic of practices in a local cultural setting (the situated culture and sustaining patterns of practical reason in a given context).

[12] Arising from the same roots as critical theories that dominated the field in the late 1980s and 1990s. Popkewitz and Brennan (1998) critique these as being characterised by a philosophy of consciousness.

7.10 Participation and Practical Reason in Developing Context

The possible dissonance between a facilitating logic for participatory practice and the more robust situated logic of intergenerational social practice can become obscured by conflating circularities and ideological closure, as in such emerging democratic genres of participatory environmental education for fostering social change. It occurs in participatory curriculum projects and participatory professional development courses made up of scaffolded and democratically mediated activity centred on ideals of cooperative learning. The recent trend towards this ideal and a reliance on technologies of participation in environmental education is currently characterised by:

- Participatory technologies and democratic approaches that become 'doxa' (representing an orthodox sense of reality) (e.g. PRA methodologies are almost synonymous with development practice and choreographed activities in workshops are almost synonymous with professional development)
- Individualisation within a dialectic process which promotes alienating inversions that foster group solidarity within a developing movement (in the examples described above which focus on personal expression and feel good outcomes through group interaction)
- Establishing system models which become self-referential (e.g. conventional wisdoms surrounding participatory practice become self-referential, as shown in Price 2004)
- Revised differentiation which creates changed patterns of inclusion and exclusion (e.g. inclusion of own perspectives, and exclusion of prior histories and experiences, as described above)
- Institutional governmentalities that download their functions to the citizenry with education as the means of democratic self-governance (seen in governance trends that focus on capacity development rather than on delivery)

Many of these concerns are evident in recent trends towards participation in education. They are tending to inscribe particular (often limited) possibilities for participation and action (as illustrated by the lesson planning example above). In clarifying these concerns, we have found the work of Bourdieu (1998) and his theory of action (practical reason) useful. This has provided a lens through which to begin exploring 'a practice that is reasoned' in a reflexive context of action, situated culture and risk (as a co-engaging clarification and situated response in the cultural context of socio-ecological risk). To use the example of lesson planning again, a different scenario to the one painted above would be involving teachers and learners in investigative activities of local environmental concerns as a basis for the lesson planning (rather than an attempted mapping of a choreographed idealisation within facilitated activity). The scenario would change to involve local deliberations in context, lesson planning would derive from an engagement with existing situated cultures, and reflexive actions would be undertaken by learners and educators.

This places participatory pedagogies that have a preordained culture of 'reasoned practice' with propositions for driving action and changing culture through 'inscribed technologies of participation' in stark relief. It allows one to begin to question and probe the implicit features of the forms of participatory pedagogical intervention as we have begun to do above. The vantage point of situated culture and practical reason allows us to begin the difficult task of exploring the troubling paradoxes inherent in many contemporary orientations to participation in education.

The clarification of these questions is part of a broader debate on the relationship between material, objective structures, and an individual agent's mental activity, which, according to Grenfell and James (1998:12) is 'part of the social theoretical tradition'. Bourdieu's theory of action (1998) considers a 'science of dialectical relations between objective structures ... and the subjective dispositions within which these structures are actualised and which tend to reproduce them' (Grenfell and James 1998:3). This interplay between structure and agency seeks to emphasise the importance of the productive activity of consciousness in socioecological contexts. Bourdieu's theory of practice emphasises the relationships between social agents and the structural, generative schemes that orientate agency and social practice. Practice or rather, praxis, according to Bourdieu, is structured and tends to partly reproduce the structures[13] of which it is a part (Grenfell and James 1998). In this regard, Archer (2000) argues for emergence, in which structural elaboration takes place through agential interactions in structural (socioecological) context. In this context, change comes about through a dynamic reproduction and elaboration of human activity in developing context, reflecting a method which Bourdieu eventually refers to as a 'structural constructivism' (Grenfell and James 1998:14), and which Archer refers to as a 'morphogenetic approach'. Here, constructivism pertains to the 'dynamic reproduction of human activity in ever-changing contexts; structuralist (refers to) the relations of those involved' (Grenfell and James 1998:13), and agency is associated with social interactions in context. This epistemology or outline of a theory of practice therefore emphasises some of the challenges in participation in education within deliberative processes of reasoning in and out of situated, socioecological contexts. It presents an argument for the foregrounding of situated culture and engagement in socioecological context in/as participatory learning processes.

7.11 Concluding Synthesis

This chapter has traced the emergence of an interest in participation in South African environmental education. It has outlined the close relationship between participation and emerging trajectories towards democracy, noting that earlier

[13] Note that we use the notion of structure here in the way in which Bourdieu refers to it. Not a structuralism that is foundational and transcendent, transcending time and cultures and innate to the human mind. Bourdieu reacts against this kind of structuralism. For him, structure is dynamic and dialectical, is manifest in links at and between the objective and subjective levels of human contingency, links which are both structural *and* structuring (Grenfell and James 1998).

trends towards participatory practice in environmental education were heavily influenced by early nature experience perspectives and later critical theories originated in the Frankfurt School. The latter gave rise to participation and an associated communicative rationality that informed a set of facilitating pedagogies for mediated socioecological change. This influence shaped a focus on empowerment of 'the other', a paradoxical structural functionalism, and the development of a wide array of participatory technologies that have seen a popularising of participatory practice for democratising transformation through capacity (professional) development.

The chapter has also opened up an emerging discontent with simplistic assumptions and technologies of participation, and reviewed developing critiques of participatory practice in South African environmental education. We have documented emerging critiques of participatory practice; notably, a latent conservatism in many participatory approaches (Price 2004); an increased institutionalising appropriation of participatory practice, with an associated change in what participation implies (Lotz-Sisitka and Burt 2004); and a concern for situated culture and participatory practice in/through indigenous knowledge research (O'Donoghue 2003; O'Donoghue and Neluvhalani 2002).

In drawing these 'threads' of critique together and analysing how participatory practices are playing out in democratic professional development contexts, we have argued that the current imperative to scaffold and choreograph activity for mediated participatory engagement towards changed practice remains flawed. Bourdieu has argued the need to explore 'a practice that is reasoned' in a reflexive context of action, situated culture and risk, and we consider Archer's work on critical realist social theory a helpful source of analytical tools for probing situated culture, structure, and agency.

Current patterns of practice in participatory education favour individualised meaning-making approaches that are disembedded from the realities of everyday life. Our analysis points to a need to probe the roles of habitus, culture, structural conditioning, emergence, and power relations in this, as key dimensions to 'reality-congruent' participatory processes and of the development and research agenda in this field.

In undertaking this review of participation in education our hope has been to open up and critically examine how emergent patterns of participatory pedagogical practice in South Africa are proving troubling as democratic practice and for effectively engaging the realms of local environmental concerns as reflexive pedagogical practice. The current evidence suggests that although some insight and research evidence is emerging, clarifying and developing participatory education activities to foster democratic ideals of change is not an easy matter.

Acknowledgments We wish to acknowledge the work of three doctoral researchers working alongside us to engage these issues in education, notably Jusin Lupele, Jane Burt, and Edgar Neluvhalani. The work of other doctoral scholars participating in our research programme, notably Leigh Price, Glenda Raven, and Daniel Babikwa, has also contributed to this chapter.

References

Archer, M. (2000) *Being Human: The Problem of Agency*. Cambridge: Cambridge University Press.

Ashwell, A. (1992) Project Water (Grahamstown): A Case Study of the Development of an Environmental Education Project. Unpublished master's thesis. Grahamstown: Rhodes University.

Atiti, B.A. (2003) Review of Development of Environmental Interpretation Resources to Foster Environmental Learning in Two Kenyan Schools. Unpublished master's thesis. Grahamstown: Rhodes University.

Babikwa, D. (2004) Tensions, contradictions and inconsistencies in community-based environmental education processes: the role of defective educational theories. *Southern African Journal of Environmental Education*, 21: 58–77.

Bauman, Z. (2001) *The Individualized Society*. Cambridge: Polity Press.

Beck, U. and Beck-Gernsheim, E. (2003) *Individualisation*. London: Sage.

Bourdieu, P. (1968) Structuralism and theory of sociological knowledge. *Social Research*, 35, 4: 681–706.

Bourdieu, P. (1998) *Practical Reason*. Cambridge: Polity Press.

Cooke, B. and Kothari, V. (eds) (2001) *Participation*: The New Tyrany? London: Zed Books.

Cross, M. (1999) *Imagery and Identity in South African Education, 1880–1990*. North Carolina: Carolina Academic Press.

De Lange, I.A. (1997) n Gevallestudie in Omgewingsopvoeding: Die Skole Waterprojek (SWAP). A case study in environmental education: the Schools Water Project (SWAP). Unpublished Masters thesis. University of Stellenbosch.

DoE (Department of Education) (2002) *Revised National Curriculum Statements Grade R-9 (schools) Policy*. Pretoria: Department of Education.

Elias, N. (1994) (1939) *The Civilising Process: The History of Manners and State Formation and Civilisation*. Oxford: Blackwell.

Fabricius, C., Koch, E., Magome, H., and Turner, S. (eds) (2004) *Rights, Resources and Rural Development: CBNRM in southern Africa*. New York/London: Earthscan.

Fien, J. (1993) *Education for the environment. Critical Curriculum Theorising and Environmental Education*. Victoria: Deakin University Press.

Freire, P. (1972) *Pedagogy of the Oppressed*. London: Penguin.

Grenfell, M. and James, D. (1998) *Bourdieu and Education: Acts of Practical Theory*. London: Falmer.

Hesselink, F., van Dempen, P.P., and Wals, A. (eds) (2000) *ESDebate: International Debate on Education for Sustainable Development, Gland and Cambridge*. IUCN Commission on Education and Communication Publications Services.

Hope, A. and Timmel, S. (1986) *Training for Transformation. A Handbook for Community Workers*. Zimbabwe: Mambo Press.

Janse van Rensburg, E. and le Roux, K. (1998) *Gold Fields Participatory Course in Environmental Education: An Evaluation in Process*. Grahamstown: Rhodes University Environmental Education and Sustainability Unit.

Janse van Rensburg, E. and Lotz-Sisitka, H. (eds) (2000) *Learning for Sustainability: An Environmental Education Professional Development Case Story Informing Education Policy and Practice*. Johannesburg: Learning for Sustainability Project.

Kristensen, J. (1999) An Evaluation of the Life Science Curriculum in Namibia and Its Development Through Stakeholders' Perceptions of 'learner centred education'. Unpublished master's thesis. University of Stellenbosch.

Lotz, H. (1996) The Development of Environmental Education Resource Materials for Junior Primary Education Through Teacher Participation: The Case of the We Care Primary Project. Unpublished doctoral thesis. University of Stellenbosch.

Lotz, H. (1999) *Developing Curriculum Frameworks: A Source on Environmental Education Amongst Adult Learners*. Howick: SADC Regional Environmental Education Centre.

Lotz-Sistka, H. and Raven, G. (2001) *Active Learning in OBE: Environmental Learning in South African Schools*. National Environmental Education Programme – General Education and Training pilot research project. Pretoria: Department of Education.

Lotz-Sisitka, H. (2002) Curriculum patterning in environmental education: a review of developments in formal education in South Africa. In: Hattingh, J., Lotz-Sisitka, H., and O'Donoghue, R. (eds), *Environmental Education, Ethics and Action in Southern Africa*. Pretoria: Environmental Education Association of Southern Africa/Human Sciences Research Council, pp. 97–120.

Lotz-Sisitka, H. (2003) Social science as a form of social praxis? A review of a selection of environmental education research in southern Africa. In: Azeiteiro, U., Goncalves, F., Fihlo, W., and Pereira, M. (eds), *World Trends in Environmental Education*. Frankfurt: Peter Lang, pp. 23–152.

Lotz-Sisitka, H. and Burt, J. (2004) *A Critical Review of Participatory Practice in Integrated Water Resource Management (IWRM)*. International review. Rhodes University. Project No: K5/1434 (http://campus.ru.ac.za/download.php?actionarg = 1307). Accessed: 19 January 2005.

Lotz-Sisitka, H. and Olivier, C. (2000) Teachers as curriculum developers: insight from the learning for sustainability project. In: Janse van Rensburg, E. and Lotz-Sisitika, H. (eds), *Learning for Sustainability: An Environmental Education Professional Development Case Story Informing Education Policy and Practice*. Johannesburg: Learning for Sustainability Project, pp. 69–102.

Lupele, J. (2003) Participatory materials development in rural Zambia. *Southern African Journal of Environmental Education*, 19: 85–96.

Masuku, L. (1999) The Role of Indigenous Knowledge in/for Environmental Education: The Case of an Nguni Story in the Schools Water Action Project. Unpublished Masters thesis. Grahamstown: Rhodes University.

Mbambisa, N.P. (2005) Strategies Used by Subject Advisors and Facilitators to Support Lesson Planning with an Environmental Learning Focus: A Case Study of the Eastern Cape NEEP–GET Cluster. Unpublished master's thesis. Grahamstown: Rhodes University.

Mbanjwa, S. (2002) The Use of Environmental Education Learning Support Materials in OBE: The Case of the Creative Solutions to Waste Project. Unpublished master's thesis. Grahamstown: Rhodes University.

Moll, I. (2002) Clarifying constructivism in a context of curriculum change. *Journal of Education*, 27: 5–32.

Motsa, E.M. (2003) Praxis-based Assignments for Professional Development in the Rhodes University/Sawziland Particpatory Course in Environmental Education: The Case of Assignment Four. Unpublished master's thesis. Grahamstown: Rhodes University.

Mukute, M. (2001) An Evaluative Case Study of Curriculum Development and Implementation in PELUM College. Unpublished master's thesis. Grahamstown: Rhodes University.

Naidoo, P. (1992) Collaborative Teacher Participation in Curriculum Development: A Case Study in Junior Secondary General Science. Unpublished master's thesis. Grahamstown: Rhodes University.

NEEP-GET (2004) *Cluster Based Professional Development Booklet*. National Environmental Education Programme-General Education and Training. Howick: Share-Net.

O'Donoghue, R. (1986) Environmental education and evaluation: a eleventh hour reconciliation. *Southern African Journal of Environmental Education*, 3: 18–21.

O'Donoghue, R. (1996) Detached Harmonies: A study in/on Developing Social Processes of Environmental Education in Eastern Southern Africa. Doctoral thesis. Grahamstown: Rhodes University.

O'Donoghue, R. (1999) Participation. An under-theorised icon in research and curriculum development. *Southern African Journal of Environmental Education*, 19: 14–27.

O'Donoghue, R. (2003) Indigenous knowledge: towards learning materials and methodologies that respond to social processes of marginalisation and appropriation in eastern southern Africa. *Australian Journal of Environmental Education*, 19: 57–68.

O'Donoghue, R. (2006) Locating the environmental in environmental education research: a review of research on nature's nature, its inscription in language and recent memory work on relating to the natural world. *Environmental Education Research*, 12, 3/4: 345–357.

O'Donoghue, R. and McNaught, C. (1991) Environmental education: the development of a curriculum through 'grass-roots' reconstructive action. *International Journal of Science Education*, 13, 4: 291–404.

O'Donoghue, R. and Neluvhalani, E. (2002) Indigenous knowledge and the school curriculum: a review of developing methods and methodological perspectives. In: Janse van Rensburg, E., Hattingh, J., Lotz-Sisitka, H., and O'Donoghue, R. (eds), *EEASA Monograph: Environmental Education, Ethics and Action in Southern Africa*. Pretoria: EEASA/HSRC, pp. 121–134.

O'Donoghue, R. and Taylor, J. (1988) Towards participant-centred resource development for environmental education. *Southern African Journal of Environmental Education*, 7: 3–6.

O'Donoghue, R. and Russo, V. (2004) Emerging patterns of abstraction in environmental education: a review of materials, methods and professional development perspectives. *Environmental Education Research*, 10, 3: 331–351.

Popkewitz, T.S. and Brennan, M. (eds) (1998) *Foucault's Challenge: Discourse, Knowledge, and Power in Education*. New York: Teachers College Press.

Popkewitz, T.S. (2000) Reform as the social administration of the child: globalization of knowledge and power. *Globalization and Education*. London: Routledge.

Pretty, J. (2002) *Agriculture: Reconnecting People, Land and Nature*. London: Earthscan.

Price, L. (2004) Participatory curriculum development: lessons drawn from teaching environmental education to industry in Zimbabwe. *Environmental Education Research*, 10, 3: 401–407.

Republic of South Africa (RSA) (1998a) *National Environmental Management Act. No. 107* (www.environment.gov.co.za). Accessed: 19 January 2005.

Republic of South Africa (RSA) (1998b) *National Water Act. No. 36* (www.dwaf.gov.za). Accessed: 19 January 2005).

Robottom, I. (1987) Two paradigms of professional development in environmental education. *The Environmentalist*, 7, 4: 291–298.

Robottom, I. and Hart, P. (1993) Towards a meta-research agenda in science and environmental education. *International Journal of Science Education*, 15, 5: 591–606.

Sugerman, D.A., Doherty, K.L., Garvey, D.E., and Gass, M.A. (2000) *Reflective Learning: Theory and Practice*. Dubuque, IA: Kendall/Hunt.

Taylor, N. and Vinjevold, P. (1999) *Getting Learning Right: Report on the President's Education Initiative Research Project*. Johannesburg: Joint Education Trust.

Wagiet, F. (1996) *Teaching the Principles of Ecology in the Urban Environment: An Investigation into the Development of Resource Materials*. Unpublished master's thesis. Grahamstown: Rhodes University.

Chapter 8
From Practice to Theory: Participation as Learning in the Context of Sustainable Development Projects

Paul Vare

Keywords learning, participation, development, communication, activity theory, practice

8.1 Introduction

This chapter starts with an autobiographical account of my experience of creating communication programmes within sustainable development projects in rural Africa. It charts the evolution of a participatory approach before turning to investigations into the way people learn in such settings. I then apply the resulting view of learning as a complex process of dialogue primarily relying on known and trusted sources to other rural communication programmes. Noting that accounts of these projects have largely relied on empirical evidence, the second part of the chapter represents a search for appropriate theoretical underpinning. I show that the concepts of *zone of proximal development* (ZPD) and *legitimate peripheral participation* are relevant to the project experience, while possible limitations in these approaches are tackled by the application of *cultural-historical activity theory* (CHAT). In conclusion, I note that the principles of situated learning and activity theory resonate strongly with real-world examples of education for sustainable development and suggest that although hitherto separate, they might become more closely aligned – albeit with a few words of caution.

8.2 Background

I never liked 'theory'. I wanted to make a difference, so I went into the world to *do* something. Of course, one wants to do things well and like so many practitioners before me, I found myself *thinking* about what we were doing. And so, from being

Learning South West, Bishops Hull House, Bishops Hull, Taunton, Somerset, TA51EP, UK, Learning4L@aol.com

A. Reid et al. (eds.), *Participation and Learning*, 128–143.

stuck on a rainforest track, scratching mosquito bites, I found myself starting to wade into theory, barely noticing the sleet on the library windows.

In writing this chapter, I have tried to make sense of these two worlds. Happily, the subject matter itself has helped me to understand that these are not two worlds at all. Starting points are always problematic; rather than beginning with a theoretical tradition, I have chosen to start my story in a time and place where I was most isolated from literature, whilst developing rural communication programmes in an increasingly participatory manner. After a foray into the world of research, the story returns to 'the field' where more ideas are put into action. The chapter concludes with an attempt to identify concepts that underpin these experiences of participation and learning. I have deliberately chosen to write in a style that reflects my perspective as 'practitioner looking at theory', rather than that of an academic researching practice.

8.3 Uganda 1991: Common Sense(s)

The Mount Elgon Conservation and Development Project was a joint venture of the Government of Uganda and IUCN, The World Conservation Union. They hired me (with a background in environmental education) as the Education/Extension Advisor. The project aimed to protect the threatened Forest Reserve (now a National Park) while supporting sustainable rural development among the 230,000 people living around the forest. The *Terms of Reference* for my involvement in the project focused on raising awareness, disseminating information, and training others to support the project aims.

As a trained teacher, it seemed self-evident or common sense that the best way to learn would be to encourage people to develop their own ideas and solutions rather than to present blueprints. The top-down education programme prescribed by the project document was quickly upended to become a programme of 'finding out', with the awareness-raising component shifting away from conveying information and towards placing ideas and suggestions into dialogue and debate.

Among our six strong senior team, there were conflicting views; it appeared that our common sense was not actually common to all of us. Our arguments revealed the tensions between project management with its concern for measurability and 'rigour', and the apparently haphazard human development processes taking place in our project area. Whilst we all wished to conserve biological diversity, there was resistance to the notion of a diversity of solutions arising from engagement with the numerous communities living around the protected area. This took place before the term 'participation' had found its way into every pore of the development world.

The context of the education programme

Education had the smallest budget on the project, most of which had been pre-designated to producing materials. The project area was remote and mountainous and literacy rates were low, particularly among the youth – a function of isolation and years of civil unrest. Against this, the thirst for knowledge and self-improvement was immense.

In attempting to promote sound environmental behaviour, it was striking how everyone wanted to educate everyone else. Among the local elite there was no shortage of offers to host meetings or place posters in villages. Meanwhile, in more remote settlements, we were providing people's first physical contact with media such as dramas, posters, and leaflets.

After 18 months, we evaluated the impact of our communication programme. We discovered that people were grateful for our efforts but virtually no new messages had been learned from them. Any behaviour change had been a reaction to the threat of an armed 'task force' of forest guards. However, people *were* learning about the possibility of developing drama or posters because the project had introduced them.

Handing over the tools of communication

Our next step was to reorient our programme so that the materials could be produced by the very communities who would 'consume' them; in this way, the language of 'target audiences' became redundant in favour of 'participants'.

By describing our approach as *gradually diminishing control* (a term borrowed from language instruction), we imbued an apparently haphazard, exploratory process of media development with a pseudoscientific aura. This helped to 'sell' the approach to our critical community at national and international level, that is, those who monitored our budgets.

Around this time, the education team attended a workshop facilitated by a British artist and author, Bob Linney. This gave us the practical tools and confidence to pursue our bottom-up approach. Bob differentiated between *participatory* approaches (as if people must be invited to participate in their own development) and *people-centred* approaches where people carry out communication activities according to their own agenda (Linney 1995).

By taking media production to communities, we were able to tap the goodwill of volunteers while giving groups of local people opportunities to explore their own issues, albeit in the context of conservation and development. The thinking was simple: as any advertising agency knows, to distil a complex issue into a snappy slogan requires a great deal of analysis. Why confine this learning to the educated project staff? Surely, the analysis had better be undertaken within the community so that local people can appreciate the complexity of the issue.

Posters were designed and screen-printed in remote villages and when an official complained that they were meaningless outside the project area, we remarked that our glossy English posters, which graced the Government offices in Kampala, were meaningless inside the project area.

Drama shows, telling local (sometimes subversive) stories, were developed and performed widely, and radio programmes, with carefully timed schedules of songs, interviews, and stories, were recorded and broadcast on local language services.

The people involved gained a deeper understanding of key issues through the process of 'reducing' them, and became more media savvy in the process. Others living in the project area experienced a media campaign that reflected their own lives. A further spin-off was that the project learned about local perspectives on conservation and development issues.

Initially, this participatory approach was at odds with the project's top-down conservation work. This changed when an external evaluation mission noted the disconnected nature of the project's activities; good participatory practice was being encouraged elsewhere (Chambers 1983) and our project model should follow suit. A *collaborative forest management* approach was suggested and the project was redesigned with the assistance of experts in participatory rural appraisal. Participation was official.

8.4 Looking into Learning

Returning to England, I directed a European Union-funded research project called *ECoSA: Education and Communication for Sustainability in Africa* (Vare 1998). The *Terms of Reference* called for 'an investigation of needs', but rather than simply canvass environmental educators for their opinions (which we did), we also conducted qualitative research in Uganda, Mali, and Mozambique with the aim of gaining insights into 'environmental learning' which might help organisations to communicate environmental issues in a more meaningful manner.

In a separate piece of research (Vare 1997), I conducted interviews in the UK and compared this with the ECoSA data. This led to the identification of four categories of learning sources:

- Own observation
- Known and trusted people
- Other human beings
- Mass media

The study highlighted the extent to which people 'own' their learning and remember experience over instruction. Thus one's *own observation* appears to be most significant followed by *known and trusted people* (often friends or family, people who share many of the 'filters' of class, culture, prejudice, etc. with the subject). The category of *other human beings* ranges from teachers to self-help groups to chance encounters. Perhaps surprisingly, *mass media* appears least effective. On reflection, I would term this *remote media*, because the common feature of this category is that communication is always one-way. My review of learning theories at this time went only as far as cognitive theories such as Kolb's learning cycle and Honey and Mumford's learning styles (Rogers 1986). Interesting though these are, they do not explain *social* learning processes, or reveal why one-way communication can be so ineffective.

In all this research, the impact of schooling is surprisingly hard to locate; this should be of great concern in settings where scarce family resources are spent on school fees. The 'crisis of relevance' in terms of many African curricula has been documented (Pennycuick 1993) but the ECoSA data suggest that schooling, with its emphasis on instruction and rote learning, interrupts the *processes* through which people traditionally gain their skills and knowledge.

With the demise of processes such as participation in everyday tasks and story-telling around the fire (Tobayiwa 1988), it is small wonder that one of my key observations was that *environmental education is the education you miss by going to school.*[1] The ECoSA survey put it thus (Vare 1998:11):

> [T]o some extent school has failed to *replace* many useful elements of the traditional learning which it has helped to *displace*.

This observation provides an interesting re-interpretation of the term: 'participation in schooling'. This is not a question of school attendance, but one of balancing opportunities to encounter new ideas with participation in the child's 'real world'. I shall return to this point later.

8.5 What's it All About? Part 1. Looking for Theories

In conducting these investigations, I considered a number of potentially binding concepts: these included *power* (Foucault 1984), *conscientization* (Freire 1968) and *participation* in relation to both of these (Chambers 1997). I also encountered Rahnema's (1992) critique of the various motives for groups appropriating the term 'participation'.[2] This chimed with personal observations of participation being applied in a mechanistic manner on conservation projects. This is 'participation as funding requirement' – a mode that is alive and well today.

To understand the extent to which people 'do' participation, I used a typology developed for the rural development context by Jules Pretty (1993); this has parallels with Hart's 'ladder of participation' (Hart 1997; see Chapter 2 by Roger Hart, this volume). Pretty's stages range from *passive participation*, where people are told what will happen, to *self-mobilisation* where people take initiatives independent of external institutions. This 'highest level' is not participation in the project sense; people constantly make decisions for themselves anyway. I would consider the highest level of participation to be *negotiation*, where power is balanced to the extent that no party can succeed in meeting its own objectives without achieving consensus, or at least the willing compliance of the other parties. In terms of my work on the project in Uganda, this was unlikely to be achieved as the Government agencies responsible for the National Park were by far the dominant party.

---

[1] This does not mean that I would advocate a return to some pre-industrialised idyll. Keeping children at home as domestic helpers while their mothers labour in the fields does not maximise human potential.

[2] This included the Rahnema quotation that was circulated to the contributors of this volume as an initial stimulus, noted on p.v of the preface.

Without further theorising, this period of research had at least led me to a view of learning characterised by two principles:

- Regardless of what is taught, people will learn in an idiosyncratic and serendipitous way and remember the aspects that uniquely have meaning for them – it is not that they *want* to learn only about things which are relevant or over which they have some agency – but these are the things that they *will* learn best.
- Learning is achieved through *dialogue*, whether with human or non-human entities – again, the more familiar/relevant the learning source, the more effective the learning episode.

8.6 Ideas into Action

Armed with these principles, I developed rural communication programmes in Ghana and Tanzania that relied on hundreds of conversations among friends and neighbours initiated by volunteers. The issues raised by these conversations were 'harvested' regularly and responded to through local meetings with project workers who had a community liaison function (Vare 2001).

In Ghana, the volunteers were called *field walkers* and in Tanzania we used the term *washirika* (a Swahili word that describes those whom we trust and with whom we cooperate). The washirika approach included 'trust mapping', through which groups formed and selected their volunteers rather than simply relying on geographical proximity to define groups as we had done in Ghana.

Both of these programmes were developed during short consultancies to conservation projects, and neither was sustained beyond two years because they fell foul of the projects' imperatives to:

- Produce materials that clearly disseminate conservation messages
- Resolve conservation/development conflicts in a project-limited time-span
- Match the activity to budget lines that had been determined before the project started

It is unusual to find a project in the natural resource management field that is prepared to invest in anything other than 'instrumental participation', that is, participation with a specific end-point in mind, such as the establishment of a community-based management committee, in a limited time frame. The rush for 'product over process' is endemic in projects, largely as a result of rigid funding cycles – 'the chains that bind' (Marsden *et al.* 1994). Where project personnel do not advocate flexibility, adherence to pre-determined objectives can stifle the innovation and creativity that meaningful participation could engender.

Most of these consultancies were conducted while I worked for the Living Earth Foundation. Living Earth is a non-governmental organisation, whose own project teams are encouraged not to view people as objects of change, but to assist them in becoming agents of their own change. Rather than specifying particular behavioural outcomes for its projects, Living Earth defines its goals in terms of *action competence* (Jensen and Schnack 1997) (see also Chapter 4 by Simovska, Chapter 10 by Breiting, and Chapter 11 by Schnack, this volume). This concept defines *action* carefully as the result of the actor's own decision, made with a change perspective. Consequently, Living Earth aims to develop in people the skills, knowledge, motivation, and self-confidence necessary to take decisions in a given context.

I witnessed a striking example of this when visiting teacher-led projects inspired by Fundación Tierra Viva, Living Earth's partner in Venezuela. At one school, teachers explained how they had transformed a rubbish tip into a playground; this had stopped vandalism and provided a valuable community resource. When asked what they personally had learned from the project, they replied, 'We can't say exactly what we have learned, but right now, we feel anything is possible.' They had action competence.

Despite many successes, organisations like Living Earth have their own sustainability problem. Institutional donors and corporate partners do not normally fund open-ended learning processes – they may expect 'participation', but their funding is tied to specific economic, social, and environmental objectives. The challenge lies in achieving these objectives while building a durable legacy in terms of learning. Analysing this kind of development is critical because donors tend to fund what they can measure. The participatory activities, I have discussed, have largely relied on empirical evidence; to date they have not benefited from a sound theoretical underpinning. Securing a rigorous basis from which to analyse project learning and participatory activity is not just a theoretical issue – it is a means of survival.

8.7 What's it All About? Part 2. Finding Theories

From the preceding section, one could expect links to many theoretical traditions and alternative views of learning. An analysis of *approaches to learning* by the Joint Information Systems Committee (JISC 2004) provides an overview of some options:

- *The associative perspective*: views learning as acquiring competence [not necessarily action competence]
- *The constructive perspective* (*individual focus*): views learning as achieving understanding – learners actively construct new ideas by building and testing hypotheses [resembles Living Earth's application of action competence]
- *The constructive perspective* (*social focus*): views learning as achieving understanding – learners actively construct new ideas through collaborative activities and/or through dialogue [parallels with *participatory learning and action*, Pretty *et al.* 1995]
- *The situative perspective*: views learning as social practice – learners develop their identity through participation in specific communities and practices (source: JISC 2004:13) [my notes added in square brackets]

ESD will involve a combination of all these approaches, although in my experience, the dominant discourse on learning in ESD is that of *learning as acquisition* viewed as an individual cognitive function. The first two perspectives in the typology are most commonly encountered therefore, even among projects focusing on social change. However, the last two approaches in the typology offer promising, socially constructed views of knowledge that relate to participatory development goals. *Situative learning* is singled out for investigation later because it has particular resonance with the issues raised in my preceding account of project experience.

8.8 Learning as Dialogue: The Zone of Proximal Development

The conceptual roots of situated learning lie in Vygotsky's concept of the ZPD (Rieber and Carton 1987:209). This could be summed up as the distance between that which we *already* know and that which we *could* know given collaboration with a more experienced other. Collaboration, or dialogue, is therefore seen as an essential component of learning. This explains the strength of the field-walker/washirika communication programmes that are founded on the notion of dialogue with known and trusted people. By giving volunteers an induction and regular access to project personnel, they inevitably became better informed individuals who could develop increased understanding of issues among their peers.

Lave and Wenger (1991) recognise that there are various interpretations of the ZPD but they see it as a zone of *social* rather than simply individual development, preferring Engeström's definition of the zone as (Lave and Wenger 1991:49):

> [The] distance between the everyday actions of individuals and the historically new form of the societal activity that can be collectively generated as a solution to the double bind potentially embedded in everyday actions.

That which we know as a society, grows from that which we already know as individuals, much of which we learned through our interaction with society. This also helps to solve the 'learning paradox' faced by an acquisition-view of learning, i.e. 'how can we want to acquire a knowledge of something that is not yet known to us?' (Sfard 1998:7).

In the field-walker/washirika programmes, there is an acceptance, even an expectation, that what is learned will be modified through the dynamic interactions among residents, volunteer field walkers, and the project staff. This supports Vygotsky's notion of socially constructed knowledge (Tryphon and Voneche 1996), but the ZPD concept does not fully address the deeply complex nature of this process unless we delve deeper into a view of the zone as one of social development.

8.9 Learning As Practice: Legitimate Peripheral Participation

Lave and Wenger (1991:15) emphasise the situated nature of learning over any consideration of instruction:

> Learning is a process that takes place in a participation framework, not in an individual mind.

They view agent, activity, and world, as mutually constitutive, hence learning is a process of participation and participation is learning. Learning takes place in a *community of practice* while pedagogy is not seen as some privileged form of learning (ibid.:93):

> The effectiveness of the circulation of information among peers suggests … that engagement in practice, rather than being its object, may well be a condition for the effectiveness of learning.

Lave and Wenger describe this engagement in social practice as *legitimate peripheral participation*, an engagement that entails learning as an integral constituent (for example, the act of participation *is* learning because it makes a difference to our capacity to participate). It is not a question of *whether* we participate, but whether we recognise exactly what it is that we are participating in.

Being *peripheral* is not a negative term. Whilst the periphery cannot be central, it is located *in* the social world and one's location will change as one learns (ibid.:36):

> Changing locations and perspectives are part of actors' learning trajectories, developing identities, and forms of membership.

Even where participation is not encouraged or allowed, it is through participation in some unintended form that learning takes place – so the content of learning may also be unintended.

Earlier in this chapter, I raised the need for schools to balance exposure to new ideas with participation in the child's context, a concern arising from the extent to which people do not remember their school learning. From the perspective of legitimate peripheral participation, this 'forgetfulness' is explained by the fact that the people we interviewed were no longer in school. What they *did* learn at school was 'how to do school'; that was their community of practice. Once they left that community they no longer needed that learning, hence it was forgotten. While we are in school, we are legitimately peripheral but we are not participating in society, so much of the learning is irrelevant beyond the confines of this *sequestration* (Lave and Wenger's term). There is support here for those who consider it inappropriate to 'teach' citizenship in schools. During our early years and adolescence, we either become good at 'doing school', or we may be disaffected from it; neither position is particularly well linked to the society that we later join. This is particularly true of schools that fail to offer opportunities for learning beyond the classroom (cf. Chapters 18, 19 and 20, this volume).

Lave and Wenger cite the role of apprenticeship as a classic example of legitimate peripheral participation. The apprentice begins with simple, repetitive tasks and gradually acquires the skills of the master. A key point to note is that the motivation for learning is that of gaining *identity*. In the case of apprenticeship, the control of the master is gradually diminished as the apprentice achieves a sense of identity as a master practitioner. In achieving this status however, apprentices create a problem for their masters because they develop ideas of their own and may wish to change the very practice that they have learned, as Lave and Wenger (1991:116) put it:

> Granting legitimate participation to newcomers with their own viewpoints introduces into any community of practice all the tensions of the continuity-displacement contradiction.

This may go some way to explain why weak forms of participation are frequently deployed on development projects, for in this way they avoid conflicts with the elites. Likewise, there are strong parallels between *legitimate peripheral participation* and the notion of *gradually diminishing control* that we used in Uganda. At first glance, they come from opposite perspectives – Lave and Wenger's term is clearly learner-orientated as it is the learner who is participating peripherally, while it is the educator who is diminishing control. But there are problems here; I wish to highlight three for the purposes of this chapter.

Three concerns with Lave and Wenger's approach

First, the extent to which we achieve mastery appears to depend, to some degree, on our overcoming the 'continuity-displacement contradiction'. Rather than view this as a contradiction, it may be more helpful to examine the complex processes by which policies and practices *do* change, with or without the blessing of their respective elites.

The process of 'structuration' suggested by sociologist Anthony Giddens (Cassell 1993), is helpful here. Giddens notes how patterns of social practice are 'structured' by rules, resources, and power. Structure is not external – agents bring structure into being and structure produces the possibility of agency, so, in an echo of 'learning as participation', we see that to be an agent is to participate in restructuring. This duality of structure 'consists in structure's two-sided existence – as both the medium and unintended outcome of social practices' (Giddens, in Cassell 1993:12).

Giddens cites the example of language. By using language and following its rules, we both communicate and (unwittingly) perpetuate the spoken word. As agents use the rules freely, the modification of the rules is an ever-present possibility. 'At each point of structural reproduction there is the potential for change' (Cassell 1993:12). This is the basis of Giddens' concept of *structuration* (Giddens, in Bryant and Jary 2001:12):

> To examine the structuration of a social system is to examine the modes whereby that system, through the application of generative rules and resources, is produced and reproduced in social interaction.

So while Lave and Wenger's legitimate peripheral participation provides us with a 'generative' process, Giddens' concept of structuration helps us see that 'continuity-displacement' is not a contradiction but a 'two-sided existence'.

My second concern is that there are instances where the practice in which we are engaged is not something that may ultimately 'be achieved' because there is no agreed or delimited body of practice. How do we know, for example, at which point we have achieved mastery in sustainable development, a field characterised by uncertainty and risk (Scott and Gough 2003)?

Even the localised examples of field walkers and washirika are not following an established practice in their respective settings. The question of mastery, therefore, does not arise. The programme should operate successfully for some years, perhaps then it can be defined. Similarly, when developing posters in Uganda, the subject matter of the poster was unknown at the start of each workshop, there was no master version or replicable discussion in which residents would participate with decreasing peripherality. The issue appears to be one of analysing learning, participation, and context when the context itself is not fully known. The advice of Lyotard (1984), to work without rules in order to 'discover the rules of what you have done', would best describe this situation, but a rigorous analysis is required to avoid unmanageable wooliness.

Finally, Lave and Wenger's strong claims for legitimate peripheral participation appear to deny that we actually *acquire* knowledge as individuals, yet our own experience informs us that each person's store of experience and knowledge is unique. What we learn in one context is not forgotten; we can and we do carry a great deal over with us to other situations and communities of practice. In this way, we may enter new communities as 'experts', at least in some aspects of that new system. Adherence to legitimate peripheral participation as dogma would certainly leave serious gaps in any analysis of learning in a complex social system. Thus, ultimately, I turn to a theoretical analysis of activity systems themselves.

8.10 Learning as Outcome of Activity: Cultural–Historical Activity Theory

Activity theory appears to address the above concerns through its analysis of human interactions within all the objects, manners, and meanings that characterise social systems. The unifying concept of the cultural-historical school is *activity* that comprises social action and individual conduct. Yrjö Engeström (1999), a leading exponent of CHAT, points out that the theory has its roots in Marxist thought. Marx recognises that change is not brought about from above or simply by self-change in individuals (Engeström 1999:9):

> The key is "revolutionary practice", which is not to be understood in narrowly political terms but as joint "practical-critical activity", potentially embedded in any mundane everyday practice.

So this is far more than watching a group of reflective practitioners at work. All our actions or social practices are imbued with cultural, social, and historical meanings and are the result of individual and social learning. To analyse this, CHAT takes as its unit of analysis (ibid.:9):

> [T]he concept of *object-oriented, collective,* and *culturally mediated human activity* or *activity system.* Minimum elements of this system include the object, subject, mediating artefacts (signs and tools), rules, community and division of labour.

The activity analyst looks down on an activity system from above but also engages with subjects within the system. Thus the analysis combines internal and external perspectives, systemic and partisan views, and (ibid.:10):

> The study of an activity system becomes a collective, multivoiced construction of its past, present, and future zones of proximal development.

Such an analysis would reveal the apparently idiosyncratic and chance encounters that tell the story of any system, including sustainable development projects, and may prove rigorous enough to overcome the conflict between exploratory development processes and rigid quality assurance.

To begin to apply this depth of analysis to something as mundane as project management, I will return to the example of the six senior project managers that I mentioned near the start of this chapter, each with conflicting versions of what we would have called 'common sense'. This should be unsurprising when we consider that we were all new to the project, came from three different countries, spoke five different mother tongues and had diverse professional backgrounds. It is also relevant that the term 'participation' (as a binding concept) had not yet entered our individual consciousness. Indeed, it could be claimed, that with no shared conceptual framework or body of experience, we lacked *common sense*.

Yet we were qualified professionals who had all been hired to work on the same project. We could all read English very well and would have no trouble understanding the project's aims, which we all supported. So why should common sense get in the way? A CHAT analysis would examine our internal perceptions, reveal the rules and regulations that bound us, and explore the contradictory histories behind the value systems under which we operated (or the rules that we all thought we shared). In order to reflect on what common sense is, I turn to Antonio Gramsci, whose work is analysed by the Italian psychologist and activity theorist, Francesco Paolo Colucci (1999).

Gramsci does not contrast common sense with the abstruseness of philosophers. Rather he recognises that all people are philosophers, capable of *spontaneous philosophy*. This is not to confuse common sense with folklore, the distinction being that common sense is not rigid but is continuously transformed, whereas folklore consists of belief systems, religions, superstitions, and opinions (and thus constitutes a part of our common sense).

In Gramsci's view, common sense originates from several sources, including language itself, but most importantly from philosophical–scientific thought (Colucci 1999:152):

> There is a continuity of passage, which implies a change of meaning, from philosophical and scientific to common language.

So, as common sense grows and becomes established, it draws its authority from philosophical–scientific thought and from other authoritative sources. When the project team formed in Uganda, we had no common pool of authoritative sources beyond the project document. Our individual sources of authority, our literature, and language, were all different. The project was an 'integrated conservation and development project', but without common practice, integration was difficult. This caused a false start in the project, something that went beyond Bruce Tuckman's (1965) well-known model of team development: *forming, storming, norming, performing*. An analysis using CHAT would not only have helped the team track its development of rules and social mores, it would have informed an emerging project world view because we were aiming to achieve our objectives through changes in social practice.

This development of common sense by drawing on authoritative sources is a process that Gramsci terms *historicity* because authorities are only adhered to by the masses (and thus infuse common sense) if their ideas 'correspond to the needs of a complex and organic historical period' (Colucci 1999:152). Common sense is therefore 'critical' in nature while conformism is only meaningful 'if we bear in mind its dialectical relation with this critical aspect' (ibid.). The idea that collective action requires individual critical thinkers who can make sense of their role within it is an empowering concept.

In this way, we can integrate collective activity and individual action, something that Alexei N. Leont'ev achieved when he first summarised activity theory. Engeström (1999) cites Leont'ev's famous example of the beater in a primitive hunt: the motive for his actions might be skin or meat but simply fulfilling his task of frightening the game cannot meet these needs. Instead he is involved in an activity with others that might eventually lead to his needs being met, thus (Engeström 1999:3):

> The beater's *activity* is the hunt, and the frightening of the game is his *action*.[3]

In this way, Leont'ev builds on Marxist analysis; he recognises that there is a division of labour as exemplified by the hunt, but sees this as a fundamental historical process that requires critically engaged individuals. Thus with Gramsci, *praxis* is identified as the interaction between theory and practical action, the process through which common sense changes and a more critical common sense can develop. The elites therefore need the masses (and their practical experience) as much as the masses need elites, and given that we are *all* philosophers, different groups at different times will develop their own form of common sense that works for them (even integrated conservation and development projects).

The role of all responsible adults (and Gramsci sees teachers as needing to be particularly responsible) is to ensure that people, especially those in marginalised groups, are able to articulate their own stories or world views while interacting with the philosophical and scientific views of the elites. Bearing in mind the significance of situated learning, informal educators such as community development workers also have a particular responsibility to bear.

---

[3] This distinction between *action* and *activity* is of a different order to the distinction made between these two terms in the action competence approach. Action competence defines an *activity* as something which is designed solely as a counterweight to academic tuition *or* something that a student is pushed to do. An *action*, on the other hand, has the dual characteristic of having a problem solving or change perspective *and* involving the student in deciding what to do (Jensen and Schnack 1997). In activity theory, an *action* is simply a component of the broader *activity*.

8.11 Conclusion, Two Further Questions and a Health Warning

It is often the case that the closer we work to everyday life, the more impossibly complex our systems appear. Reducing these processes to 'scientific' cognitive rules does not explain many of the (often obscure) motivations that are being played out in any human activity system. Indeed, discoveries of *neuromodulation*, revealing a virtually infinite range of neural responses to stimuli (Greenfield 1997), supports the argument that brains are not at all like computers. Participatory activities therefore, with their opportunities for social engagement, challenge, and reinforcement, would better help us achieve common, meaningful interpretations of the world than disseminating instructions, however clearly.

CHAT is potentially of great value to sustainable development projects, not only as a means of analysing complex systems but because its approach characterises learning as an *outcome* of activity within a system. This both informs and justifies the argument that sustainable development is a learning process (Foster 2002; Scott and Gough 2003). Indeed, situated learning perceives *any* sort of social development as a learning process.

The complex nature of situated learning and participation also undermines dominant western notions of linear development. The specific outcomes of participation will depend more on the potential of the participants, their cultural-historical background and a wide range of other factors (as identified by Engeström), than on a specific project intervention. The learning that takes place cannot be a predetermined development that is 'done to' people, but will be an unfolding or an emergence that requires close analysis if it is to be understood. Such a perspective resonates so well with systemic views of education for sustainable development (Sterling 2001) that it seems inevitable that these hitherto separate traditions will become more closely aligned.

This leads to a question that merits further investigation. Situated learning suggests that to invite people to 'participate in their own development' is nonsense because they are *always already* participants. The challenge for external change agents is to participate in the lives (or activity systems) of the people they would assist. So what are the implications for the staff, volunteers, and organisations who facilitate learning? To what extent can both parties recognise where participation on both sides is realistic, sensible, and helpful? Is it enough for the change agent to simply participate in other people's lives and see what happens? (It might be a powerful stimulus). Would anyone fund this as a mode of development rather than a mode of research?

At the institutional level, Lyotard's suggestion of 'working without rules' is unlikely to impress those charged with managing public funds, yet CHAT is just such an exploratory process. To what extent will policy makers accept investigation of outcomes over traditional, target-driven project management? This question applies to national policy and at project level (virtually all major donors insist on projects having predetermined logical frameworks).

There is some hope here. The UK Government commissioned a major study of the operation of the new Children's Fund (Edwards *et al.* 2006). The Children's Fund aims to support new cross-sector partnerships (e.g. across social services, education, health, police), in order to deliver services to children. It is appropriate therefore that the study is based on CHAT, an inherently interdisciplinary approach.

Having said this, we should avoid getting carried away. Activity theory is only one possible approach. Whilst for the academic there is a danger of becoming ghettoised in an exclusive club of activity theorists, for the practitioner, there is the danger that this becomes another trend to be applied mechanistically to all we do. As Sfard (1998:10) warns us:

> When a theory is translated into instructional prescription, exclusivity becomes the worst enemy of success.

CHAT may answer the need for rigorous evaluation of an open-ended development process, but it had better not be overplayed lest donors and institutions are encouraged to apply this as an even more intrusive tool of surveillance. Rather than failing to record unforeseen outcomes, *this* quality assurance tool leaves nothing to chance. Yet some of our most creative moments occur in the hidden cracks between regulated activities often *because* we are safely out of sight from analysts. Can we trust our authorities to balance serendipity with quality assurance? Would chance not be a fine thing?

Acknowledgments This chapter is based on experiences gained while working with the Living Earth Foundation, IUCN, The World Conservation Union, and the International Centre for Conservation Education. Some of these reflections were made while being interviewed by Alan Reid of the University of Bath and I would like to acknowledge Alan's input and his insistence on my writing this at all. Above all, I must thank Natalie, Asha, and Bathsheba for tolerating me while I wrote this.

References

Bryant, C. and Jary, D. (eds) (2001) *The Contemporary Giddens: Social Theory in a Globalizing Age*. Basingstoke: Palgrave.

Cassell, P. (ed) (1993) *The Giddens Reader*. London: Macmillan.

Chambers, R. (1983) *Rural Development: Putting the Last First*. Harlow: Longman.

Chambers, R. (1997) *Whose Reality Counts? Putting the First Last*. London: IT Publications.

Colucci, F.P. (1999) The Relevance to psychology of Antonio Gramsci's ideas on activity and common sense. In: Engeström, Y. *et al.* (eds), *Perspectives on Activity Theory*. Cambridge: Cambridge University Press, pp. 147–161.

Edwards, A, Barnes, M., Plewis, I., and Morris, K. (2006) *Working to Prevent the Social Exclusion of Children and Young People: Final Lessons from the national Evaluation of the Children's Fund*. London, DfES Research Report 734.

Engeström, Y. *et al.* (eds) (1999) *Perspectives on Activity Theory*. Cambridge: Cambridge University Press.

Foster, J. (2002) Sustainability, higher education and the learning society. *Environmental Education Research*, 8, 1: 35–41.

Foucault, M. (Rabinow, P. (ed)) (1984) *The Foucault Reader*. London: Penguin.

Freire, P. (1968) *Pedagogy of the Oppressed*. London: Penguin.

Greenfield, S.A. (1997) *The Human Brain: A Guided Tour*. London: Weidenfeld & Nicolson.

Hart, R. (1997) *Children's Participation: The Theory and Practice of Involving Young Citizens in Community Development and Environmental Care*. London: Earthscan.

Jensen B.B. and Schnack K. (1997) The action competence approach in environmental education. *Environmental Education Research*, 3, 2: 163–178.

Joint Information Systems Committee (2004) *Effective Practice with e-Learning: A Good Practice Guide in Designing for Learning [online]*. Available from: http://www.jisc.ac.uk/uploaded_documents/jisc%20effective%20practice3.pdf [Accessed: April 2005]

Lave, J. and Wenger, E. (1991) *Situated Learning: Legitimate Peripheral Participation*. Cambridge: Cambridge University Press.

Linney, R. (1995) *Pictures, People and Power: People-centred Visual Aids for Development*. London: Macmillan Education.

Lyotard, J.-F. (1984) *The Postmodern Condition: A Report on Knowledge*. Manchester: Manchester University Press.

Marsden, D., Oakley, P., and Pratt, B. (1994) *Measuring the Process: Guidelines for Evaluating Social Development*. Oxford: INTRAC.

Pennycuick, D. (1993) *School Effectiveness in Developing Countries: A Summary of the Research Evidence*. London: ODA.

Pretty, J.N. (1993) *Participatory Enquiry for Sustainable Agriculture*. London: International Institute for Environment and Development.

Pretty, J., *et al.* (1995) *A Trainer's Guide for Participatory Learning and Action*. London: International Institute for Environment and Development.

Rahnema, M. (1992) Participation. In: Sachs, W. (ed), *The Development Dictionary*. New York: Zed Books, pp. 116–131.

Rieber, R.W. and Carton, A.S. (1987) *The Collected Works of L. S. Vygotsky. Volume 1 The Problems of General Psychology*. New York/London: Plenum Press.

Rogers, A. (1986) *Teaching Adults*. Milton Keynes: Open University Press.

Scott, W. and Gough, S. (2003) *Sustainable Development and Learning: Framing the Issues*. London: Routledge/Falmer.

Sfard, A. (1998) On two metaphors for learning and the dangers of choosing just one. *Educational Researcher*, 27, 2: 4–13.

Sterling, S. (2001) *Sustainable Education: Re-visioning Learning and Change*. Totnes: Green Books.

Tobayiwa, C. (1988) The midnight classroom. In: Briceño, S. and Pitt, D.C. (eds), *New Ideas in Environmental Education*. London: Croom Helm, pp. 137–141.

Tryphon, A. and Voneche, J. (eds) (1996) *Piaget – Vygotsky: The Social Genesis of Thought*. Hove: Psychology Press.

Tuckman, B. (1965) Developmental sequence in small groups. *Psychological Bulletin*, 63, 384–399.

Vare, P. (1998) EcoSA: a report on a pan-African environmental education survey. *Environmental Education Research*, 4, 1: 5–24.

Vare, P. (2001) 'Washirika': raising awareness without mass media. *PLA Notes*, September, London: International Institute for Environment and Development.

Vare, P. (1997) *Serendipity or Schooling? Identifying the Process of Environmental Learning*. Unpublished MPhil thesis. Graduate School of Education, University of Bristol.

Chapter 9
Participation and Sustainable Development: The Role and Challenges of Mediating Agents

Jeppe Læssøe

Keywords democracy, sustainability, mediators, dilemmas, green guides

9.1 Introduction

This chapter proposes a transgression of narrow fixations on top-down versus bottom-up approaches to citizen participation. Even though there is good reason to attack top-down approaches, I argue that when, for example, environmental sociologists turn their criticism of the top-down into an idealisation of the bottom-up, they are stuck with a simplistic model. Moreover, when confronted with the quest for sustainable development, an emancipatory conception of participation can neglect the tension between participation as a defence of democratic rights, and participation as a tool for promoting learning processes that aims to replace narrow interests with a collective responsibility for a sustainable future. In this context, this chapter takes a closer look at a new kind of professional agent faced with this tension and charged with implementing participative processes on sustainable development. These professional agents are conceptualised as 'mediators', and their role is differentiated into 'networkers', 'interpreters', and 'facilitators'. Based on examples from Denmark, this chapter outlines four dilemmas with which these kinds of mediating change agents have to cope and the qualitative differences between them. The dilemmas are:

1. Populism versus paternalism
2. Local settings versus global scope
3. 'Environment-centredness' versus cultural orientation
4. Independence versus involvement

Instead of just opposing top-down participation with an ideal of bottom-up processes, there is a need for the high-level education of mediators in order to qualify them to cope with these dilemmas.

DPU, Research Programme for Environmental and Health Education, Danish School of Education, University of Aarhus Tuborgvej 164, DK–2400 Copenhagen NV. Denmark, jepl@dpu.dk

A. Reid et al. (eds.), *Participation and Learning*, 144–158.

9.2 Background

Participation and sustainable development are both buzzwords – it is hard to find anyone who does not agree on the importance of ensuring a sustainable world and the need for involvement of ordinary citizens in processes towards sustainability. However, behind the rhetoric, different conceptions of participation and sustainable development are at play. For this very reason, we need to clarify key concepts and the relationships between them. It does not, however, make much sense to do this at an abstract, general level. They are living concepts woven into historical processes and specific contexts, which means they are reinterpreted and ascribed with new meanings over time and from place to place. Right now, there are good reasons for reflecting on these concepts and their relationships again. As an introduction to the focus of this chapter, I will do this from my position as a researcher on citizen participation in sustainable development in Denmark.

9.3 The Harmonious Visions of Participation and Sustainability

A story of general relevance

In 2003, the Danish section of Friends of the Earth (an organisation called NOAH), invited me to join an internal discussion about their core values. By means of a 'brainstorming' activity they produced a list of keywords expressing these values. Among the keywords, I found both 'participation' and 'sustainability', and to provoke them, I asked whether they fitted together well. At first they hesitated a bit, but then one of them said: 'No, I have just been to another seminar and there one of the contributors actually stressed that public participation is the most important obstacle to sustainable development!' This statement really sent a shock through the group. Is it possible to proceed on the basis that: 'Yes, our demands for sustainability are really in conflict with the values and the visions of the population', and – on the other hand – 'No, we have always believed in participation as the only reasonable way to achieving a sustainable world'? A key issue raised by this event is whether this example of cognitive dissonance – the uncomfortable tension arising from holding two conflicting thoughts at the same time – reflects the grounds for an incipient shift in the way the relationship between participation and sustainable development is conceived and approached. In the following sections, I will briefly sketch the dominant conceptions of this relationship as they have developed in political discourse and environmental sociology since the 1970s, before I elaborate on the tension between democracy and sustainable development and propose a way to cope with it.

Emancipation and ecological modernisation

When environmental discourse emerged in Denmark in the late 1960s, there was a collective neglect of environmental problems among established societal institutions and little suggestion that lay people should be involved in addressing the rapid transformation of nature and society. In response to this neglect, a new environmentalism was formulated and organised by students who made it a part of ongoing student and youth activism. In this context, participation became part of a social emancipation process. The mobilisation of the populace in local environmental groups became the dominant strategy. Thus, in this context, there was no contradiction between citizen participation and the struggle for environmental improvements (Læssøe 1990).

From the late 1970s and during the 1980s, the public debate about environmental issues became professionalised. Experts from scientific institutions, public administration, and big companies took over responsibility for the environment. Instead of trying to fight against this professionalisation by means of grass-roots mobilisation, the environmental movement adapted to the new conditions and became themselves more or less professional non-governmental organisations (NGOs). For this reason, ordinary citizens were reduced to the status of a 'worried audience'. However, in the late 1980s, inspired by the UN Brundtland report, the Danish government took over the initiative by supporting the participation of citizens and consumers in projects that aimed to promote sustainable forms of development at the local level. Indeed, in this context, participation was not seen as an integrated part of a social emancipative process; rather, it became a part of the rising ecological modernisation discourse claiming a consensus between all agents about sustainable development – it is for the common good and everybody has to contribute and cooperate (Hajer 1995).

There are obvious differences between the bottom-up emancipatory approaches of the 1970s and the top-down approaches to participation in sustainable development during the 1990s. It is however remarkable that they have both stressed an accordance between participation and sustainable development. No one has talked about problems in combining these two ideals.

Top-down versus bottom-up

The emancipatory conception of participation has not disappeared. For good reasons, researchers in the field of environmental sociology have attacked dominant top-down approaches to the involvement of lay people in environmental risk issues. They do not only ignore the uncertainties of scientific knowledge, they marginalise the experience-based local knowledge of citizens as well (Irwin 1995; Wynne 1996; Fischer 2000). However, even though this criticism is unquestionably important, it is limited in the sense that it maintains a focus on participation and sustainable development as a matter of top-down suppression versus support of bottom-up emancipation. The bottom-up ideal is reproduced as a contrast to top-down approaches, but only with a marginal focus on empirical knowledge of trials to implement the bottom-up ideal. As shown by researchers of 'development', there are good reasons for being critical regarding what actually happens during participatory processes. Democratic ideals do not eliminate a gradual construction of power relations during participative projects – experts will snatch up and tap the local discourse to promote their own ideas. The local discourse is not pure experiential knowledge, but is constructed en route through local power struggles, alliances, and the speculative use of prospects for support. These very real processes are masked at the surface layer of participative processes (Cooke and Kothari 2001). Furthermore, the emancipatory lens narrows the focus on participation to a matter of either a top-down or a bottom-up model, even though this – as I will return to later – seems too simplistic compared with recent trends in participation. Last but not least, the emancipatory reaction to the top-down approach maintains a harmonistic view on the relationship between participation and sustainable development. Citizens are positioned as simply pro-sustainable development agents. There is no room for opposite interests and ambivalences in this conception.

The challenge of sustainable development

The relationship between participation and sustainable development is not only a matter of democracy and power. As with the story from the NOAH meeting at the beginning of this chapter, participation contains an inherent tension between individual and common interests. The democratic argument is that every citizen should have the opportunity to influence their own life conditions (see Nelkin 1975:37; Fischer 2000:2). The sustainable development argument is that, in doing so, they should take care of the common good – a core topic in political theory on democracy (Dobson 2003). However, in relation to sustainable development, the meaning of 'the common good' is radicalised. As the Danish environmentalist, Heinberg (2003:2),[1] puts it:

> The core of it is that if you want to put sustainability on the agenda, then it implies a radically new kind of policy. A kind of policy where the basic consideration has to deal with the fact that the consequences of a given practice here and now will hit someone at another time and place. Thus: What I am doing here to take care of my self, my family, my time, my world, has consequences – not for me but for someone else, not here but at a another place, not now but in a generation.

Thus, the relationship between participation and sustainable development is not just about the democratic rights and real possibilities of ordinary people to gain influence on political decisions. It is also a matter of promoting reflections, learning, and problem-solving regarding the tensions between 'what I would like here and now' and 'what we should do to take care of the other, at another place in the future'. Thus, instead of a

[1] Translated from the Danish by the author.

naive belief in a bottom-up strategy for sustainable development or in a manipulative top-down approach, a participatory strategy for sustainable development ought to expose the ambivalences and conflicts, and facilitate the involvement of the populace in the democratic process of developing answers to these challenging problems. These points have serious consequences for the following discussion of the role of the new professional organisers of participatory processes regarding sustainable development.

9.4 The Mediating Agents

The forms of social agency involved in the construction of participative processes are undertheorised and rarely the subject of explicit investigations that try to unpack their different qualities and consequences. The distinction between top-down versus bottom-up has worked as a reasonably simple and relevant way of discerning different approaches to organising participative processes, and also the differences between the positions of the various agents involved – they were either associated with 'the top' or 'the bottom'. As I will demonstrate, there is a need for a more richly faceted, multidimensional identification of the positions and roles of key players.

The new professional third-party mediators

Since the new environmentalism rose in the late 1960s, the agents who have tried to promote citizen participation in sustainable development have typically been rooted in either environmental NGOs or public administration. However, during the 1990s, a third and striking trend emerged, so that in Denmark today we find mediation professionals working to promote civic participation for a sustainable future from a distinctly third-party position; that is, from a position in which they are not personally associated with a local stakeholder. Instead, as independent professional agents they have the role of facilitating civic participation in local sustainability initiatives. Occasionally such third-party mediators are employed by planning advisory firms or by opinion poll agencies hired to run the participative process. In other cases, they are associated with national government bodies to promote 'empowerment for democracy'[2] or with adult education associations. In these cases government

[2] Since the introduction of democracy in the mid-1800s, Denmark has had a tradition of allocating state funds for institutions and activities focusing on 'empowerment for democracy'. The Danish term for this is 'folkeoplysning', a value-loaded, condensed concept that defies direct translation into English (Borish 1991). Originally, it was a formal kind of adult education that aimed at promoting a democratic 'bildung' of the populace. Today, there is still governmental support for 'folkehøjskoler' (people's high schools), but since the 1980s, more modern versions of 'folkeoplysning' have emerged as well. They are concerned with the facilitation of non-formal community projects – typically related to social welfare and/or local environment.

grants fund the involvement of the third-party mediators in local participative processes. In a third, though less frequent model, teachers and researchers from schools and universities engage in local participative processes for a sustainable future as part of their action- and experience-based teaching or research (Kofoed *et al.* 1994; Aagaard Nielsen 1996).

Aspects of mediation

I will describe this kind of agency by means of the concept of *mediation* and, as I will explain by means of three concepts expressing differing aspects of mediation: *networking*, *interpretation*, and *facilitation*.[3]

Mediation is the most general term and can be thought of as the explicit process of connecting otherwise separated individuals or groups. Mediators are agents of social interaction, the people whose task it is to mobilise participants in projects of public participation. In a recent collection of articles entitled, 'Mediating Sustainability', the term is characterised as 'a process of "coming between" different social interests with a view to finding a way forward from what is, or is in danger of becoming, a cul-de-sac of conflict or inertia' (Blauert and Zadek 1998:10). In this sense, mediation is a form of agency that is meant to break through barriers, be they institutional, social, political, or cultural.

In relation to participatory activities, the mediators can play the part of those who are bringing people together (networking), transferring ideas and practices from one place to another (interpretation), and/or organising communicative or action-oriented processes in order to promote learning and sociocultural change (facilitation).[4] Thus, *networking* represents the *organisational* aspect of mediation, *interpretation* the *substantiating* aspect, and *facilitation* the *process* aspect. Even though all aspects will be present in one way or another in the practice of mediators, there will also be differences regarding which aspect they will emphasise. For this reason, it sometimes makes sense to characterise a mediator as, more or less, a 'networker', 'interpreter', or 'facilitator'.

These new professional mediators are furthermore characterised by their *third-party position* in relation to the involved agents. They are not a part of the specific case but, as facilitators, are supposed to guide the process of the involved agents (Rasmussen 2003:310f). Facilitation, then, is about the communicative practice of mobilising participants and managing the dialogue.

---

[3] This conceptualisation is inspired by discussions with Andrew Jamison, Karsten Schnack, and Bjarne Bruun Jensen.

[4] Inspired by Jamison (2004).

The need for a closer look

These new mediators are interesting for several reasons. As relatively autonomous third parties, they bridge the simplistic split between 'top-down' and 'bottom-up'. They have mediation and participation as their professional task, which can be seen as a way of potentially strengthening democracy in complex and time-pressured 'network societies' (Castells 2000). But then again that in itself should also raise critical questions as to how such third-party mediators actually influence the construction of participative processes and how their influence impinges on aspects such as power, powerlessness, marginalisation, manipulation, legitimisation, and so on. In other words there is every reason to take a closer look at what goes on in actual practice and to develop a rather more differentiated understanding of third-party mediation in its various forms. I will present my own suggestions regarding ways to distinguish between various approaches among mediators who are involved in the promotion of participation in sustainable development.

9.5 Four Dilemmas of Third-Party Mediators

In this section I outline four dilemmas associated with the third-party mediation of participative-sustainable development processes. I use the term 'dilemma' because each case involves contradictions for which no unequivocal and easy resolutions can be given. The relevant third-party mediators are not necessarily aware of these dilemmas but they definitely encounter them in practice and respond to them. The interesting part is how their responses differ. Unlike a definition of participative processes that is either top-down or bottom-up, third-party mediators approach the four dilemmas in ways that provide us with a basis for a more detailed description of participative processes in terms of the differences between their design, processes, and outcome.

However, before presenting these dilemmas I introduce the empirical examples to which reference will be made.[5]

Introduction to empirical examples

The empirical basis of my description of the four dilemmas is primarily twofold:

- Interactive research and process evaluation of a government-funded local sustainability advisory scheme, the Danish Green Guides scheme, which I carried out from 1996–2001.

[5] The four dilemmas are derived from a number of empirical studies on attempts to promote citizen participation in sustainable development in Denmark that I have carried out since 1991 (Læssøe 1992, 1995, 2000, 2001).

- Material from my participation in the background group of a project featuring citizens' hearings on a sustainable future, conducted in 2002 by the Danish Board of Technology (a government funded national body), in four local authority districts.

The Green Guide scheme was launched by the Danish government in 1996 and ran until 2005. It has encouraged local stakeholders to join forces to promote local sustainable development. For this purpose, they could hire a 'Green Guide' for a three-year period and have 70% of the cost covered by a government grant (an extension period was also possible). In total, a little more than 100 Green Guide projects have been supported. The Green Guides have functioned as local environmental advisors, whose job – apart from providing information – has been to be inspirational, action-oriented, stimulating, and coordinating. As the scheme has provided considerable latitude for Green Guide projects, a wide range of initiatives to promote civic participation for a sustainable future has evolved within the framework. Compared with other experiments that support citizen participation, the interesting part of this scheme is that government funding has made it possible to hire people with professional expertise on a full rather than part-time basis in supporting civic participation in local sustainable development, and over a protracted period.

Unlike the Green Guides scheme, the civic hearings of the Danish Board of Technology exemplify an approach in which mediators, as a third-party agent, undertake 'flying visits' to a local community to run a participatory event. The Danish Board of Technology is a government-funded body with the mission of facilitating public debate on technology development, and in that capacity, offering advice to the Danish government. Over a number of years, its secretariat has evolved a professional competency in participative methods. One is the consensus conference method, today used by several countries (Klüver 1995). In this example, civic hearings on a sustainable future were prepared jointly with a team of external experts and then offered to four different types of local authorities. In each district, they drew attendance from some 100 out of about 1,000 randomly invited residents. Before the event, the participants received a pamphlet featuring extracts from interviews on sustainable development with a number of experts. The event itself was mainly in the form of a dialogue workshop, that is, a structured sequence of group sessions during which participants generated and selected their preferred suggestions to promote sustainability. At the conclusion of the event their ideas were presented to local councillors. The Danish Board of Technology rounded off its engagement in the local district processes with a follow-up session in each local district, during which, proposed local initiatives to continue the process were debated. Finally, the entire process of the four civic hearings concluded with a public presentation at which a compiled catalogue of ideas for promoting sustainability was handed over to the Danish Minister of the Environment and to members of parliament.

My description of the following four dilemmas will relate experience gained from the Green Guides scheme and from the civic hearings on a sustainable future conducted by the Danish Board of Technology.

Dilemma A: Populism Versus Paternalism

At the beginning of this chapter, I discussed this dilemma in terms of two different motivations behind the promotion of participation in relation to sustainable development. On the one hand, there is an emancipatory, democratic interest in making it possible for citizens to influence societal decisions regarding sustainable development. On the other, at its core, sustainable development seeks the transcendence of narrow interests regarding what is good for me, here and now. Thus, participatory approaches in relation to this topic may also imply an interest in 'bildung', that is, promoting reflection and learning about the common good and developing the competencies of the populace to take responsibility for each other, in another place and time.

This tension between listening to the voice(s) of the people versus trying to motivate them to transcend their own interests is not just a dissonance in the head of the mediator but a concrete methodological challenge when it comes to how to involve citizens in sustainable development. On the one hand, it is a fundamental rule that motivation and learning presuppose a process that springs from a participant's current perspective and experience. On the other, the starting point is also necessarily the case at hand, the issue that participants are supposed to relate to. The extremes can be presented as a laissez-faire approach, in which participants are left to define the issue and control their own process, and conversely, an authoritarian pedagogy, in which targets and substance are defined before dealing with the issue of motivating participants (c.f. Chapter 7, by Lotz-Sisitka and O'Donoghue, this volume).

When the Green Guide scheme was constructed there was just one demand regarding the content of the projects: that in some way or another they should address issues of environment and sustainable development. Yet there were a number of process requirements attached, implying that the key objectives were to address citizens in a broad sense and to establish broad partnerships across local stakeholders. The Guides experienced considerable pressure in terms of proving themselves by activating citizen participation. As a result they typically took the lead by launching a broad spectrum of activities. En route they realised that the residents' spontaneous interest could be aroused most easily by advice regarding simple actions that individuals could integrate directly into their everyday lives. As for their mission – promoting a sustainable future – it is remarkable that with few exceptions they showed little interest or skill in drawing attention to other implicit sustainability issues, for example, individual ambivalences and socio-political conflicts, attitudinal and economic barriers, or difficult choices and prioritisations in risk issues. Generally the Green Guides chose a laissez-faire pedagogy to facilitate the process. They kept within the consensus paradigm of the ecological modernisation discourse (Hajer 1995) instead of trying to stimulate other forms of involvement by, for instance, engaging residents in deliberations about conceptual or material opposites and their possibly productive interplays.

Overall, the Danish Board of Technology has been successful in taking up current, conflict-ridden issues and in ensuring an unbiased dialogue about them between different stakeholders. However, with the civic hearings on a sustainable future they departed from this position. Firstly, they invited several environmentalists and professional experts to offer their inputs to the planning group concerning the current challenges regarding sustainability. Based on these inputs they formulated the topic of the civic hearings as 'sustainable growth – how?' Like the 'inspiration pamphlet' featuring expert suggestions for a sustainable future, this theme holds an obvious tension – are sustainability and continuing growth compatible? Furthermore, how are ideals to be translated into day-to-day practice? However, the design of the civic hearings failed to highlight these tensions. Participants were left free to define the issues they were to work on (and therefore fell back on familiar environmental issues). Moreover, they were pressured to produce quick conclusions and consensus on their proposed improvements. In addition, they were told that their suggestions should be as constructive and concrete as possible. Thus this part of the process also ended up avoiding fundamental tensions and conflicts by adopting a laissez-faire approach and foregrounding consensus as the endgame.

Dilemma B: Local Settings Versus Global Scope

On paper, the potential conflict between the global scope of sustainable development and the local settings of citizens is easily resolved – *think global, act local!* Things are rather more difficult in practice. The challenge is not just about bringing global and abstract perspectives into local reality, but that a local reality has narrow confines for solutions. Elements such as production, distribution, and consumption are difficult to change within a local community. Moreover, there is a parallel cultural trend towards globalisation and individualisation, both of which weaken the local arena. As with the other dilemmas, mediators have various options in coping with the tension between global versus local.

A basic requirement of the Green Guide scheme was for projects to be local and action oriented. As concluded in the evaluation report of a previous nationwide campaign for a sustainable future, also focusing on action orientation, the result was typically that people have 'thought local and acted local' (Gram *et al.* 1992). Global perspectives were either left out altogether or referred to sporadically, as the background of local efforts. With a local frame of reference, action prescriptions were bound to address mainly individual citizens or neighbourhoods, their main focus being housing, consumption, and housekeeping. In a few cases, Green Guides went beyond the local setting, by taking up transport issues, for example, exemplifying a problem that should be solved regionally or nationally rather than locally.

The hearings of the Danish Board of Technology embraced both poles. On the one hand their introductory presentations for the hearings focused mainly on the challenges of creating a globally sustainable future. On the other the hearings were run as local events. In the same vein, citizens' proposals were communicated to local politicians, as inputs to Local Agenda 21 work. However, there were no attempts to try to reflect on the local proposals in a global perspective.

Dilemma C: 'Environment-Centeredness' Versus Cultural Orientation

There are different discourses, and hence strategies, for sustainable development. Most particularly, an *environment-centred* perspective – in which environmental problems are investigated and 'solved' technoscientifically, after which attempts are made to implement these solutions in society as such – differs from a *socio-cultural* perspective, where the main emphasis is to understand the circumstances within a society that will create/construct environmental problems, set the conditions, and hold resources relevant to possible solutions (Læssøe 2000:225). The two perspectives are complementary and will thus engender profoundly different views on how to mediate participative processes for sustainable development.

Most of the Green Guides had an environment-centred approach as their starting point. As a consequence, and especially in the beginning, they spent much of their time staging different public events, during which they would inform people about environmentally friendly products and behaviour. However, in the process, many Guides found this to be a sluggish path and that having to take every initiative on one's own was a back-breaking exercise. They found that a negotiable way to solve the problem was to gradually get integrated, via their numerous contacts, with the existing social networks of the local community. By doing so, they decentred, allowing themselves and the environmental issue to become integrated with local social life. In these ways, the process shows a shift towards a rather more sociocultural approach.

The civic hearings of the Danish Board of Technology took a broad approach to the sustainability issue, also thematising both economic and sociocultural aspects. From the above description of other dilemmas, it could be assumed that the expert presentations for a sustainable development debate had little impact on the hearing itself. However, that is not entirely true. It was evident that expert contributions on sustainable development, understood as lifestyle development based on both ecological and social motives, had a fair impact on the priority that citizens gave to the aspects of sustainable development they considered important. However, when encouraged to offer their concrete suggestions, the visions of the citizens to that effect resorted to rather more pedestrian, commonplace environmental advice.

Dilemma D: Independence Versus Involvement

This is actually not one, but two dilemmas. However, they are both about the involvement versus the independence of the third-party mediator. In the first case, the subject of this dilemma is about the relationship between the third-party mediator and the participants; while in the second case, it is about the mediator's professional distance versus engagement in sustainable development. Put another way, in the first case the dilemma is related to the networking and facilitation aspects, while in the second it is dealing with a potential conflict between facilitation and interpretation. I will present them separately.

Independence versus involvement with the stakeholders

As described earlier, a third-party mediating agent is ideally characterised by being independent of the stakeholders. Yet these agents cannot just go about their own business. In order to fill their networking and facilitating roles they have to involve themselves and build relations with the stakeholders. This can potentially compromise their autonomous position. They risk being unable to mediate and facilitate because their involvement could cause local stakeholders to 'label' them as either being parties to the case or as actually serving the objectives of one or another local stakeholder. Thus the core of the dilemma is that they can compromise their mediating mission by being overly autonomous or too biased in their involvement. We could say that mediators cannot stay beyond the top-down versus bottom-up issue. However, a local community will typically have a number of different agents, so in real-life terms relations are often more complex than just 'top' and 'bottom'. Still, for the sake of clarity, I will confine myself to that type of relationship between local authorities and residents.

Distinctly different 'answers' to that dilemma can be identified among the Green Guides. They arrived as new agents in the local arena who had to be able to function quickly and effectively vis-à-vis local administration and politicians. Though local authorities initially had a sceptical or hesitant attitude towards the Guides, positive outcomes were seen in more than 75% of cases. Typically, the Guides succeeded in building productive, personal contacts within the local administrative system as materialised in cooperation on actual tasks. These in turn led to more continuing contacts and cooperations involving more and more stakeholders within the local administrative system. The Guides' close contacts and rapport with citizens implied that local authorities could benefit from their knowledge and also utilise them as their 'extended arms', in pesticide campaigns and guidance on composting, for instance. Still, not all local authorities were equally active on the environmental front. That again means that there were several cases where Green Guides were acting as the local authority's critical opponents – they took on the role as 'advocates of the citizens and the environment', pressing local authorities to upgrade their Local Agenda 21 efforts. Both the Guides and their cooperation partners made the point that this lets them function as 'free players' – free to work both as partners and opponents in relation to the relevant local authority and free to move between local agents, experts, laypersons, and various council departments. As a distinct position, this lets them promote synergy and cooperation, yet also a potential to work as the dialogue-building third party, who can help overcome barriers by encouraging debate and conflict-solving among the local authority, citizens, and other local agents.

Thus, the Green Guides varied in their response to the positioning dilemma. Some positioned themselves close to the local authorities, as their 'extended arm', others chose a position as a 'spokesperson' for green citizens and nature. However, no matter what their choice it did not normally leave them with their hands tied. By varying their work they were able to retain their autonomy and were respected as 'free players'.

Due to the transient nature of the civic hearings on sustainable futures conducted by the Danish Board of Technology, they were in less danger of 'contaminating' their independent status, owing to their contact with local players. Some green citizens were critical because the Board of Technology entered the arena from outside and planned a civic hearing with the local authority without having acquainted themselves with previous local initiatives in the field. However, their misgivings waned once it became apparent that the civic hearings mobilised many new residents and set the local process going. The role of the Board of Technology was easily identified as an external and impartial process catalyst. Their authority and financial contribution made local stakeholders open up to these possibilities for participation, and since they were only involved for a short period they posed no major positioning problems vis-à-vis the local players.

Professionalism versus engagement in relation to sustainable development

Mediators have to be good at facilitating a process that will enable local players to make progress in their work. At the same time, to allow mediators to assess the proper progress of the matter, they need both qualified knowledge of, and commitment to, the issue at hand. That implies a double risk. On the one side their process facilitation can end up as a purely technical exercise. On the other a knowledgeable and committed facilitator can cause his/her actual views and values to curtail and control the participants' process.

Overall, the Green Guides belong to the latter category. That is, they have been quite knowledgeable about and committed to environmental matters, and they have been extremely eager to communicate their expertise and views to others in the hope of making them as active and conscious as they are themselves. This creates facilitation problems since a 'dark green' image could make it difficult for facilitators to win general support and because when it comes to facilitating participative processes their basic skills were rather limited and conventional. On the other hand, the second shortcoming was addressed in part since Green Guides were given supplementary training. Moreover, due to their prolonged local presence they gradually became well-integrated members of local social networks which in turn allowed them to reach beyond the ranks of those who are already 'green'.

Staff at the Danish Board of Technology do have comprehensive environmental expertise, commitment, and considerable technical skills in facilitating participative processes. However, making a point of their own neutral status is an inseparable part of their professional approach to participation, understandably so, since they are an advisory board to the Danish government and because (as mentioned earlier) personal engagement could compromise the process. Yet that fact also influenced the outcomes of the civic hearing process on sustainable futures. As mentioned earlier, they started by mailing a pamphlet to the participants, featuring interviews with sustainability experts. However, this expert knowledge was not brought into dialogue with citizen knowledge and attitudes, either during the dialogue workshops or later. In the absence of this dialogue, the citizens' proposals for a sustainable future ended up as a mix of largely abstract attitudes and concrete, trivial, and familiar advice. To be fair, that does not apply to all approaches used by the Danish Board of Technology. For the present case, however, the consequence of their low profile was that the end product was not qualified by professional interplay and counterplay.

9.6 Conclusion

Third-party mediators are not robots. They are people with particular backgrounds and social contexts. As such, they are always already tainted by, and taint discourses on sustainable development, participation, mediation, and learning. In the field of education for sustainable development (ESD), different mixtures of these discourses are offered. According to my own experiences with mediators in Denmark, they may be influenced by these approaches but not in a directly reflected manner. Their approaches are to a large extent self-made, developed through practice and through dialogue with other equally self-made mediators. Researchers and practitioners working with ESD can definitely learn from their experiences. However, their knowledge interests are typically technical, i.e. 'how do we do this or that?' When it comes to basic questions like the relationship between participation and sustainable development, or the other dilemmas mentioned in this chapter, there seems to be a lack of explicit reflection and dialogue. The consequences are significant given their considerable influence on how participative processes for sustainable development are planned. It may cause the resulting processes to be profoundly different, be it in terms of substance or formal set-up and process. With the four dilemmas, I have tried to conceptualise a few aspects that can be used to differentiate between different approaches and to evaluate strengths and weaknesses in each of them. At present, this is still incomplete but should be understood as an opening in relation to the rather more stereotyped idealisations of top-down versus bottom-up or, for that matter, third-party mediated participative processes. There is a need for further research to conceptualise the different types and approaches of professional mediators – their interests, influence, approaches to social change and learning, methods, competencies, and ethics. There is also need for high-level education to transform knowledge gained from research on these issues into materials and processes that can support the reflections, and thus qualifications of these new catalysts of ESD.

References

Aagaard Nielsen, K. (1996) *Arbejdets Sociale Orienteering*. København: Forlaget Sociologi.
Borish, S.M. (1991) *The Land of the Living*. Nevada City, CA: Blue Dolphin.

Blauert, J. and Zadek, S. (eds) (1998) *Mediating Sustainability: Growing Policy from the Grassroots*. West Hartford, CN: Kumarian Press.

Castells, M. (2000) *The Rise of the Network Society*. 2nd edn., Oxford: Blackwell.

Cooke, B. and Kothari, U. (eds) (2001) *Participation – The New Tyranny?* London/New York: Zed Books.

Dobson, A. (2003) *Citizenship and the Environment*. Oxford: Oxford University Press.

Fischer, F. (2000) *Citizens, Experts, and the Environment – The Politics of Local Knowledge*. Durham, NC/London: Duke University Press.

Gram, S., Jeppesen, T., and Sørensen, T.P. (1992) *Evaluering af Brundtlandkampagnen Vor Fælles Fremtid*. København: CASA.

Hajer, M.A. (1995) *The Politics of Environmental Discourse*. Oxford: Clarendon Press.

Heinberg, C. (2003) Det begynder med hinanden. *Nyhedsbladet Øko-Net*. 55, December.

Irwin, A. (1995) *Citizen Science – A Study of People, Expertise and Sustainable Development*. London/New York: Routledge.

Jamison, A., Eyerman, R., and Cramer, J. (with Læssøe, J.) (1990) *The Making of the New Environmental Consciousness*. Edinburgh: Edinburgh University Press.

Jamison, A. (2004) Participation and agency: hybrid identities in the making of green knowledge. In: Ribeiro, G., Lund, S., and Mullins, M. (eds), *Debating Participation – Actors Shaping Sustainable Urban Development*. Copenhagen: Royal Danish Academy of Fine Arts, School of Architecture, pp. 21–35.

Klüver, L. (1995) Consensus conferences at the Danish Board of Technology. In: Joss, S. and Durant, J. (eds), *Public Participation in Science: The Role of Consensus Conferences in Europe*. London: Science Museum with the Support of the European Commission Directorate General XII, pp. 42–53.

Kofoed, J., Vognsen, C., and Jensen, B.B. (1994) *Quark – et samarbejde om udvikling af miljøundervisning og byøkologiske teknologier i lokalsamfundet*. København: Danmarks Lærerhøjskole.

Læssøe, J. (1990) The making of the new environmentalism in Denmark. In: Jamison, A., Eyerman, R., Cramer, J. with Læssøe, J. (eds), *The Making of the New Environmental Consciousness – A Comparative Study of the Environmental Movements in Sweden, Denmark and the Netherlands*. Edinburgh: Edinburgh University Press, pp. 66–120.

Læssøe, J. (1992) *Folkeoplysningens opgave i indsatsen for en bæredygtig udvikling*. Lyngby: Tværfagligt Center, Danmarks Tekniske Højskole.

Læssøe, J. (1995) Borgerinddragelse: Mellem åbenhed og styring. In: Lübcke, P. (ed), *Miljøet, markedet og velfærdsstaten*. Copenhagen: Fremad, pp. 334–363.

Læssøe, J. (2000) Folkelig deltagelse i bæredygtig udvikling. In: Holten-Andersen, J. *et al.* (eds), *Dansk Naturpolitik i bæredygtighedens perspektiv*. København: Naturrådet. pp. 224–236.

Læssøe, J. (2001) *Evaluering af Grøn Guide ordningen 1997–2000*. Rapport udgivet af Den Grønne Fond/Miljøstyrelsen.

Nelkin, D. (1975) The political impact of technical expertise. *Social Studies of Science*, 5: 35–54.

Rasmussen, L.B. (2003) The facilitation of groups and networks: capabilities to shape creative cooperation. *AI and Society*, 17: 307–321.

Wynne, B. (1996) May the sheep safely graze? A reflexive view of the expert-lay knowledge divide. In: Lash, S., Szerszynski, B., and Wynne, B. (eds), *Risk, Environment and Modernity: Towards a New Ecology*. London: Sage, pp. 44–83.

Chapter 10
Mental Ownership and Participation for Innovation in Environmental Education and Education for Sustainable Development

Søren Breiting

Keywords mental ownership, psychological ownership, environmental education, action-competence models

10.1 Introduction

What specific qualities make participation *genuine*, and how do they link with a democratic view of education, cooperation, and empowerment? This chapter explores these themes, linking reflections on existing theoretical perspectives on the 'ownership of participation' with professional experience on educational development initiatives in a number of countries around the world (mainly Thailand, Namibia, and Denmark). In particular, the chapter argues that while the process of developing 'mental ownership' is a neglected aspect of successful participatory approaches in development efforts as well as in education, there is good reason to consider the hypothesis that the level of mental ownership that a participatory initiative is able to generate among participants, corresponds with the experienced quality of the participatory approach. In so doing, the chapter discusses the potential value of generating mental ownership in participation for improving the quality of different approaches to innovation in education, and illustrates wider debates on the need to democratise environmental education and Education for Sustainable Development (ESD), and whether educational goals should be principally regarded as addressing adaptation or emancipation (e.g. Hellesnes 1982).

10.2 Background

According to Kelly (2003:1): 'Probably the most significant problem that critiques of "participatory development" have highlighted is its potential to reinforce rather than challenge power relations, and thus to fall short of its own declared goal of

DPU, Research Programme for Environmental and Health Education, Danish School of Education, University of Aarhus, Tuborgvej 164, DK-2400 Copenhagen NV, Denmark, breiting@dpu.dk

A. Reid et al. (eds.), *Participation and Learning*, 159–180.

encouraging "empowerment."' Kelly's analysis draws attention to the importance of the wider context to participation – the depoliticisation of participation that accompanies the valorisation of the personal, the local and the 'community' – and the dynamics of particular instances of participation that reinforce existing privileges and discourage an articulation of subordinate perspectives (see, e.g. Chapter 13 by Barrett, this volume).

As Reid and Nikel show in Chapter 3, and Lotz-Sisitka and O'Donoghue in Chapter 7, participation has become a 'buzzword' in development projects and school innovation, particularly in seeking to foster contributions to sustainable development. From the early days of environmental education, but also in more recent documents in relation to ESD, participation has been positively linked to the resolution of environmental problems (e.g. Belgrade Charter 1976; Tbilisi Declaration 1977; UNESCO 1992), and to new forms of public governance (e.g. United Nations 1993; UNESCO 2005 – the UN's Decade for ESD 2005–2014). Across these documents and policy statements, participation is emphasised as a need rather than an option, with participatory approaches being typically regarded as implying a more 'sustainable' model for achieving *long-term changes* in society as well as in school development projects. The direct challenge of the title of a recent Oxfam publication, 'From "donorship" to ownership?' (Lawson 2004), typifies the desired shift in the *modus operandi*, and it is seen to apply as much to societal innovation as it does to how school development is preferably conceived and practised by students, teachers, support staff, and other members of the school community, as Schnack goes on to show in Chapter 11.

However, while the need for participation is argued in different ways, a key focus has now become a shared concern over determining *the kinds and levels of participation* that an individual experiences and enacts (see the summary in Chapter 4 by Simovska, this volume). This implies taking different stakeholders and participants into consideration and in so doing, addressing the possibility of uneven distributions of power and responsibility in their roles and activities. Thus, while the distinction between stakeholders and participants is often slight, these partly overlapping categories are distinguishable in terms of their potential 'mental ownership' of the process and outcomes. They differ, in that participants are those people who are involved in an activity, while stakeholders can be those who only have a direct interest in its outcome.

In the *Research Programme for Environmental and Health Education* at the Danish School of Education, the quest for participation as part of the democratisation of learning processes in environmental and health education is regarded as crucial within pedagogy as well as for the whole approach to innovation. The preferred approach has been to ensure that all 'democratic approaches' are based on participation respecting different participants and stakeholders' points of view and expectations (see Breiting 1996; Breiting *et al.*, 1997; Jensen 2004). The focus has been on the learner's development of action competence (Jensen and Schnack 1997) and the necessity of their participation as part of their learning experience in school, community, and society (e.g. Uzzell *et al.* 1997; Breiting and Janniche 1995; Breiting *et al.* 1999; Christensen 1994; Clift and Jensen 2005).

However, even if the need for participatory approaches in environmental education and ESD is emphasised, the actual level of participation in the action competence approaches mentioned earlier is not necessarily genuine. *This might be considered as a deficit* in heuristics and tools available to determining *the kinds and levels of participation* an individual experiences within a participatory setting. There are a number of implications, one of the most crucial of which is their limiting effects in the sustainability of educational and development innovations and impacts.

The need for a more explicit focus on the conditions for developing a genuine sense of ownership in participation became clear to me during a review project in Namibia in 1997. An international team was brought together to review the mid-term progress of the *Life Science Project* in Namibia, at national and regional levels. The review report identified a short list of aspects in the project that were important for the sustainability of the outcomes of the project (cf. Breiting *et al.* 1997). The key aspects for stimulating ownership among participants in Namibia are listed in Figure 10.1.

These ideas, emerging from a series of visits to the project, resonate with many other instances from my professional experience, although they appear to be discussed and theorised only to a limited degree in the research literature on genuine participation in development and education. In the following discussion, I will address this from two perspectives:

- First, the preconditions for the development of ownership are explored through a search for concepts and heuristics that capture and illustrate these important processes (including from disciplines other than education). I introduce the concept of 'mental ownership' and interpret its nature as a phenomenon and concept, drawing on examples from everyday experience. This is complemented by a presentation of 'psychological ownership' as a relevant parallel concept developed in the fields of psychology related to management perspectives.

The development of a sense of ownership is stimulated:

- If all involved participate in the goal setting or strategy formulation, etc.
- If all concerned are regarded as "equal" partners in the process.
- If all have a direct interest in the changes.
- If those involved give input to the process.
- If they can find their "fingerprint" in the final outcome.
- If they receive some form of recognition for their contribution to the process.

Figure 10.1 A summary of specific aspects that support the development of a sense of ownership among stakeholders in educational development (after Breiting *et al.* 1997)

- Second, the argumentation draws on experiences and knowledge about 'what works' within projects aiming at democratising schools and education. This is achieved through exploring aspects of genuine participation and mental ownership based on experiences and reflections from the project in Namibia in 1997 and an educational development project carried out during 2001–2004 in Thailand. From these cases, I introduce a set of graphical heuristics illustrating the basic relationships between mental ownership and participation which identify focus points for future exploration and inspiration.

Finally a proposal for the innovative power of mental ownership for education is discussed and summarised.

10.3 The Concept of Mental Ownership

Mental ownership is understood here as a concept referring to a mental disposition combining affective and cognitive domains as these relate to a specific situation or to certain achievements. Its etymological roots make it a derivate of the forms of ownership that relate to physical commodities or intellectual properties that have an economic value. Thus, ownership is commonly understood to refer to someone's legal or established right to be regarded as the owner of some object or other. In short, to continue with a financial analogy, mental ownership develops in situations where you 'invest' your mental energies in an activity, for example, when you are aiming for some kind of change in that situation.

A few examples illustrate the crux of the main ideas. If you help your neighbour to build a bicycle shack, your mental ownership of the shack is higher compared to a situation when the neighbour builds it on his or her own. Through your involvement you have the feeling that the shack partly exists because of you; resulting in a sense of mental ownership. Furthermore, your 'investment' in the shack does not need to require physical work. If you have been invited to discuss the location and the design before the construction you are likely to develop some mental ownership, too.

In another example, consider offering help to a work colleague via commenting on a manuscript he or she is preparing. Consequently, your mental ownership to the final document is higher compared to a situation where you were not involved in the drafting of the manuscript. The more you know your comments have been taking into account the more you will tend to feel mental ownership to the manuscript too. This may be understood in the way that you have been able to invest a part of your 'self' in the manuscript, such that it is also 'a part of you'.

The point being made here is that the more involvement and effort to achieve a certain change, process, or outcome in a situation, the higher the level of mental ownership possible for those involved. Theoretically, we can imagine a full continuum of levels of mental ownership, from very low by not having being involved at all, to very high by being the sole person that has decided everything and done everything alone.

Furthermore, it is assumed that the level of mental ownership influences a person's future engagement and motivation in situations involving the 'thing(s)' or situations, to which the person feels ownership. For example, if your neighbour's shack is neglected, you might feel sorry about the situation and may have to say to yourself 'never mind', if you are not able or willing to do something about it. The important point is that you are affected by the situation. Similarly, if your colleague's document receives positive feedback, it is likely that you may gain a sense of satisfaction as well.

Besides the value of the concept of mental ownership in explaining daily experiences, we can also note that it has been used as a psychological mechanism in marketing and commerce. A classic example is a car trader inviting a customer to sit down in a car or to take it for a test drive, to increase their interest in the vehicle. This activity aims at developing the 'mental ownership' of someone else, the customer, in order to lead to a positive effect on their willingness to purchase the car. The aim of the trader is that the car becomes – so to speak – a real and not just potential extension of the other's 'self'. A teacher might express similar goals – wanting students 'to buy into' their expectations for a lesson, understand or enjoy their subject area in similar ways to them, and so forth.

However, while the mechanisms related to mental ownership are considered to be well known, the potential of 'mental ownership' as an important element to the conceptualisation of participation, and as a heuristic to facilitate consideration about learning outcomes for the participants and the sustainability of the results of an intervention, are yet not understood in a systematic and theoretical way. Clearly, 'feeling mental ownership for something' is closely related to other concepts like commitment, engagement, involvement, showing interest, sense of belonging, or responsibility with or for something.

For the purposes of this discussion, we can note that while mental ownership as a concept in an educational context is currently an under-researched area, research on work-place management has fostered a growing interest in another related concept that addresses many of these areas, namely 'psychological ownership'.

10.4 Psychological Ownership

Psychological ownership is a cognitive–affective state that characterises the human condition related to property, possession, and ownership, with the notion of 'mine' being an important indicator of its presence. The titles of recent reports indicate some of the interests in this research: 'Employees that think and act like owners: Effects of ownership beliefs and behaviours on organizational effectiveness' (Wagner *et al.* 2003), and 'Possessive expression at work: "Those machines are mine"' (Kindred 2005). In its basic form, psychological ownership is related to objects, and the subjective aspects of ownership have been researched from different viewpoints. A review by Pierce *et al.* (2003:87) provides a conceptual perspective on psychological ownership and clarifies how it relates to legal ownership:

> [L]egal ownership is recognised foremost by society, and hence the rights that come with ownership are specified and protected by the legal system. In contrast, psychological ownership is recognised foremost by the individual who holds this feeling. Consequently, it is the individual who manifests the "felt" rights associated with psychological ownership. Furthermore, psychological ownership can exist in the absence of legal ownership.

The authors primarily ground their analysis of relevant literature in the assumption that the concept of 'possession' is a recurring interest to the Western cultural tradition. Political endeavours such as 'the ownership society' argued for by US President George W. Bush along the lines of "'making every citizen an agent of his or her own destiny' (The Heritage Foundation 2005) are a recent and powerful contemporary version of this tradition. (While such endeavours are embedded in an ideology of individualism, the present motivation in this chapter is to examine different ends: how ownership might enhance social responsibility, empowerment, and action competence.)

Pierce *et al.* (2003) focus their analysis of pyschological ownership on questions such as:

- Why do individuals come to feel ownership?
- What factors cause individuals to experience these feelings, i.e. the human experiences that result in the emergence of psychological ownership?
- How is this psychological state (of psychological ownership) achieved?

Concerning the motivation for psychological ownership, they conclude that achieving it fulfils three basic human motives ('*roots*'): *efficacy and effectance* (the motivation to be efficacious in relation to one's environment), *self-identity* (through being attached to objects), and *having one's own place* (coming to feel at home).

The mechanisms and paths ('*routes*') in generating this feeling of psychological ownership are identified as integrative of three kinds of experiences in which the individual: (i) controls the ownership target (object), (ii) comes to know the target intimately, and (iii) invests 'the self' in the target. They cite several authors to support the concluding assumption that the more one experiences control over an object, the more the object is experienced as part of 'the self' and the higher the level of psychological ownership. In contrast, those objects that are controlled by others or out of one's control are less likely to become part of the individual's sense of self. Pierce *et al.* (2003:87) describe this as follows:

> [P]sychological ownership is grounded, in part, in the motivation to be efficacious in relation to one's environment. Because of the innate need for feelings of efficacy and competence, individuals are propelled to explore and manipulate their environment. These person–environment interactions may result in the exercise of control and subsequent feelings of personal efficacy and competence.

In other words, interpreted in relation to participation, one might conclude that 'People come to find themselves psychologically tied to things as a result of their active participation or association with those things.'

The idea and importance of investing the self in the target can be traced back in Western culture to, for example, John Locke in 1690, and to later writers:

- '[W]e own our labour and ourselves, and therefore we are likely to feel that we own that which we create, shape, or produce. Through our labour, we invest not

only our time and physical effort but also our psychic energy into the product of that labour' (Locke (1690) after Pierce *et al.* 2003:93).
- 'The most obvious and perhaps the most powerful means by which an individual invests himself or herself into an object is to create it. Creation involves investing time, energy, and even one's values and identity. "Things" are attached to the person who created them because they are his or her product; they derive their being and form from his or her efforts. Thus, the individual who created them owns them in much the same way as he or she owns himself or herself' (Durkheim 1950, 1957; after Pierce *et al.* 2003:93).
- 'The investment of an individual's self into objects causes the self to become one with the object and to develop feelings of ownership toward the object'.... 'In other vocations, individuals may feel ownership for the products they create through scholarly pursuits (academics), the organizations they establish (entrepreneurs), or the bills they draft (politicians). Investment of the self allows individuals to see their reflection in the target and to feel their own effort in its existence' (Rochberg-Halton 1980; after Pierce *et al.* 2003:93).

Pierce *et al.* (2003) suggest that the process of developing psychological ownership takes place on the basis of complex interactions among the elements: the roots, the routes, the target factors, and individual factors. Furthermore, it is assumed that the more the properties of the object target are able to satisfy the three motives ('roots') and enable 'routes' for enaction with the target object, the more likely it is that psychological ownership will develop. Among the individual factors, differences linked to gender and to the type of personality are suggested, while a *time* factor is also expected; suggesting that psychological ownership is not likely to be consistent across people, or persist forever once developed.

Obviously, there are similarities between mental ownership and psychological ownership as both appear to pursue the same mental or psychological mechanisms related to the subjective elements of developing a feeling of 'mine' to something. However, while the concept of psychological ownership is rooted and derived from management studies and workplace or organisational matters, the concept of mental ownership appears to be more closely linked to engagement in issues and problems, in decisions, innovations, and the sustainability of interventions.

While the preceding discussion has introduced the two concepts of mental ownership and psychological ownership, the next section reflects on recent empirical projects to gain further insights into the development of mental ownership from participatory involvement of individuals in environmental education and ESD.

10.5 Strengthening Environmental Education in Thailand – Some Preconditions

Strengthening Environmental Education in Thailand (*SEET Project* 2001–2004) was funded by the Danish Development Agency (Danida 2002) and coordinated by the Thai Ministry of Education to develop an 'integrated learner centred environmental education based on an action competence approach carried out in interaction with the communities' (Danida 2002:9). The SEET project covered all levels of the Thai general educational system: ministries, provincial authorities, educational advisors, school administration, teachers, and their classroom practices and interactions. The focus was on decentralisation, development of a school-based curriculum, learner-centred teaching methods, teachers' and advisors' professional development, whole-school approaches, and building links with the local communities. Throughout all stages of the project (initiation, pre-project period, and implementing phases), SEET had the modernisation of Thai schools and teaching as the target, and environmental education as the point of departure.

The project team (Danish–Thai cooperation) included five environmental education non-governmental organisations (NGOs), a big zoological garden, universities, and teacher training institutions. Already from the pre-project auditing work, it was clear that many aspects of the project were in line with the points highlighted in the Namibian experience in Figure 10.1. Later in the project implementation, efforts were made to keep these points in mind and take them into account when this was feasible and acceptable to the other stakeholders in the project-management group.

The complexity and goals of SEET made it possible to explore participation and mental ownership in relation to a number of issues:

- Project development and achieved innovations
- Conceptual development of environmental education (e.g. links between environmental education and ESD)
- Professional development
- Approaches to teacher training
- Action research
- The idea of 'greening schools' (whole school approaches)

In addition, we could also explore mental ownership in relation to the wider context of international and cooperation, and democratisation and civil society.

Initial findings

By way of background, the Thai educational system mirrors the wider culture in that it is strongly hierarchical and authoritarian in spirit and in distribution of power. Not surprisingly, we found that the schoolteachers as well as the advisors from the Thai Ministry of Education brought little by way of training and expertise with them in enhancing students' ability to think for themselves; an important prerequisite for participation beyond the most basic level of attendance.

During the initial phases of the project, it became evident that one of the most important aspects to address was enhancing the ability of educational professionals 'to think for themselves' instead of just looking for answers and instruction further up in the hierarchy. It was assumed that without developing these kinds of abilities among teachers, how could they encourage and help pupils to do so too? Similarly, how could a supervisor encourage and help teachers in this endeavour if he or she did not have adequate first-hand experiences of 'thinking for oneself'?

The lack of training and skills related to 'thinking for themselves', is obviously a general problem in schools. It is manifest, for example, in the dominating style of questioning in which the teacher expects the class to answer immediately in chorus. The challenge to change the traditional authoritarian Thai style of teaching towards a more democratic and participatory approach created an excellent opportunity for investigating mechanisms involved in the development of mental ownership to the project ideas among the SEET participants.

Another issue arose in relation to local perspectives on the requirements for developing a New Knowledge Economy. The Thai government and the prime minister had emphasised on several occasions that the future of Thailand in a globalised world would depend on the Thai population's ability to be creative and innovative. An expression, like 'Thais are too often copycats', captures something of the criticism in this regard. At the same time the government has been keen to support the development of the Thais' self-esteem and the use of local products and ancient traditions in handicraft. These efforts lead to the integration of 'local wisdom' in the school curriculum and classroom teaching through, for example, inviting older village people to teach traditional handicraft or prepare local food specialities. However, while these efforts are highly valued, the way learning is organised militates against mental ownership here, reducing 'creative work' to following the instructions of a teacher or resource person to copy a model of handcrafts or following a recipe, leaving little space for developing one's own creativity.

As a result of the general situation in Thailand, the SEET project faced a major challenge in the school system's culture in relation to genuine participation. It was assumed that participation was a precondition for the engagement of formal stakeholders (representatives from ministries, regional authorities, educational advisor, school administrator, teachers, and students) in issues of sustainability in a way that hopefully resulted in an increased level of mental ownership.

10.6 The Action Research System of 'SEET'

The *a priori* intention of the action research system within SEET, was to provide professionals in the school system with a 'tool' to help them to support the development and practice of environmental education in Thailand. Furthermore, it was recognised that first and foremost it was important to encourage the stakeholders to 'think for themselves'. Consequently, the focus of the action research system was to support practitioners in reflecting on and developing their own professional thinking and performance as it related to environmental education. This involved a 'reflective practitioner' approach (Schön 1983).

The action research system offered to selected primary and secondary school-teachers, headmasters, and advisors was structured as follows:

1. The action researcher (i.e. the teacher, head teacher, or advisor) was supported by a coach who helped with researching one's own professional thinking and performance related to environmental education.
2. The action researcher had to formulate his/her own problem to investigate and develop solutions.
3. The coach provided support to help the action researcher to overcome obstacles and to analyse motives and possibilities, but without imposing solutions or 'the solution'.
4. The action researchers were advised to focus on making small achievable changes in their daily practice. This was grounded in the assumption that small steps in a good direction would be more sustainable and positive than highly ambitious plans.
5. All action researchers had to keep a personal ('secret') diary containing their thoughts, intentions, observations, analysis, and reflections.
6. The action researchers were invited to regular workshops to exchange, explore, and deepen their reflections on their experience (mainly based on their diary).
7. The coaching team initiated and considered their involvement as a professional development project, involving regular seminars and action research, reflecting and discussing their performance as coaches.

This action research system turned out to be surprisingly effective in terms of the stated intentions. The processes are well documented in Thai (Suwannaketnikom *et al.* 2004). A simple way to understand the impact of the approach is to draw on participants' self-reported changes. These were communicated during a number of seminars for the action researchers and later documented as the action researchers wrote their stories as 'interesting cases' (ibid.). Members of the coaching team reported that their involvement had dramatically changed their own teaching at their university or teacher training institution. Some of the school principals stated that the programme completely changed their views on their teachers' abilities, for example, in terms of their capacities and contributions in identifying solutions to a problem.

For instance, previously, a school principal may have often felt that teachers demonstrated low levels of responsibility and commitment, resulting in a lack of ownership in taking part in the resolution of problems. One school principal reported that this had changed dramatically, as she became more competent herself through her own reflections, in participating in problem solving with her staff.

Teachers reported similar eye-opening impacts in relation to the perceived abilities of their students, as demonstrated by the following quotation from a teacher: 'I had no idea that they were able to think about so much and to come up with so many ideas.' At deeper and more extensive levels, the findings of the action researchers support the general and concrete outcomes of the SEET activities as documented and distributed at the end of the project, in the final reports (Ministry of Education 2004).

10.7 Interviewing 'SEET' Participants Regarding Mental Ownership

In 2005, ten months after the SEET project was officially completed, a qualitative follow-up study investigated the impact of participation in the project, with a specific focus on participants' sense of ownership. Interviews took place with participants from all levels (ranging from the top management people of SEET at the Ministry of Education, to teachers in school). The study was not intended to be used directly as part of the overall SEET project evaluation; in contrast, it aimed to create a distance from the aims and activities of the SEET project, to enable a space to investigate mental ownership and related processes in general. Thus the intention was to search for experiences and statements that might deepen the understanding of the list of specific aspects supporting the development of a sense of ownership identified during the midterm review of The Life Science Project in Namibia (see Figure 10.1).

Twelve individual interviews took place, each one lasting 45–60 minutes (two teachers, four members of senior management from Ministry of Education, and six supervisors from the Ministry of Education, two of whom had served as coaches). The participants were volunteers, although it might be assumed that expectations from their surroundings and their involvement in working teams were likely to have influenced their decision to participate to some extent.

In two cases the recorded interviews were held in English while the others required the help of a Thai–English translator. For each question, the interviewee gave a long answer and the translator took notes and translated the answer. Trust in the quality of the translation was assumed because the translator had been co-operating with the author in Thailand for 3 ½ years with excellent results. Nevertheless, the main data from the interviews have to be understood as interpretations and some times they were summaries by the translator of the interviewee's opinion.

While the terms ownership, mental ownership, and participation were not explicitly used in the semi-structured interviews, questions addressing related aspects of the development of mental ownership were, exploring the concepts from different angles. The questions addressed:

- The experience and motivation ('eagerness') of the interviewee for disseminating the project's ideas (e.g. teaching approaches) and achievements, after the end of the project
- Personal aspects of having been part of the SEET project such as perceptions about having been influential towards project outcomes and its overall success, expressions of the project as 'mine', and reactions from other people about their involvement
- The lessons learned concerning how to initiate engagement and involvement among participants in a potential new project

The interviewees were all positive about participating in the interviews, and provided rich details of their experiences and reflections. It became apparent in the interviews that while the answers were strongly embedded in direct SEET experiences, the ten months afterwards had also led to substantial additional experiences.

Overall findings

All interviewees demonstrated a high level of engagement in the work related to the SEET project and expressed a positive attitude towards the experience in general. But a closer look revealed differences. While there was a common agreement about a high burden of work, most of them did not focus on providing compensation of different kinds when they explained their views on what makes people engage in the ideas and outcome of a project. Only one (a teacher) had the opinion and the experience that compensation by money is important for many participants (teacher colleagues) to make sure they would commit themselves to a new project (interpreted by the author as a sign of low levels of mental ownership).

The teachers and supervisors described their engagement with the ideas and approaches of the project as characterised by the following aspects:

- Learning new approaches and techniques for their teaching that had been satisfying to use, i.e. the teachers had received a positive response from the students.
- Improving their ability 'to think for themselves' and to help students (or teachers) to do likewise. This aspect was especially valued by participants who had been conducting action research on their own professional performance in teaching or supervision.
- Gaining self-esteem through the project was perceived as a positive aspect that strengthened their engagement in the project elements and outcomes.
- Being offered sufficient time to understand the ideas and the new approaches to feel confident in using them. (Many projects of shorter duration, e.g. one year's duration in Thailand, were said to face problems around this issue.)
- 'Surviving' the challenges of a pressing curriculum reform and with a quest for child-centred learning because the SEET experience provided useful preparation for the challenges.

In what follows, the main findings from the study are presented that provide further insights into the prerequisites for developing mental ownership.

Findings on mental ownership

First of all, it was obvious that teachers who had signed up voluntarily appeared to develop a deeper and more persistent engagement in the project than teachers that were more or less forced or convinced by their head teacher to be involved (in some schools the head teacher decided that all teachers should be included in SEET).

Second, mental ownership was experienced or became obvious to some participants in the case of a change of manager or supervisor during the project period. The participants recognised the difference in impact between an engaged boss who supported the project, and a new boss with a lack of interest in the project. Interestingly, in cases where this happened, all interviewees demonstrated continued commitment to the SEET ideas despite a less supportive environment. A similar mechanism was reported when managers or supervisors were moved to another region and their efforts to integrate SEET ideas continued within their new place or field of work.

All interviewees, except one overburdened teacher, expressed a strong sense of personal mental ownership towards the project. For the persons in charge of the central organisation of SEET, the mental ownership appeared to be in relation to the project as a whole, while others expressed a sense of ownership more narrowly towards aspects of the project in their region (for supervisors), or for the project part at their own school (teachers). This can be interpreted in relation to the development of mental ownership to the part of the organisation and innovation that the interviewee had immediately been influencing and involved in.

The interviewees also reported incidences during the project intervention in which different viewpoints clashed and unavoidable conflicts emerged. However, all interviewed key persons, i.e. persons that had some kind of key function as an organiser on a school or in a supervision unit, expressed that despite this there was an overall sense of the project as 'mine' but shared with others. Some referred directly to their influence ('my hope was, that…'), or to the spirit of or their intentions for the project to illustrate this.

However, interviewees also pointed to the influence of recognition at higher level in the hierarchy (such as the Ministry) as an important factor influencing commitment to the project and responsibility.

Findings on external recognition supporting mental ownership

Some interviewees expressed that SEET had generated more engagement than other projects because participants felt they were recognised for their work and could develop their own ideas. The findings point out that:

- The local community (leaders as well as the public) were a main source for social recognition as they apparently realised the usefulness of the school's efforts, related to address environmental problems or other issues.
- Participants experienced official recognition through the Thai promotion system as participants (at all levels) could make use of documentation from their SEET achievements to support their annual request for promotion, which is an integrated part of their professional conditions.

- Several of the interviewees experienced recognition by being invited to participate as a speaker or resource person in other schools or regions after the closing of the project.
- Participants gained an additional feeling of achievement and recognition through the publication and wide availability of small reports which they had individually written up as 'interesting cases' and which were perceived as project achievements.

To conclude this section, the interviews supported the preliminary list of aspects that enhance participants' mental ownership in project innovations developed from the Namibia project (Figure 10.1). We can also add the following aspects to that list:

- To have responded to outsider's expectations regarding one's engagement in innovations and ideas
- To have experienced being part of some kind of new social positive relationship with other partners as established during an innovation
- To have declared some kind of commitment to a project or to ideas
- To have achieved a certain level of master of new ideas and approaches as part of an innovation and to feel confident about them

These aspects too seem to fit into mechanisms of psychological ownership if this concept is extended into project innovations.

10.8 Relations between Participation and Ownership

To begin formalising some of these observations, this next section introduces a series of diagrams to illustrate basic features of the relationship between participation and mental ownership. The development of the diagrams is informed by both the conceptual commentary at the beginning of this chapter, as well as the findings from the SEET follow-up study. In their current state they represent initial visualisations only, and will be further developed and shaped through analysis and experiences in future areas of research and innovation.

Figure 10.2 graphically illustrates a direct relationship between the level of participation and the level of mental ownership. In its simplest form, the relationship can be displayed as a linear graph. Helping the neighbour with the building of the shack or providing feedback on a colleague's draft document, are both examples here. The graph illustrates the simple fact that, the more the individual recognises that outcomes derived from his or her participation, the higher the level of mental ownership.

The graph in and of itself cannot explain a causal relationship; it only suggests that the relationship is a proportional one. Further, the graph does not indicate a time perspective.

An alternative way to explore the relationship between levels of participation and the development of mental ownership is through a focus on sequence: for

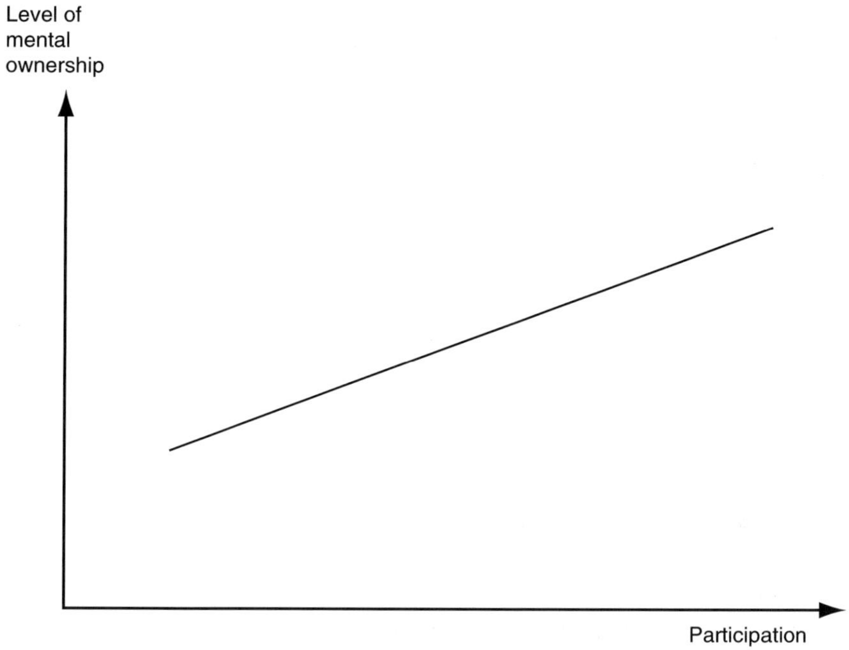

Figure 10.2 Level of mental ownership and participation. A hypothetical way of showing the relationship between the levels of participation and the level of mental ownership in any kind of change action

example, without any kind of involvement in the SEET project participants would not have a chance to develop mental ownership to its ideas but the amount of mental ownership developed depended on the level and quality of the participatory involvement. In this case, participation generated ownership. On the other hand a certain level of pre-existing ownership might help to improve the quality and level of participation.

The main point is that the process has a potential to be self-reinforcing where ownership and participation mutually support and develop the other. While these abstractions might be oversimplified, they have value in terms of highlighting a perhaps trivial but important point, that 'engagement' in an issue can be interpreted as the combination of participation and mental ownership.

Figure 10.3 focuses on the importance of the quality of the participation (e.g. the level of influence during participation and the mechanism of how rewarding it is) on the development of mental ownership.

In Figure 10.3, at the lower level of participation, in the interval besides A, little mental ownership has been generated. However, if the quality of participation was higher (e.g. through being more and equally involved in decision-making), it could generate a much higher level of mental ownership, without changing the quantity

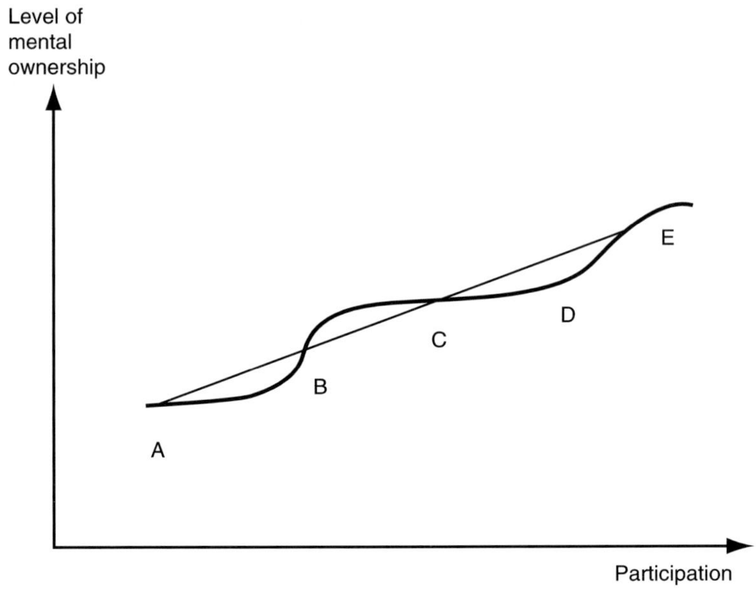

Figure 10.3 How mental ownership may vary with the quality and quantity of participation. The bold line illustrates a hypothetical way of showing relationships between levels of participation related to quality as well as quantity and the level of achieved mental ownership. The slim line represents just the oversimplified relationship of Figure 10.2. Time is not included in the graph. See text for further explanation

of participation that much, as at B. Similarly we can imagine that quite a wide range of 'quantitative' participation does not change the level of mental ownership very much (as at C). Related to the shack example, it could be to increase the time you spend helping your neighbour in building their shack from 10 to 20 hours (a quantitative increase).

On the other hand, if your neighbour makes use of your advice while you are both building, and changes, say, the construction of the roof (and thus changes the quality of your participation) that might generate a much higher level of mental ownership, even if your time invested in the project remains as before (e.g. above B). Also, on the relatively high level of participation it is possible for a qualitative improvement of the level of participation to raise the level of mental ownership considerably (comparing D and E).

Normally we would expect that improving the quality of participation would have much to do with involvement in decision-making as in the example above, and this is clearly reflected in the 'ladder of participation' (Hart 1992, see Chapter 2 by Hart, this volume) and general experience. But we could easily imagine a situation when one has been very much involved in designing the shack and shaping its form and function and that the improved quality of participation had to do with your concrete involvement with constructing the shack in reality, with 'blood, sweat and tears', but without that much extra time (i.e. increase in quantity) involved.

Finally it might be possible to lay out a relationship between participation and mental ownership where a high level of participation becomes counterproductive in generating mental ownership, see Figure 10.4.

The lower level of generated ownership in Figure 10.4 at D might be illustrated by impressions by a person (say, a student) of being left alone (too little support) with an overwhelming problem or too high a degree of responsibility, or he or she may have a negative social experience in the situation. Again the level of achieved mental ownership might be seen as a quality indicator of how well the participatory process has been working.

If this discussion represents the reality of participation in any way it lends support to the need to pay closer attention to the quality of the actual participation in activities and change processes.

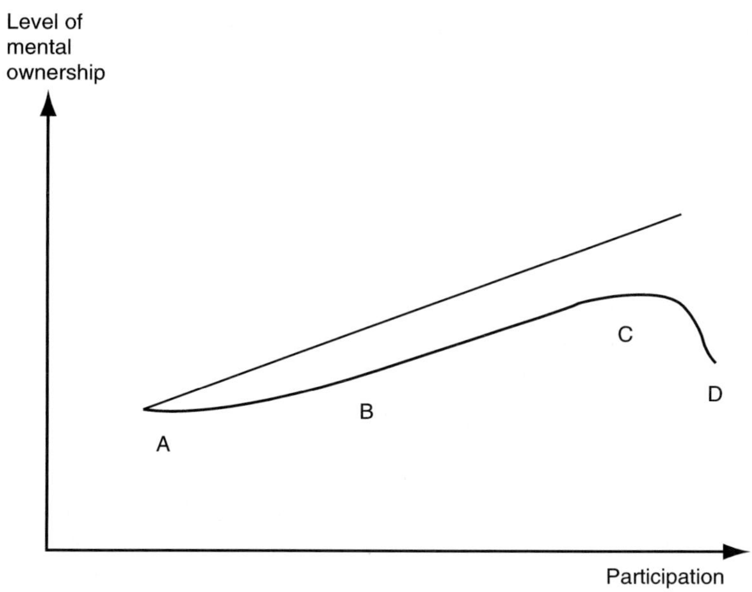

Figure 10.4 Mental ownership and low-quality participation. A hypothetical way of showing the relationship between levels of participation and the level of mental ownership in a series of scenarios with levels of low quality of participation (the bold line). The slim line refers to the over-simplified version in Figure 10.2. The general level between the bold and the slim line is not important for the point, see text for further explanation

10.9 Discussion: The Innovative Power of Ownership for Education

This chapter has introduced the concepts of 'mental ownership' and 'psychological ownership', and used findings and conclusions from studies in Namibia and Thailand to exemplify and demonstrate the potential of focusing on the notion of developing 'mental ownership' to support innovations in education linked to participation. While this perspective have been taken into account in relevant settings (see examples of approaches for improving the quality of education through participation and ownership in American Institutes for Research 2000 and Gross 1997), the full potential has yet to be explored and made more accessible through further theoretical work, research, and application in practice.

In relation to environmental education and ESD, a focus on mental ownership and its value for genuine participation could apply to a number of levels (all of which were involved in the SEET project):

- The learners' generation of ownership to what is happening during the teaching
- The learners' generation of ownership to problems and solutions to issues outside the classroom at present and in the future, e.g. environmental problems and other problems related to sustainable development
- The teacher's ownership to the development of the individual pupil
- The teacher's feeling of ownership to the curriculum
- The teacher's generation of ownership to innovations in her own teaching and to broader school development efforts and innovations, including school community cooperation
- The teacher's generation of ownership to her personal professional development

Clearly the use of participatory approaches is argued for from different perspectives. For some the importance of a participatory approach is due to its *effectiveness* compared to traditional top-down approaches. Others view participatory approaches as crucial within a general *democratic ideal* where learning opens up opportunities for empowerment and emancipation. Although the first view is often embedded in 'management thinking' it is worth recognising that the alternative is not typically a free-choice scenario with endless democratic possibilities, but a substitute, still stiff top-down approach, with very limited feedback and influence from the involved persons. So even with the (often criticised) management rationale for participation, such a participatory approach may be preferable when the alternatives are considered.

Furthermore, it is important to realise that even in educational settings where participation is grounded in a democratic philosophy, there are 'frames' and structures restricting the level of choices and influences. For example, in the SEET project in Thailand, the aim was to contribute clearly towards the democratisation of schools (and civil society in the long term). At the same time, the effort had to set some borders for the freedom of the participants, and in reality many stakeholders also had a kind of management function related to others. It might be more a personal scepticism than a real problem when traditional management philosophy changes in a so-called participatory direction, as described here by Burns (1994: unpaginated):

> Make Your Planning Process Inclusive. Ownership is a critical ingredient when implementing a strategic plan. Without ownership, individuals who are most affected by the plan may actively or passively resist their part in implementing it. The most effective means of developing ownership is to make sure there is broad participation in the planning process.

This advice is far from the situation of a pupil in the classroom but in reality the difference to an educational setting might be difficult to discern and the advice worthwhile for the teacher and student, when operating within a participatory paradigm (see Chapter 11 by Schnack, this volume, for further examples). Indeed, for a while now, a participatory approach to development has been regarded as a successful management strategy that mobilises people towards a pre-defined goal or objective. But to ensure that this approach has deeper roots than a paradigm of persuasion, it is important to focus on the outcomes of mental ownership. The level of mental ownership generated might, so to speak, be regarded as the 'litmus test' for the quality of a participatory approach.

According to Mattessich and Monsey (1992:22), there are key implications for participatory approaches if the notion of ownership is an important indicator of the quality of the participatory approach and the overall success of the innovation. They argue for:

1. Adequate time and resources being devoted to developing ownership among all participants in a collaborative effort
2. Having operating principles and procedures of a collaborative group in place, that promote among members a feeling of ownership about decisions and outcomes
3. Continuous monitoring of the ownership of a collaborative group over time, and making needed changes in process or structure in order to ensure the feeling of ownership and finally
4. Initiating and supporting interagency work groups with regular planning and monitoring meetings about the collaborative effort, to solidify ownership and ongoing commitment.

Mattessich and Monsey (1992) emphasise the importance of a high level of mental ownership of participants throughout the project where important components are the initial phase of a project as well as a continuous monitoring. A sense of ownership can relate to the nature of the involvement as well as to the final outcome. It can even be stated that being challenged in one's perceptions and thinking can also lead to mental ownership as long as it is felt as a personal achievement.

From the experiences and theoretical explorations presented thus far, it would appear to be the case that the most important outcomes of a participatory approach (that generates high levels of mental ownership to innovations) are:

- Participants are much more satisfied with their involvement and the process.
- Innovations are much more sustainable because all stakeholders will tend to care for them in the future, including being able to adapt them to future changes and feel proud about them in general.

- Innovations are much more relevant and well adapted to local concerns and circumstances and become integrated in stakeholders thinking and doing.

In addition to these most important aspects, we might expect that:

- Innovations are more diverse.
- Innovations are less complicated.
- People's empowerment in general is strengthened with a transfer effect to other matters/issues.
- People's cooperation around similar future problems will be more probable.
- Innovations strengthen local self-esteem and 'survival skills' and make people more robust towards future challenges.

In an educational setting there are limitations to the extent of participation judged from the point of view of ownership generating participation. That is probably one reason for the essential need for motivating learners as a part of every teacher's professional competence. Ever since the experience in Namibia in 1997, I have been reminded of the list in Figure 10.1 when designing new initiatives or investigating innovations in progress. It seems that the more any kind of innovation gives opportunities for as many people as possible to be involved in relation to those aspects, the more these people will experience they are a part of the innovation and support it; that is, support its sustainability, generate mental ownership, and benefit from it. The aspects are not only psychological mechanisms but add substance and quality to the actual work. Critical reflections on how the notion of participation is anticipated and brought into real life situations are important. To make use of a focus on the potential development of mental ownership comparing alternative approaches of participation seems to be a worthwhile strategy in a search for the most useful kinds of participation.

References

American Institutes for Research (2000) *Participation as a Method to Improve Education Quality: The Principles. Improving Educational Quality*. Arlington, VA: US Agency for International Development.

Breiting, S. (1996) Is there anything special about environmental education in the Nordic region? *Conflicting Interests in the Use of Natural Resources – A Nordic School Developmental Project on Environmental Education, Nordic Council of Ministers*, Nordic Publishing House, 519: 13–17.

Breiting, S. and Janniche, P.M. (1995) *MUVIN-DK. Background Information for Schools in Denmark Participating in Nordic Environmental Education in 1994–96*. Copenhagen: The Ministry of Education and The Royal Danish School of Educational Studies.

Breiting, S., Hedegaard, K., Mogensen, F., Nielsen, K., and Schnack K. (1999) *Handlekompetence, interessekonflikter og miljøundervisning - MUVIN-projektet*. Odense: Odense Universitetsforlag.

Breiting, S., Imene, S.N., and Macfarlane, I. (1997) *Life Science Project. Midterm Review*. Windhoek: Ministry of Basic Education and Culture, Namibia & Ibis.

Burns, M. (1994) Off the shelf: how to ensure that your strategic plan becomes a valued tool. lessons learned. *National Endowment for the Arts*. (http://www.nea.gov/resources/Lessons/BURNS.HTML)

Christensen, C.U. (ed) (1994) Environmental and school initiatives ENSI – phase 2. The Danish report to OECD. Proceedings from the Research Centre for Environmental and Health Education. 22. Copenhagen: The Royal Danish School of Educational Studies.
Clift, S. and Jensen, B.B. (eds) (2005) *The Health Promoting School: International Advances in Theory, Evaluation and Practice*. Copenhagen: Danish University of Education Press.
Danida (2002) *Strengthening Environmental Education in Thailand. Revised Project Document*. Bangkok: Danida, Royal Danish Ministry of Foreign Affairs. Danida Ref No. 104.Thai.1. MFS.42.
Gross, P.A. (1997) *Joint Curriculum Design: Facilitating Learner Ownership and Active Participation in Secondary Classrooms*. Mahwah, NJ: Lawrence Erlbaum.
Hart, R. (1992) *Children's Participation: From Tokenism to Citizenship* (Innocenti Essay No. 4). Florence: UNICEF ICDC.
Hellesnes, J. (1982) Education and the concept of critique. In: Bredo, E. and Feinberg, W. (eds), *Knowledge and Values in Social and Educational Research*. Philadelphia, PA: Temple University Press, pp. 355–68.
The Heritage Foundation (2005) *Education Notebook: Taking Ownership of Education*. (http://www.heritage.org/Research/Education/EdNotes6.cfm)
Jensen, B.B. and Schnack K. (1997) The action competence approach in environmental education. *Environmental Education Research*, 3, 2: 163–178.
Jensen, B.B. (2004) Environmental and health education viewed from an action-oriented perspective: a case from Denmark. *Journal of Curriculum Studies*, 36, 4: 405–425.
Kelly, U. (2003) Beyond 'The Tyranny of Safety': Reflections on a Potential Alternative. Paper prepared for the Conference 'Participation: From Tyranny to Transformation?', Manchester, 27/28 February 2003. (http://www.sed.manchester.ac.uk/idpm/research/events/participation03/Buhler-Kelly.doc)
Kindered, J. (2005) *Possessive Expression at work: "Those Machines are Mine"*, Unpublished doctoral thesis, City University of New York.
Lawson, M. (ed.) (2004) *From Donorship to Ownership?* Moving Towards PRSP. Round Two. Oxfam Briefing Paper 51. Oxford: Oxfam.
Mattessich, P.W. and Monsey, B.R. (1992) *Collaboration: What Makes it Work – A Review of Literature on Factors Influencing Successful Collaboration*. St. Paul, MN: Amherst H. Wilder Foundation, pp. 7–22.
Ministry of Education (2004) *Case Study. Experiences from Environmental Education Strengthening Environmental Education in Thailand (SEET). Inspiration and Guidance for Thai Schools*. Ministry of Education. Bangkok.
Pierce, J.L., Kostova, T., and Dirks, K.T. (2003) The state of psychological ownership: integrating and extending a century of research. *Review of General Psychology*, 7, 1: 84–107.
Schön, D.A. (1983) *The Reflective Practitioner: How Professionals Think in Action*. New York: Basic Books.
Suwannaketnikom, S. *et al.* (2004) *Action Research on Environmental Education Experience for Reflective Practitioner Development. Strengthening Environmental Education in Thailand (SEET)*. 5 volumes. Bangkok: Chulalongkorn University and Ministry of Education (In Thai).
United Nations (1993) *Earth Summit. Agenda 21. The United Nations Programme of Action from Rio*. New York: United Nations Publications.
UNESCO (1976) The Belgrade Charter: a global framework for environmental education. *Connect*, UNESCO-UNEP Environmental Education Newsletter, 1, 1, 1–2.
UNESCO (1992) *The Tbilisi Declaration: Final Report Intergovernmental Conference on Environmental Education. Organized by UNESCO in Cooperation with UNEP, Tbilisi, USSR, 14–26 October 1977*, Paris: UNESCO ED/MD/49. UN
UNESCO (2005) *UN Decade for Education for Sustainable Development (2005–2014)*. (http://portal.unesco.org/education/en/ev.php-URL_ID=23279&URL_DO=DO_TOPIC&URL_SECTION=201.html)

Uzzell, D., Davallon, D.J., Fontes, P.J., Gottesdiener, H., Jensen, B.B., Kofoed, J., Uhrenholdt, J.G., and Vognsen, C. (1997) *Children as Catalysts of Environmental Change*. In collaboration with Instituto de Promoçao Ambiental, Lisbon, Portugal.

Wagner, S.H., Parker, C.P., and Christiansen, N.D. (2003) Employees that think and act like owners: effects of ownership beliefs and behaviors on organizational effectiveness, *Personnel Psychology*, 56, 4: 847–871.

Chapter 11
Participation, Education, and Democracy: Implications for Environmental Education, Health Education, and Education for Sustainable Development

Karsten Schnack

Keywords participation, co-determination, democracy, general education, adjectival education

11.1 Introduction

This chapter focuses on participation in the context of European and Nordic debates about whether education should be understood as an end in itself, as in the humanistic tradition, or instrumentally, to bring about social change. In the case of the latter, environmental education, health education, and education for sustainable development can each be understood as denoting initiatives and efforts that are driven by a shared ideal of improving the world we live in now, and for the future. As such, they are often seen as overruling education in the former, humanistic sense, particularly in relation to its legitimate purposes and modes. Whilst acknowledging this trans-educational function, this chapter sets out to recover the relevance of general education to 'adjectival educations' like environmental education, health education, and education for sustainable development, and vice versa, that is, in relation to efforts to educate pupils in a broader, humanistic sense. It asks what, after all, is the aim of our educational efforts, and how are we to assess their success and outcomes, particularly if we are to prevent 'adjectival educations' from being reduced to instruments of ideology or policy when the overarching goal is to foster deeper and more meaningful participation in education.

11.2 The Relationship Between Education and Social Objectives

While earlier chapters in this volume have illustrated common agreement that participation has become a buzzword in both education and development studies, recent struggles and debate are characterised by asking 'where next' with the term

DPU, Research Programme for Environmental and Health Education, Danish School of Education, University of Aarhus, Tuborgvej 164, DK-2400 Copenhagen NV, Denmark, schnack@dpu.dk

A. Reid et al. (eds.), *Participation and Learning*, 181–196.

(e.g. Jensen 2000b; Hickey and Mohan 2004). Some argue that participation is better replaced by another term. For others, it is a notion that still has relevance and potential. This chapter explores the second position, given that participation remains an important notion to environmental education, health education, and education for sustainable development. Reasons for this include that:

- The concept illustrates an important overlap between how each of these adjectival educations has come to be constituted (e.g. by participatory approaches and goals), as well as in terms of the overlap between their goals and approaches and those of general education in fostering similar objectives.
- While participation is constitutive of the idea of democracy, it can easily be reduced to a token; for example, as a means for legitimating arbitrary power.
- Participation has different conceptualisations and interpretations, in that it can be understood in strong, more radical senses, but also in weaker senses, with quite different implications for education and development.
- In both education and development, the possibilities for and practice of participation pose questions about the efficiency as well as morality of current activities, and it may be advocated as a requirement for either motivational or ethical reasons.

Current educational discourse can give the impression that education and health improvement, environmental protection, and sustainable development are linked together by instrumental relationships and functions. Here, education is perceived as a means rather than an end in itself. However, to separate means and ends can be artificial and fruitless because they are usually entwined, excepting extreme situations for the sake of discussion. Even if different pedagogical methods are sometimes discussed as alternative means to a specific end, e.g. reading skills, mathematical competence, sociological imagination, or democratic attitudes, the preference for some methodologies over others is most often based on the sound assumption that the methods are not means to identical outcomes. In education, means are not neutral tools; they are in themselves alive with experiential and formative consequences that cannot meaningfully be abstracted in their totality from the definition of the outcomes. We might then consider whether a more useful way to talk about means and ends in educational endeavours is as two extremes on a continuum. At one pole, education is interpreted as specialised training with no other aim than being the means to achieving a certain predefined objective (e.g. a profession or craft or skill). An appropriate term for this exercise in instrumentalism could be 'instruction' or 'training'. At the other pole, education is understood as an end in itself. It is combined with certain fundamental values in relation to which this kind of education might be called a 'means', but here, this is understood in a more hermeneutic sense, for example, as a 'precondition', a 'dimension of', or a 'perspective for' developing or achieving some overarching goal or other. In this line of thinking, to be 'educated' is often interpreted as a precondition for becoming a 'fulfilled' self, of being able to actualise one's potentials in a balanced or harmonious way, and of being able to become an active and integrated member of a modern democratic society.

The practices of most educational programmes and settings within and outside the formal educational system are located somewhere in between the two poles of the continuum, and the evaluation of these practices differs substantially depending on the perspective taken upon it. Examining or evaluating an educational issue from the 'means' perspective can be very different from that from the 'end' perspective, especially when 'education' is considered as an instrumental or functional means for obtaining goals other than those of 'education' and the 'educated and fulfilled' person in society. For example, a programme about energy saving or battery collecting may be evaluated according to the amount of energy saved or batteries collected, *or* it may be evaluated with reference to what the students learned about society, their community, and themselves by being involved in these activities (Jensen and Schnack 2004; Breiting *et al.* 2005; Mayer and Mogensen 2005).

It is important to emphasise that this distinction is analytical rather than, say, ontological (a point we will return to later). That is, the value of making this analytical distinction lies largely in its ability to promote clarification and evaluation of the status of practices, but also in the options it presents for grounding the debate in terms of a humanistic tradition, rather than solely in terms of its role in pursuing or realising an instrumentalist agenda. In other words, the differentiation challenges us to engage discussions about what is, after all, the purpose of our educational efforts, and what is relevant when we want to assess their 'effectiveness' or 'efficiency' (e.g. who gets to define and contest these terms, and how). In what follows, the distinction will be used to inform wider reflections on participation, education, and democracy in relation to environmental education, health education, and education for sustainable development, by focusing on how they relate to education in general, alongside their role in bringing about social change.

Environmental education, health education, and education for sustainable development as education

Environmental education, health education, and education for sustainable development can each exemplify initiatives, activities, and efforts driven by a goal to better the world. In such cases, education is often seen as a means to something else than education itself.

Problems in contemporary Western society are often identified in relation to the health of people, ecological issues about their relationship to nature, and challenges connected to the lack of the sustainability of economic development and many political and sociocultural processes. Many of these problems are experienced (initially) as very local and personal; they are often manifested as individual and maybe even private difficulties. However, when the individual cases become more widespread or frequent, personal troubles are transformed in scale and interpretation, constituting examples of public issues of social structure (Mills 1959).

Take the example of a few overweight people. This situation may be considered problematic in terms of their own health and well-being, and their social surroundings. However, as the average number of overweight people rises, it may be considered a social phenomenon and thus a societal issue. In the same way, the occasional extinction of a few species of plants and animals may be regarded as a local, intermittent problem, while an acceleration of extinction rates and effects may be seen as a threat to a habitat's, ecosystem's or in some cases, the earth's biodiversity, and thus represents a serious challenge for humanity.

Besides the sheer number of cases, the characteristics of the people affected, as argued by the environmental justice movement, can also make a difference as to whether something is considered an individual/local issue or a social one. For example, the increased polluting of a watershed or the progressive deterioration of the urban or natural environment around many factories and industrial areas may be an immediate problem for the workers and the local community, but when it spreads to the areas where the owners and other better-off people live, it might start to be considered a societal issue (Galtung 1972).

Societal issues often become political issues when they draw the attention of political activists and representatives and are acted towards with political means, such as through policy, legislation, and regulation. The most difficult political challenges, which can include those related to health and the environment, may call for education to be used as a complementary tool to develop not only the necessary awareness, knowledge, and understanding of the issue, but also to support the development of the 'right' attitudes and 'responses' in terms of the behaviours, habits, and ultimately, lifestyles, of people.

This illustrates the tension in interpreting education as an instrument that can solve some societal problems, or is fundamentally of value in and of itself. In a way, both views are appropriate and possible and not necessarily in contradiction. However, the evaluative criteria for judging success differ in important ways. It makes a lot of difference if you look for indicators for success in such things as visible and measurable changes at a social level – such as decreases in the number of overweight people, a lower number of cases of drug abuse, and reduction in pollution levels or energy consumption. These examples can be contrasted with the less tangible or immediate outcomes for individual and social learning, or those that may not be amenable to indicator measures or instrumentation (Breiting *et al.* 2005; Mayer and Mogensen 2005).

Of course, it is possible to have school projects dealing with health or environmental issues reporting considerable benefits or effects in terms of 'improved health' or a 'cleaner environment', without any reflections on the broader educational value of the project activities and the learning outcomes for the participating pupils. Viewed from the other pole of the continuum and evaluated from the general educational perspective, many school-based health and environmental projects can be considered inadequate because while they may prove effective in changing a health-related habit or improving the local environment in the immediate and typically short term, they do not enhance, let alone demonstrate a measurable gain in longer-lasting pupil learning or competence applicable to changing situations (Jensen and Schnack 2004; Clift and Jensen 2005).

What is at risk here is making one's focus solely that of a shared commitment to improving the world in and through environmental education, health education, and education for sustainable development, and the ideological motivation to do so. Such a situation can easily end up reducing these 'educations' to an instrumental logic related to some kind of social engineering. An alternative is to step back and recognise the compound status of these adjectival educations, that is, to emphasise and examine the educational dimension of environmental *education*, health *education*, and *education* for sustainable development. This is the focus of the next part of the chapter, in which I discuss this status in relation to participation in general and adjective forms of education, and the possible tensions and achievements they might encounter in the life of a school.

11.3 Why Participation? Efficiency Versus Morality

To return to an earlier theme, within international development theory and practice, participatory ideas have become widespread and have almost gained the status of an imperative. The 'rise of participation' began during the 1980s with increasing critique of the shortcomings of top-down approaches to development. Since then we have witnessed an almost Kuhnean paradigm shift towards the endorsement and practice of participatory approaches to community development (see Chapter 3 by Reid and Nikel, this volume). In a recent book, Robert Chambers describes this as participation constituting the new paradigm of development (Chambers 1997), and the situation is characterised by bottom-up change, democratisation, and empowerment.

We can note here that the concept of 'empowerment' plays a particularly significant role in health education (Tones and Green 2004; Clift and Jensen 2005). The use of this particular terminology is interesting in this context owing to its built-in reference to power, which arguably lies at both the heart and reality of participation. For good reason, power and issues relating to power (inequalities, distribution, gradients, its exercise, and so on) have become the battleground for much recent discussion of participation in the development field. Cooke and Kothari (2001:4), editors of the collection on, 'Participation: The New Tyranny?' make a frontal attack on current practices of participation in the field, stating that 'tyranny is the illegitimate and/or unjust exercise of power; this book is about how participatory development facilitates this.' They also argue for an increased recognition of the tyrannical potential of participatory development, perceiving this as systemic rather than merely a matter of how a specific practitioner operates.

Most of the debate about the 'paradigm' of participatory development has focused on methodological issues and questions about the use of different tools, and this has been ongoing within the field since its early days (see, for example, Rahnema 1992). Cooke and Kothari (2001) however, have accomplished a much more fundamental ideological analysis and critique, culminating in the picturesque and provocative title of the last chapter, by Heiko Henkel and Roderick Stirrat, on: 'Participation as Spiritual Duty; Empowerment as Secular Subjection'.

While the rationales for adopting a participatory approach in both development and educational programmes vary in several ways, a useful distinction may be drawn between the arguments that reference efficiency and those that invoke morality. Participation itself is, of course, not something simply 'out there' to be discovered and analysed; on the contrary it is a very complex phenomenon and to analyse it also requires focusing on the very construction of a reality. Thus it can require asking critical questions such as whose reality is relevant when we talk about participation, and what is understood by that the very term, 'reality'. So, even if Chambers seems to take the question: *Whose reality counts?* as a rather simple and almost rhetorical confrontation with the colonial tradition, it may also be the starting point for an advanced postcolonial critical reflection on the power structure of the Western or Northern, reality, with its built-in, perhaps taken-for-granted notions of human rights, empowerment, and participation.

These questions about power can also foreground considerations of their equivalences across both development and education. For example, in both arenas participation may be a part of ethically grounded democratic 'power equalisation'. Also, in both fields, participation can be used as an efficiency-driven strategy of self-governance. However, even in the case of the ethical rationale for participation there is not an easy and straightforward answer to the question 'whose reality counts'. To give one example: when the modern conception of education (as *Bildung* in German) is understood in a democratic perspective, it automatically implies a kind of political education, in which the overall ideal is the development of 'action competence' and participation is interpreted as co-determination (Schnack 2003). Yet when democracy and education are then exported to the developing world as part of an international development project, the cultural clashes that we may also find within the Western world itself regarding these terms and goals are often magnified. For example, my Nepalese students find the whole philosophy of participation interesting and fascinating but tend to acquire a knowledge of it in a sort of parallel learning; as a Western ideology with little relevance for Hindu culture. It took me some time to realise that their apparent problems in appropriating the ideas of participation were actually due to sound cultural reasons and intentional resistance rather than learning difficulties (cf. Chapters 10, by Breiting, this volume). There is a lesson to be learnt here, and not only in relation to development and the developing world: namely, whose reality counts in education?

The idea of 'general education' (Bildung, Allgemeinbildung)

Following on from the key ideas of the Enlightenment and modern conceptions of democracy (such as the ideals of emancipation, autonomy, and equity), general education has become both more important and more 'general'. In the modern continental European tradition, such general education can be seen to be both comprehensive and extensive in relation to three key dimensions (Klafki 2000):

1. Scope: it is education for all and everybody
2. Aim: to develop a wide range of human competences (cognitive, emotional, physical, social, etc.)
3. Content: it is related to topics and issues of general rather than special interest.

The third of these dimensions presupposes a further differentiation between:

(a) Vocational education or training (in German: *Ausbildung*; in Danish: *uddannelse*), which relates to the societal differentiation of (job) functions.
(b) General or liberal education (in German: *Bildung*; in Danish: *dannelse*), which refers to what is of relevance for all of us as members in a (democratic) society, or in the world.

It is important to note that in this view, general education (*Bildung*) cannot be simply reduced to a question of effectiveness (e.g. how effective the education is in achieving predefined objectives or goals in forms of attitudes, knowledge, and skills). Education in this sense also refers to a kind of formation and open-ended development of the person. Furthermore, we could also consider how in relation to content, health education, environmental education, and education for sustainable development might be or are framed as important aspects of a general education, in theory and practice.

However, rather than unpack those themes here (though see Schnack 1995b and 1998), I will focus on how participation plays a significant role in the logic of this framework. To begin with, while democracy may be defined in many ways, it can be assumed that their key referents, such as the equal distribution of power and the participation of laypersons in debates and decisions about issues of common concern, constitute prominent and shared features. Thus, it is argued, one of the duties of general education in a democratic society must be to give all members of society access to the opportunities, structures, and activities that will build up understanding of issues that, in principle, concern all members of the society, such as questions about health, environment, and sustainable development.

This can be reframed as a curriculum principle, taking those challenges facing humanity into consideration as one of the criteria for selecting the content of education (Schnack 1995a), not forgetting that what is considered as being of concern to us all is subject to the *Zeitgeist* (foregrounding the need to be aware of historical context), but also a playground for political, cultural, and ideological positions and their influence on decision-making.

On this point, Wolfgang Klafki, a prominent representative of the German *Didaktik* tradition, argues for a curriculum that is grounded in what he calls the 'key problems/characteristic of the period' (*Schlüsselprobleme*). Klafki contends that general education must contain a historically mediated consciousness of the central issues of the age, and, as far as it is possible, those of the future. He further emphasises the need to accept that we all share the responsibility for these kinds of problems and should develop a readiness to participate in the solutions of the problems (Klafki 1985). For Klafki, during the mid-1980s the five key issues (*Schlüsselprobleme*) that concerned general education were: (i) peace; (ii) environmental issues; (iii) socially created inequalities; (iv) effects of technology, especially ICT; and (v) I-You relationships between people. Accordingly, the embedded potential tension throughout the five is between, on the one hand, an individual's claims or rights to happiness and fulfilment, and on the other, inter-human considerations and respect for others.

Participation and General Education

In relation to the *Schlüsselprobleme*, it is important to emphasise that the main point is not the precise choice and number of key issues, or for that matter, their continued or continuing relevance, but the logical link that can be made to participation, which stems from the democratic orientation of this *Didaktik* (curriculum theory).[1] A useful way to illustrate this is by using Klafki's summary of his conception of *Bildung* (general education). Here, *Bildung* is viewed as an end in itself, articulating three interrelated notions:

- *Selbst-bestimmung* (self-determination)
- *Mit-bestimmung* (co-determination/participation)
- *Solidarität* (solidarity)

To clarify, *Solidarität* is of course, not solidarity with 'everyone', but to emphasise being for and with the weak and disabled, who do not necessarily have the opportunity or equal potential to display self-determination and co-determination (in effect, an anti-fascistic point) (Klafki 1985).

Also, the dual options for *Mit-bestimmung* arise because sometimes the term is translated very literally into English as 'co-determination', or alternatively as 'participation'. This uncertainty with *Mit-bestimmung* presents an attractive double possibility for participation in an educational context, namely, distinguishing between participation as taking part, and participation as having an influence. In Danish, this is readily illustrated as the language has at least two different senses of participation, that is: (i) *deltagelse* which implies, to take part in (as opposed to being an onlooker) and (ii) *medbestemmelse* which implies, to be involved in decision-making and to have a say (see also Reid and Nikel, Chapter 3).

Both meanings are important in a democratic perspective, in general and for education. However, even though they are interrelated they are often treated separately as two different pedagogical challenges. Of course, one may argue that the second interpretation (participation as influence) is always a special instance of the first one (participation as taking part); and, therefore, that the first one is

[1] A short series of invitational seminars called 'Didaktik meets curriculum', discussed the relationship between the German/continental tradition in educational philosophy and that of Anglo-Saxon perspectives. This resulted in three books, which also contain translations of German texts (Klafki and others) into English (Hopmann and Riquarts (eds) 1995, Gundem and Hopmann (eds) 1998, and Westbury, Hopmann, and Riquarts (eds) 2000).

the most fundamental one. On the other hand, the second and stronger meaning of co-determination is the more difficult one in practice, and it is too easily neglected even if it has important democratic learning potentials, particularly as these relate to questions of ownership, responsibility, fairness, handling conflict, and so forth.

Participation interpreted in the Danish context – from co-determination to individualisation via differentiation

This ‘double’ notion of participation has played a decisive role in the Danish pedagogical debate and reflections on practices of the classroom life, and since 1975 it has been incorporated in the national education legislation. Passed in 1975, the Danish ‘School Act for the *Folkeskole*’ (primary and lower secondary schools) contained a paragraph (Paragraph 16.4) that explicitly states that together with teachers, students are to be involved as co-determinants (having a say, in a strong sense) in the organisation, pedagogies, and methodologies of the teaching–learning process in the classroom, as well as its content. Even if the very words are not used in the Danish legislation, you might say that the Act requires students and teachers together to decide on, plan, and assess the practised curriculum, and this has been its typical interpretation.

The Act coincides with a relatively strong movement towards fostering greater levels of student participation in the 1970s and 1980s, in which participation was seen as not only an ingredient in active learning, experiential learning, and constructivism, but also in the stronger sense of co-determination. Since those times, responses to the Act have been divided, some teachers stating that it is very difficult or even impossible to achieve, while others indicating that it has become part of daily life in classrooms. For the latter, co-determination is regarded as always in an imperfect state of affairs, and consequently, it is a continuous learning process. While it is fair to acknowledge that co-determination has not become a predominant trait in the classrooms in the Danish *Folkeskole*, reports show that students do feel that they can influence the teaching, and that they have a say (Daniel *et al.* 1985; Nørgård 1992).

There is an important point of clarification to be made about this Danish experience: participation as co-determination is not understood as an additional time-consuming, disturbing, or distracting factor, but as a central and important part of the learning. The decision-making processes about the curriculum are themselves considered to be part of the curriculum. The school has a special obligation here; it is not only a democratic institution but also an arena for the experiential learning of democracy. Qualified democratic participation is learned by gaining experience with participation.

Why then was this interpretation of participation as co-determination introduced and stressed in the Act? On the one hand, it was viewed as a response to the widespread problems in a lack of student motivation. In the 1970s, talk about a 'crisis' was common, and addressing 'school fatigue' became part of the vocabulary of educational debates and conversations. On the other hand, co-determination as genuine student participation was understood as a concrete expression of democratisation. In that sense it was viewed as a necessary element in fulfilling the ideal of 'education *for* democracy *with* democracy'. In this respect, a strong focus on the collective aspect was prevalent: individual students had to talk, listen, argue, and compromise in order to reach joint agreements.

The policy statements about cooperation between teachers and students are still valid in the contemporary Danish context. The Act formulated in the mid-1970s was revised in 1993; the revised version (Section 4, Paragraph 18) opens with a statement saying that:

> [T]he organisation of the teaching, including the choice of teaching and working methods, teaching materials and the selection of subject-matter, shall in each subject live up to the aims of the Folkeskole and shall be varied so that it corresponds to the needs and prerequisites of the individual pupil.

One way to interpret this revision is to focus on its strong emphasis on the differentiation of teaching, typically expressed as individualisation in terms of teaching that meets the diverse learning needs of each individual student. However, while there is some reasonable awareness of the need for this and the practical difficulties of what can be a time-consuming task, the realisation of this 'personalisation' of learning pathways in the classroom remains a slow process. Typically it shows itself more as a variation in the teaching–learning process than as a principle for classroom practice. This is subject to change now owing to a heavy political pressure, followed by national tests and demands for continuous assessment and revision of learning objectives for each individual pupil (Carlgren *et al.* 2006). Nevertheless, the notion of 'putting students at the centre' of educational efforts was, and still is, a trend in Denmark; while in recent decades, the sense of the plural in 'students' tends to be understood as a move towards the singular via an accumulation of individuals (the additive plural), in contrast to a collective sense of the plural – a community of socially interacting individuals.

In all this, 'differentiation of the teaching' or simply 'teaching differentiation', in contrast to 'pupil differentiation', has now become the key concept. As opposed to pupil differentiation, it has been understood first and foremost as an idea about integration. For this reason, there has been a continuous debate about the balance in general education between individualisation and community – and solidarity-focused learning (Nielsen 1995).

One variation of the teaching–learning process that has contributed very much to the development of teaching differentiation and participatory learning has been the 'project method', where groups of students work together with problem-oriented projects. They choose an issue and formulate a set of questions that they investigate and, finally, make a report. Introduced into higher education and adult education in the second half of the 1970s (Illeris 1974; Berthelsen *et al.* 1977; Holten-Andersen *et al.* 1980), it soon became part of the progressive experiments in the *Folkeskole* too (Daniel *et al.* 1985). In the 1993 School Act it officially entered the classroom, albeit via an ironic route through a paragraph about assessment (Paragraph, 5): 'At the 9th and 10th form levels, the pupils shall carry out an obligatory project assignment, for which the assessment shall be given in the form of a written statement and by a mark, if the pupil so wishes.' As a result, the policy-level status of the 'project method' helped to legitimise different versions of it, and led to its more widespread use, even in lower grades (as exams and assessment forms always influence the preceding teaching).

Given such developments over the last 20 years in Denmark and the influence of Didaktik on the Nordic education system, the final part discusses Nordic examples of attaching great importance to participation and democracy in the context of environmental education, health education, and education for sustainable development.

11.4 Adjectival Educations – Democracy – Participation: Nordic Examples

As stated in the opening part of this chapter, participation is a key notion in these three adjectival educations. It can be viewed as binding them together and each one with general education. It can also emphasise the relationship between education and democracy, given the view that democracy without educated people is an 'empty shell'; and, education with no democratic dimension is little more than training or a tool for indoctrination.

The first generation of environmental education (Breiting 1993) in Denmark, as in other countries, was almost solely construed as conservation education and aimed at changing the behaviour of pupils. In reaction to this, members of the Research Centre for Environmental and Health Education in Copenhagen opposed this viewpoint by stating that environmental issues are better understood as problems generated in society and culture, rather than in nature. This position emphasises that at the heart of environmental problems lie conflicting human interests in relation to nature (Schnack 1984 and 1998). Consequently, a central feature of this perspective is that conflicts or problems are not in nature or between people and nature – they are between (groups of) human beings.

From this point of view, conceptions of the environment are always value loaded and by extension, often political issues, with questions of participation and democratic power distribution as crucial to this approach to environmental education as they are to general education; a position that has been examined in several development and research studies, conceptually and empirically, for example, the large-scale Nordic developmental and action research project MUVIN (Environmental Education in the Nordic Countries) 1992–1996 (Schnack 1995b, 1998; Breiting 1996; Breiting *et al.* 1999).

Similar developments were apparent in health education and health promotion in schools in Denmark over that period. The same Research Centre, now called the Research Programme for Environmental and Health Education (RPEHE), worked with a distinction between two paradigms in health education: the moralistic and the democratic (Jensen 2000a), to emphasise the shift of focus of health education from behaviour modification with respect to health, to participatory, and action-orientated pedagogical approaches, aiming at enabling students to reflect upon and address the social determinants of health. As part of this work, the democratic paradigm has also been applied and evaluated conceptually and empirically, within the terms and experience of 'The Health Promoting School' initiative (Clift and Jensen 2005). In place of behaviour modification and referring to a broad and open conception of health and the aim of developing students' action competence, the democratic paradigm focuses on a democratic, participatory pedagogical approach.

The final example in this section concerns the overlaps between environmental education and education for sustainable development (ESD) in Swedish schools. Here, three 'selective traditions' have been identified: the 'fact-based tradition', the 'normative tradition', and the 'ESD' or 'pluralistic tradition' (Sandell, Öhman, and Östman 2003; Öhman 2004). All three can be related to democracy, though in different ways. While the fact-based tradition may prepare for democratic participation *after* education, the normative one presupposes a kind of democratic consensus process *before* education. For Sandell *et al.* (2003:177) in the ESD tradition:

> [T]he democratic process is an integral part of the education process and is situated *in* education – the critical discussion on different alternatives and their implications is an essential part of education itself.

Without calling the first two 'undemocratic', the third approach may be viewed as the only one that represents a 'democratic education' in theory as well as practice, in that it is the only one that tries to educate *for* democracy *with* democracy.

In summary, if environmental education, health education, and education for sustainability are to be 'democratic education', the concept of participation has to be an integral part of the pedagogical philosophy – not only for efficiency reasons but also for ethical and educational reasons. At the same time the key concepts in these three areas, such as sustainable development, health, and democracy, have to be dealt with as open notions and consequently made the object of discussion and (re)construction in the teaching–learning processes. In a democratic, participatory education you cannot *teach* health, sustainable development, or democracy. The meanings of the concepts are open and contested and with an expression of John Huckle (1996:3), a key function of education is 'to help people reflect and act on these meanings and so realise alternative futures in more informed and democratic ways'.

11.5 Key Issues for Research

Among the many educational questions and challenges discussed in this chapter, four key issues can be identified for further research:

Methodological problems related to empirical study of the decisions made in classrooms

When research participants (teachers and students) are asked about their experience by researchers, depending on how the questions are formulated, the responses will differ. In one study (Christensen and Schnack 1992) students were asked the following: *Who has decided on the topic or issue you are working with?* Participants most often responded that they did not know or did not remember. It is as if the process of co-determination – when it is not a special project but quite normal life in the classroom – is a subtle process, and researchers and evaluators should be very cautious so as not to misinterpret what is going on. It might, however, for sound educational, research, and development reasons, be a good idea to make the process more explicit.

Forms of participation in different phases of the teaching–learning process

Thus it might also be a beneficial to be more conscious and explicit about the different aspects or elements of the participatory process. Inspired by the analysis of the original concept of 'work', RPEHE researchers have looked at the phases of identifying a need (or a problem), getting an idea or designing a vision, sharing ideas, creating a plan, doing something according to the decided plan, assessing the results and reconstructing the plan, and so on. Jensen (1983) has argued that broadening the landscape of student participation to more phases than those typically experienced will be an essential aspect of democratisation, and more studies would be relevant in this area. One such attempt has looked into the gradations of participation and different varieties of student participation (Jensen 2000b). This suggests that student participation can differ across phases – for good and bad reasons – and sometimes without any apparent reason at all.

Tensions between standardised, objectified curriculum, and participation

Worldwide policy trends towards more specified objectives and standardised assessments are developing at the same time as increased expectations of child-centred pedagogy and learning for citizenship. The potential tensions between these two sets of interests may be actualised in different ways. From the point of view of democratisation, investigations into this field will be very important. Conceptual clarification, comparative studies, as well as participatory action research or dialogue research will be very relevant.

Assessment and success criteria

As we live in an era of assessment and evaluation, the focus tends to be more on educational outcomes than processes. This is a problem when participation and democratisation are primarily understood in terms of their status as a means rather than an end. It is very difficult and time consuming to evaluate genuine participation, and it is almost impossible to generate end point quantitative criteria of any use. More research has to be done in this field to qualify the discussion of what can sensibly be meant by 'evidence-based knowledge' or 'evidence-based practice' in relation to participation and democratisation in education. For example, when it comes to evaluation, it often seems too easy to change the focus from education to observable or spectacular outcomes. In the area of 'green schools' or 'eco-schools' this is a well-known issue as the problem presents itself in what counts as the criteria for awards and celebrations (Ward and Schnack 2003). Further research is needed not only to qualify the (self-)evaluation process, but also to study how educational measures or quality criteria (Breiting, Mayer, and Mogensen 2005; Mayer and Mogensen 2005) work and influence the participatory pedagogies and processes of the schools.

References

Berthelsen, J., Illeris, K., and Poulsen, S.C. (1977) *Projektarbejde – erfaringer og praktisk vejledning* [Project work – experience and practical guidance, in Danish]. Holstebro: Borgen.

Breiting, S. (1993) The new generation of environmental education focus on democracy as part of an alternative paradigm. In: Mrazek, R. (ed), *Alternative Paradigms in Environmental Education Research*. The North American Association of Environmental Education, pp. 199–202.

Breiting, S. (1996) MUVIN-DK. Innovation and research in environmental education. In: Breiting, S. and Nielsen, K. (eds), *Environmental Education Research in the Nordic Countries*. Proceedings from the Research Centre for Environmental and Health Education 33. Copenhagen: The Royal Danish School of Educational Studies, pp. 75–96.

Breiting, S., Hedegaard, K., Mogensen, F., Nielsen, K., and Schnack, K. (1999) *Handlekompetence, interessekonflikter og miljøundervisning – MUVIN-projektet* [Action competence, conflicts of interests, and environmental education – the MUVIN project, in Danish]. Odense: Odense Universitetsforlag.

Breiting, S., Mayer, M., and Mogensen, F. (2005) *Quality Criteria for ESD-schools: Guidelines to Enhance the Quality of Education for Sustainable Development*. Vienna: Austrian Federal Ministry of Education, Science and Culture, Department V/11c, Environmental Education Affairs.

Carlgren, I., Klette, K., Myrdal, S., Schnack, K., and Simola, H. (2006) Changes in Nordic teaching practices: from individualized teaching to the teaching of individuals. *Scandinavian Journal of Educational Research*, 50, 3: 301–326.

Chambers, R. (1997) *Whose Reality Counts? Putting the First Last*. London: IT Publications.

Christensen, C.U. and Schnack, K. (1992) Nature and environment. In: Nørgaard, E. (ed), *School Development and Innovation. Abstracts from 14 Research Reports of Evaluation of the Innovation Programmes of the Danish Folkeskole 1987–91*. Copenhagen: The Royal Danish School of Educational Studies, pp. 31–42.

Clift, S. and Jensen, B.B. (eds) (2005) *The Health Promoting School: International Advances in Theory, Evaluation and Practice*. Danish University of Education Press.

Cooke, B. and Kothari, U. (eds) (2001) *Participation: The New Tyranny?* London/New York: Zed Books.

Daniel, J., Krogstrup, P., and Pollas, B. (eds) (1985) *Projektarbejde – kan man det I folkeskolen?* [Project work – is that possible in the primary and lower secondary school? in Danish]. Copenhagen: Statens Pædagogiske Forsøgscenter and Folkeskolens Forsøgsråd.

Galtung, J. (1972) *Økologi og klassepolitik* [Ecology and class-politics, in Norwegian]. Copenhagen: Christian Ejlers' Forlag.

Gundem, B.B. and Hopmann, S. (eds) (1998) *Didaktik and/or Curriculum: An International Dialogue*. New York: Peter Lang.

Hickey, S. and Mohan, G. (eds) (2004) *Participation: From Tyranny to Transformation? Exploring New Approaches to Participation in Development*. London: Zed Books.

Holten-Andersen, C., Schnack, K., and Wahlgren, B. (1980) *Invitation til projektarbejde – en problematiserende fremstilling* [Invitation to project work, in Danish]. Copenhagen: Gyldendal.

Hopmann, S. and Riquarts, K. (eds) (1995) *Didaktik and/or Curriculum*. Kiel: Institut für die Pädagogik der Naturwissenschaften an der Universität Kiel.

Huckle, J. (1996) Realizing sustainability in changing times. In: Huckle, J. and Sterling S. (eds), *Education for Sustainability*. London: Earthscan, pp. 3–17.

Illeris, K. (1974) *Problemorientering og deltagerstyring – oplæg til en alternativ didaktik* [Problem-orientation learning run by the participants, in Danish]. Copenhagen: Munksgaard.

Jensen, B.B. (2000a) Health knowledge and health education in the democratic health-promoting school. *Health Education*, 100, 4: 146–153.

Jensen, B.B. (2000b) Participation, commitment and knowledge as components of pupils' action competence. In: Jensen, B.B., Schnack, K. and Simovska, V. (eds), *Critical Environmental and Health Education: Research Issues and Challenges*. Copenhagen: The Danish University of Education, pp. 219–238.

Jensen, B.B. and Schnack, K. (2004) Assessing action competence? In: Scott, W. and Gough, S. (eds), *Key Issues in Sustainable Development and Learning*. London: RoutledgeFalmer, pp. 164–167.

Jensen K. (1983) *Arbejde – undervisning og demokratisering* [Work – teaching and democratisation, in Danish]. København: Danmarks Lærerhøjskole; Copenhagen: The Royal Danish School of Educational Studies.

Klafki, W. (1985) *Neue Studien zur Bildungstheorie und Didaktik* [New studies in educational theory and Didaktik, in German]. Beltz Verlag.

Klafki, W. (2000) The significance of classical theories of Bildung for a contemporary concept of Allgemeinbildung. In: Westbury, I., Hopmann, S., and Riquarts, K. (eds), *Teaching as a Reflective Practice. The German Didaktik Tradition*. Mahwah, NJ: Lawrence Erlbaum, pp. 85–107.

Mayer, M. and Mogensen, F. (2005) Evaluation in environmental education and the use of quality criteria. In: Mogensen, F. and Mayer, M. (eds), *ECO-schools: Trends and Divergences*. Vienna: Austrian Federal Ministry of Education, Science and Culture, pp. 26–41.

Mills, C.W. (1959) *The Sociological Imagination*. Oxford: Oxford University Press, reprint 1973.

Nielsen, V.O. (1995) *Folkeskolen i 90'erne – loven og samfundet* [The Folkeskole in the nineties – the education act and the society, in Danish]. Vejle: Kroghs Forlag.

Nørgård, E. (ed) (1992) *School Development and Innovation. Abstracts from 14 Research Reports of Evaluation of the Innovation Programmes of the Danish Folkeskole 1987–91*. Copenhagen: The Royal Danish School of Educational Studies.

Öhman, J. (2004) Moral perspectives in selective traditions of environmental education – conditions for environmental moral meaning-making and students' constitution as democratic citizens. In: Wickenberg, P. *et al.* (eds), *Learning to Change Our World?* Lund: Studentlitteratur, pp. 33–57.

Rahnema, M. (1992) Participation. In: Sachs, W. (ed), *The Development Dictionary*. London: Zed Books, pp. 116–131.

Sandell, K., Öhman, J., and Östman, L. (2003) *Education for Sustainable Development. Nature, School and Democracy*. Lund: Studentlitteratur.

Schnack, K. (1984) Miljøundervisning i folkeskolen [Environmental education in the primary and lower secondary school, in Danish]. In: Bjørn, M. (ed), *Miljøundervisning i folkeskolen – en konferencerapport*. København: Miljøministeriet, pp. 47–55.

Schnack, K. (1995a) The didactics of challenge. In: Hopmann, S. and Riquarts, K. (eds), *Didaktik and/or Curriculum*. Kiel: Institut für die Pädagogik der Naturwissenschaften an der Universität Kiel, pp. 407–416.

Schnack, K. (1995b) Environmental education as political education in a democratic society. In: Jensen, B.B. (ed), *Research in Environmental and Health Education*. Copenhagen: The Royal Danish School of Educational Studies, pp. 7–20.

Schnack, K. (1998) Why focus on conflicting interests in environmental education? In: Ahlberg, M. and Filho, W.L. (eds), *Environmental Education for Sustainability: Good Environment, Good Life*. Frankfurt am Main: Peter Lang, pp. 83–96.

Schnack, K. (2003) Action competence as an educational ideal. In: Trueit, D., Doll, W.E., Wang, H., and Pinar, W.F. (eds), *The Internationalization of Curriculum Studies*. New York: Peter Lang, pp. 271–291.

Tones, K. and Green, J. (2004) *Health Promotion: Planning and Strategies*. London: Sage Publications.

Ward, K. and Schnack, K. (2003) Perspectives on the eco-schools programme: an environment/education dialogue. *Southern African Journal of Environmental Education*, 20: 142–148.

Westbury, I., Hopmann, S., and Riquarts, K. (eds) (2000) *Teaching as a Reflective Practice: The German Didaktik Tradition*. Mahwah, NJ: Lawrence Erlbaum.

Chapter 12
What Comes Before Participation? Searching for Meaning in Teachers' Constructions of Participatory Learning in Environmental Education

Paul Hart

Keywords environmental education, participation, learning, teacher's thinking, intentionality

12.1 Introduction

This chapter contributes to the debate on participatory education through examining some of the intentions and dispositions that underwrite teachers' ways of thinking and acting in educational settings. Rather than looking at participation directly, the study on which the chapter is based attempted to generate understandings of how certain teachers become disposed towards the participatory interests that are almost synonymous with environment-related education. My interest is in the assumptions and predispositions that underlie teachers' notions of what counts as participatory learning within their genealogies of context, and so this chapter explores, the aims and methods for elucidating teachers' stories of their actions within their theories of context.

12.2 Background

> Beneath the veneer of civilization…lies…the human in us who knows what is right and necessary for becoming fully human:…a rich nonhuman environment…the discipline of natural history…the cultivation of metaphorical significance of natural phenomena of all kinds…. There is a secret person undamaged in each of us, aware of the validity of these conditions, sensitive to their right moments in our lives. (Shepard 1982:39.)

Environmental education is more a philosophy than a curriculum area such as science or social studies. Teachers who have somehow acquired the 'ethic' always seem to find ways to incorporate environmental education into their programmes

Faculty of Education, University of Regina, Regina, Saskatchewan, S4S 0A2, Canada, hartpa@uregina.ca

A. Reid et al. (eds.), *Participation and Learning*, 197–211.

of study. Contrary to official stories of the teacher-proof curriculum, environmental education occurs because teachers believe it is important to the education of young people in their care (Hart 2003). Our research group[1] has traced teacher's participation in environmental education to certain core values, such as respect, caring, and responsibility, and these seem to be associated with early-life history. Some teachers reflect them in their personal and social values to the extent that they incorporate environmental ethics as part of their personal practical theory; in other words, this philosophical identity is manifest in their educational practice.

In attempting to understand the precise nature of the environmentalist part of teachers' discourses-practices, we have adapted our earlier narrative-based inquiries in pursuit of a more nuanced genealogy of teachers' experiences within their more immediate historical contexts. Given recent developments in research on learning that implicate its socio-cultural dimensions (Cole 1996), our focus now incorporates activities that include the generation of memory maps, autobiographical accounts, and sense of place encounters. These have proved useful in helping participants link their personal experiences to their pedagogical beliefs and practices. This narrative-based adaptation to our work represents an attempt to explore teachers' theory/practice connections as genealogical tracings of those pedagogical ideas that appear to have helped them to reflect on the origins of their participatory practices in environmental education, as the socio-cultural dimensions of their own learning (see Gutiérrez and Rogoff 2003). This chapter elaborates how and why we reframed our focus, and discusses a short piece of textual evidence that illustrates this shift.

12.3 Conceptual/Contextual Frames

To be an environmental educator is to understand what it means to know based on a participative relationship with the world. Although educators often discuss knowledge as cognition – as something that can be acquired by individuals and exists 'inside the head' – ideas about knowledge have changed in line with the evolution of social theories, initially based within a philosophy of knowledge grounded in orthodox science. For example, Reason (1988) has identified three interrelated aspects of an ontological shift in thinking about education that implicate participation, critical subjectivity, and knowing in action, as constitutive of a postmodern view of knowledge. Whilst not achieving the status of a new orthodoxy, such a shift has created openings for ways of knowing based on broadened notions of consciousness and a world view that encompasses a variety of less certain and more dense webs of knowing, in representing human social interaction. Sterling (2003) captures the spirit of this shift in forms of thinking by distinguishing between 'learning within education' and 'learning through education'. This implies an epistemological change towards more participatory modes of knowledge generation,

[1] I use the words 'we' and 'our' to refer to our team of research associates that included Ann Camozzi, Susan Gesner, Christine Robertson, Loraine Thompson, and Judith McPhie at various stages of the research. The Social Sciences and Humanities Research Council for Canada funded the work over a period of approximately ten years.

and is tune with the participatory learning found within the historical rhetoric of much environmental education.

It may be difficult for many of us living in industrial societies and wondering what it means to be entering a postmodern age to imagine how we can experience participatory, empathetic, responsible, and primary meanings to our lived experience. Heron (1996: 181) argues that such meanings are 'both prior to and consciously underneath and within our use of language', where language is 'secondary meaning', and remains a 'partial and incomplete transformation of our primary meaning'. Such 'participatory consciousness', set aside for a hundred years for ideological and political reasons (Berman 1981), remains a fundamental part of human intersubjectivity with nature (see Abram 1996). The term captures the work of 20th-century philosophers and educators who seek to redress a scientifically narrow and outdated Cartesian epistemology and to engender embodied ways of knowing that reflect new understandings of human learning and meaning-making. In Heshusius' (1994) terms, participatory consciousness is an awareness of a deeper level of kinship, a way of being with others, and the world, facilitating more profound understandings of others and the landscapes within which they reside.

To a large extent, this consciousness has permeated the work of environmental educators over the last decade. The idea that there is more to teaching than observable behaviours is based on assumptions other than those relating teachers' competence and effectiveness to individual performance. Our work with Canadian teachers who incorporate environment-related activity within their elementary school classes has led us to attend to the relationships between their actions and their perceptions, understandings, beliefs, and values. We think it is reasonable to view teachers' conceptions of their professional work in terms of what they believe is good for 'their' children, as well as what seems to them to be educationally effective. So we have questioned how their beliefs, attitudes, and conceptions are connected to their identities as teachers (Sfard and Prusak 2005). Because identity can be thought of as created and recreated in interactions with people, we explored the role of relationships in shaping the dynamics of their ideas and practices (see Bauman 1996; Holland and Lave 2001; Roth 2004). Thus, for complex reasons amid the tensions of structure (as social 'givens') and agency (not in a unified sense but as something life history gives them to make sense of themselves as multiple subjectivities), we have explored how, in essence, environmental education theories have become pedagogical practices.

Evidence of a fundamental agentive dimension to teaching is grounded in research on teacher's thinking (see Day *et al.* 1993) and in life history research (see Goodson 1992). As a shift in focus from the search for greater 'skill' within educational research to the search for 'will' (Sarason 1983), understanding how the discursive production of teachers' identities provides frames for their practices accrues value, as it can expand the research frame beyond largely anthropocentric interests rooted in multiple levels of social relationships to recognise other agents and agencies, e.g. from the 'more-than-human' world. Using Clandinin and Connelly's (2000) narrative investigation of teachers' implicit theories, we have explored how images constructed from childhood experiences could be viewed as culturally endorsed, implicit, taken-for-granted values and beliefs that connect directly and dialectically with teachers' pedagogical practices. However, studies of personal, professional knowledge (Elbaz 1983, 1991), professional development (Louden 1991), and attempts to characterise teaching as moral or creative (Olson 1992; Woods and Jeffrey 1996) provide evidence for what Bruner and Haste (1987) describe as a quiet revolution occurring within learning research and developmental psychology. So we have now incorporated into our work ideas that make a case for investigating teacher's learning through interactions located in historical and sociocultural contexts as a way of making sense of social processes. This is a more complex conceptualisation of identity beyond intrapsychic and social dimensions.

The idea that, in certain situations, learning was more likely to result from participation in social learning experiences, such as young girls learning 'weaving' skills from close observation and intent participation (Rogoff *et al.* 2003), has begun to transform learning theory (see Bransford *et al.* 2000) in ways that directly tie participation to the development of social consciousness and social responsibility (see Moore *et al.* 1985; Wenger 1998). However, this recent propensity to implicate personal qualities in teaching decisions/practice with the social participatory nature of learning is troubled by social theory that provides explanations of social determination of personal identity that leave little room for agency. For example, Foucault (1981) and Bourdieu and Wacquant (1992) show in their different ways that personal identity and action are inevitably conditioned by the operation of power in society. Although issues of identity and structure are far from being settled, we found that our interpretation of environmental education practice was strengthened by our attempt to envision the development and contestation of teacher's identities in terms of issues of power.

So, while we began by exploring 'reasons' for teachers' pro-environmental professional actions in terms of personal qualities as 'latent' awakenings of significant life experiences, our current interest is in the actions of teachers in terms of what MacIntyre (1981) might call 'traditions' or Bourdieu (1984) fields of choice within relational (social) space. Viewing teachers as being aware of arrays of possible choices (and actively choosing participatory environmental education activities) and, in essence, rejecting other possibilities, we can look for the missing links by which they affirm their affiliations and make distinctions that come to constitute their professional identities. The process of understanding such complexity is messy because the meanings of such choices may change over time, they may even contradict themselves on occasion and are often tacit or unconscious parts of teachers' thinking. Thus, inquiry into reasons for particular actions being situated and historical is tenuous at best. However, assuming that actions and choices are not completely random and whimsical, it seemed reasonable, following Taylor (1989) and others (see Noblit and Dempsey 1996), to look at tendencies historically and as embodied and generated by 'moral sources' and other resources (personal and social) that inform a teacher's creative articulation of such choices within the structural/power constraints of the profession.

Our interest in the historical agentive dimensions of teaching environmental education as participation attends to the fact that people are the way they are because of their past, but it is also more than this. For example, our work has been informed more recently by well-constructed arguments about learning through participation in meaningful experiences that are identity-forming, meaning-driven, and socially situated (see, for example, Solomon *et al.* 1992; Rogoff *et al.* 2003). Developmental learning research affirms earlier findings in disparate areas of social consciousness, political socialisation, and moral education, suggesting that participation is grounded in one's sense of larger meaning and purpose, one's identification with morality, and that these emerge early in childhood (see Thomashow 1995; Berman 1997; Hutchison 1998). If, however, we begin to think of biography as a mixture of personal biography and social biography, we can then examine life-history research (see Louden 1991; Goodson 1992) in terms of how biography affects the practice and professional identities of a teacher in both senses. Thus, Poulson's (2001) concern that researchers focus on teacher's learning as a means of understanding teacher's actions led us to find ways of looking at where in teachers' learning experiences their implicit theories have been generated and developed throughout their lives, as well as at how their professional choices were exercised in relation to relatively intractable social constructs, including cognitive and sociocultural (affective) frames, and institutional structures (i.e. relational positioning in social space – see Bourdieu and Wacquant 1992).

Sociocultural frames have also been useful in our examination of the influences of professional mentors and colleagues across the generations because they provided a focus on the social nature of learning (see Cole 1985; Smagorinsky 1995; Ball 1997). A neo-Vygotskian perspective suggests that learning involves the use of cultural and symbolic 'tools' such as language, texts, and experiences to develop understandings of subjects (such as environmental education), while engaged in particular activities within particular environments. Our focus in talking to teachers, as students of educational professionals and as mentors for their own school students, emphasised this generative social/contextual nature of learning in producing new understandings that themselves acted to reshape the original symbols (images, concepts of environment, issues) as organising frames for their own thinking (i.e. beyond a seemingly intractable moral grounding). Thus, teachers' learning may be morally grounded but it is also contextually bound, shaped by meaning through their participation in experiences past and present, in particular situations (i.e. natural or urban settings), and as configured by significant people (i.e. mentor-types) in the setting/location/space.

In our genealogical tracings, Russell's (1997) notion of 'activity networks' has also helped conceptualise ongoing, meaning-directed, yet historically conditioned tool-mediated interactions between teachers and their students. In our inquiries, these interactions are centred around exploring certain experiences in post-secondary courses of study and in professional encounters, often in school and university contexts (though not always located on campus), within the contexts of school and classroom cultures of pre/in-service teachers. Such contexts can be thought of as networks or systems of activity or participation (see Sterling 2003), where certain beliefs, ethics, values, and assumptions are assumed and enacted. Accordingly, our teacher-participants in the research came to think about teaching and learning in relation to participation networks in which they found themselves as learners, learning 'through' social interaction where cognitive dimensions of learning are intimately interconnected and dependent on the society. Frames of sociocultural theory (Vygotsky 1978; Cole 1985; Wertsch 1991; Smagorinsky 1995; Wertsch *et al.* 1995) thus informed both our data collection and 'analysis/ communication'.

12.4 Researching Indentity via Narrative

Let us suppose for a moment that a university-based teacher-educator and a small group of pre-service (or in-service) teachers share a predisposition, perhaps an ethic, which can be characterised as ecophilosophical, ecological, or environmentally sensitive. Suppose that such a shared ethic, although messy, full of contradictions and only partially coherent and certainly not unitary or comprehensive, consists of some ideas, images, and predispositions that can be used as tools to investigate learning. As Grumet (1987) says, to investigate this kind of interactive process we begin with narrative. We do so because we are asking people to 'tell identities'. These tellings may be identified as a set of reifying, significant, and endorsable stories about a person. According to Sfard and Prusak (2005), such stories, even if individually told, are products of collective storytelling. We were therefore led to consider a more complex process of autobiographical excavation as a way into understanding the 'landscapes of learning' (Wilson 2002).

Let us now further suppose that if we view integration between teacher and students as a learning process, and if learning implies becoming a different person, then it involves the construction of identity (Lave and Wenger 1991; Wenger 1998). Such broadening of perspectives about learning might implicate educational practices in new ways of learning/knowing based, for example, on different considerations of what counts as learning, legitimate experiences, or educational settings. Such a scenario, if appropriately framed in terms of narrative identity, for instance, gives new meaning to the value of narrative inquiry as a conceptual bridge between identity and sociocultural dimensions of learning. Sharing an ethic means sharing some agency – through the experience of identity as a way of being in the world (Wenger 1998:151).

Viewing our narrative work with teachers as something more than their own stories of who they are, we attempted to link teachers' identities as environmental educators (even if only part of their identity) to communicative activity beyond self-dialogue. We attempted to get beyond claims about teachers' stories as mere windows into teacher's thinking or representing their essences, to a view of their thoughts and actions as discursive constructs that are collectively shaped (even if individually told), and although they may change according to the authors' and recipients' perceptions and needs, accessible to researchers. This is not to say identities do not originate in daily experiences (see Wenger 1998), but that they operate as discursive counterparts of one's lived experiences – that is, as visions and memories of those experiences rather than the experiences themselves (see Sfard and Prusak 2005).

Although we remain uncertain as researchers about what we can learn from what might be termed genealogical tracings of ideas, trajectories of relationships, or even a sociological flow of ideas, we think that we can draw on what is known about how narratives may interact with one another beyond personal stories, narratives of experience, and personal life histories (see Clandinin and Connelly 2000). We were looking for social/cultural learning connections as if they could be found. We believed that the meanings transfered through such symbolic and cultural tools as coursework experiences, writings, drawings, stories, and images shaped by purposes, situations, or contexts (see Kagan 1992; Richardson 1996), were somewhere to be found in the social relationships, the relational spaces and places of interactions, and the conversations over time about the environment, sustainability, or ethics. As Sfard and Prusak (2005) say, identities are the product of discursive diffusion: our proclivity to recycle ideas of others even if we become unaware of the origins of these texts.

In spite of our doubts about the possibility of understanding the personal/social construction of meaning through relationship, we began by creating family case studies centred around active environmental education, teacher education programme components, working to interview professors, their own mentors if possible, and several teachers who indicated through their subsequent teaching work that their environmental education experiences had helped to develop, nurture or awaken their desire to teach environmental education. The idea for this strategy was based on the notion that those stories that make it into a teacher's identity depend on the significance that teachers/students attach to the voices of others. Significant narrators, as the most influential voices, are carriers of those cultural messages that will likely have the greatest impact on one's actions (Sfard and Prusak 2005).

Our interviews were conversational, interactive, of a historical (life history), autobiographical and genealogical nature. We pursued teacher-generated stories, anecdotes and ideas through conversations and publications of mentors that served to articulate their philosophical bases. We assumed teachers' discursive practices were traceable to many sources, particularly in teachers' education where many teachers had identified significant mentors in previous studies. We viewed our mentors as the 'heroes' of these narratives because they engage routinely in the elaboration of thought about discourses-practices in environmental education that may be influential in the generation of another's thinking, by actively, interactively, penetrating existing ideas. We assumed that these utterances and words would not have been effective were it not for the power of relationships with significant others to contribute to teacher/student narratives about themselves (perhaps as tacit co-authors of their own identities). In other words, a story was never just a story but a statement of belief that could take on a life of its own beyond itself in the work of others, or as Goffman (1959) puts it, as 'presentations of self in everyday life'.

It was anticipated that the narrative evidence provided here could capture aspects of our attempt to access narrative identities and penetrate social consciousness across two or three generations. Our tracings were in search of what Bakhtin (1981, 1999) calls 'internally persuasive discourse' – those ideas about thinking and acting that have engaged us from within and become part of us, part of our being. We think that some of these ideas may have been products of a discursive diffusion or narrative osmosis (Gee 2001) with enough power to contribute to one's sense of identity, that is, as stories that somehow speak to us so powerfully that we cannot continue our work without them. We also recognise that not all such ideas or stories are simply accepted as elegant or immaculate, but come to us as we critique, modify, adapt, or reject the ideas of our teachers. Nor do most of us propose our ideas to students in ways that anticipate anything other than engagement in them as complex possibilities, sensitive to differences of many kinds. It is not unreasonable to conjecture that learning occurs where shared knowledge is exchanged interactively as one's values are exposed, critiqued, and changed within the relative safety net of collegial relations. Critical stories, those that shake our sense of identity, could even make one feel as if one's whole identity has been changed (Sfard and Prusak 2005).

On the one hand then, our inquiry strategy was epistemologically individualist, as one that regarded teachers' ideas/practices as theory-guided, experience-based, and critically reflexive, as participants themselves struggled to learn about their own tacit 'drivers' (Carr and Kemmis 1986; Kemmis 1998; Kemmis and McTaggart 2000). On the other hand, our method was epistemologically interactionist, acknowledging personal, practical theory and experience as valid ways of knowing as they implicate hermeneutic, participatory processes. Blending narrative autobiography with genealogy represented our attempt to link narrative notions of identity to the activity of communication as discursive or reflexive. It also represented an attempt to address the role of narrative diffusion in learning. Framed ontologically, our study was based on the philosophical premise that people who come to know interactively enter a kind of hermeneutic cycle of learning. That is, they come to understand complexities in their own thinking through dialogue, but particularly the sort of dialogue that maintains space for uncertainties, inconsistencies, and indeterminacies across multiple subjectivities (see Beyer 1988; Lave and Wenger 1991; Krechevsky and Stork 2000).

There is always a tension, it seems, between teachers' intuitive, remembered ways of knowing manifest in their agency in classrooms, and those discourses of culture and education that form the context of teaching (Stevenson 1987; Gilmore 2001). Our inquiry was more about (re)storying and identity-building through interpretation of philosophies of environmental education teaching as participative, than the social dimensions of a larger learning process. How else can one search for those ideas that mattered most – as a sociological flow of lasting impressions now manifest as 'can't not do' practices? How do researchers help teachers 'see' differences they can make in the lives of their students? Narratives that constitute one's identity, being an important factor in shaping one's actions, seem useful in research even if they communicate one's experiences only as well as human words can tell (Sfard and Prusak 2005).

12.5 Tracing Teacher's Identities as Stories Coming from Different Narrators

In a case study involving a university professor (Cynthia[2]) and three of her former students (Whinnie, Eunice, and Neeson), we explored how their interactions seemed to have generated some lasting meanings for each of their identified lives

[2] Pseudonyms have been used when participants are referred to by name.

as teachers. We have illustrated, as excerpts from this case study, how the narratives of one person appear to have woven through personal narratives of another. Because this sense of connection in relationship was difficult to represent textually, narratives were focused on the significance of various kinds of interactions where participants derived meaning for their orientation to and participation in environmental education. These passages capture small portions of text that seem to represent connections to both Cynthia's mentors and her students. Obviously, the narratives, as life stories, are more complex than these testimonials can express, as we strived to achieve as proximate a relation as possible to what really is the case of that discourse (Chawla 1994).

We begin with one teacher's (Whinnie) description of how her parents influenced her ideas about individual responsibility and social justice through their own participation in their Canadian community.

> They (my parents) were very vocal – very interested in their community, in politics. Always they would tell us to stand up for what you believe in...the whole family would attend protests and marches in the streets. In one such incident, Dad used his umbrella to tear down a sign he didn't like. His picture was in the newspaper and he showed it to us proudly.

Neeson related incidents of his participation in the TREK programme – an outdoor programme of integrated environment-focused studies becoming increasingly popular in Canadian high schools.

> There was lots of outdoor experience – nature was experienced and appreciated as a place of refuge, as an important protector of human life...it can be very empowering...but it needs an ongoing component so students don't get dropped back into that other life without support.

> There was a special teacher. He always raised issues of social justice...and the importance of spiritual development, which I ignored when I was a high school student. He engaged his students in social projects and got them involved in the community.

In university, Eunice said she 'ended up' in geography because it had a landscape and an outdoors component. She grew up on a hobby farm with parents who were 'wonderful models of love of the outdoors'. She was politically active, she says, particularly in the area of social justice, but returned from South Africa (where she went after completing her geography degree) because:

> [W]orking on justice issues while not working on environmental issues seemed to be problematic – missing a huge piece of the fit...without clean air and water all struggles to lessen poverty, racism, would be for naught in terms of our human ability to survive.

Each of these teachers gravitated toward Masters-level studies in environmental education – 'it was a moral imperative' for Eunice. Neeson emigrated from community education and Whinnie had enrolled in a Masters programme when she took an environmental education unit and met Cynthia there, her professor and subsequently one of the significant influences in her life as a teacher, or so it seems.

Cynthia's early landscape was an industrial town, but she remembers the 'Clean up America' campaign.

> My father loved nature, loved to be outdoors and would take us hiking. Though he wasn't an environmentalist, he bought me a copy of Thoreau; he affirmed my own passion…somehow I equated the notion of citizenship with looking after…environment.

One of Cynthia's mentors was a social studies teacher who organised a student club to fight pollution. He was able to affirm her early sensitivities to environmental degradation and to provide her with some of the tools for social action. She recognised him as her inspiration for the commitment and effort required to make a difference – something she has taken seriously in her life as a professor.

From her own educational experiences, both formal and informal, she developed a strong sense of personal responsibility. This principle guides her personal participation as well as her professional sense of commitment. Arising in childhood, Cynthia attributes her active moral stance to her philosophy professor who challenged dominant social values, and to authors such as David Orr (1994), who advocates responsible citizen participation as environmental stewardship.

We ask, in our role as researchers, whether it is possible to assert that an influential teacher has made a difference in the moral and ethical stance taken by a student? We ask whether this is possible, without coercion or indoctrination, but rather through creating conditions for them to find their own voice? We also ask, in this role, whether our descriptions of these influences can capture such ephemeral connections, certainly not as explanations, but at least as plausible interrelations of interactional learning?

Always aware and wary of the power embedded in her role as a teacher, Cynthia's postmodern sensibility can be seen repeatedly in her struggle with the notion of 'empowering' her students. She challenges herself and her students to ask some hard questions of themselves. She urges her students to develop their awareness and understanding of their personal landscape; she speaks of citizenship, of connections between personal awareness and personal commitment, but there is no question that it is her view that environmental education is foundational to all learning and all subject matter. She states this view many times in the course of our conversations:

> I would like to see teachers modelling and teaching stewardship. Get the kids outdoors – have them looking, drawing, writing, becoming aware and appreciative…to feel a connection to place, to what causes peace and what causes disjunction…to understand that the way we do science is weighted with the values of society.

Her students describe Cynthia in terms of her intellect, her caring, and her emulation of mutual respect.

> She would share her opinions and values but never made you feel that they were the only ones…just hers.
>
> She encouraged us to know our own voice, to explore our individual beliefs and values. Change happens inside first.

Cynthia's themes include the need to construct knowledge of one's personal landscape in order to understand the social construction of knowledge on a path to responsible citizenship (which involves environmental stewardship and a measure of humility toward nature).

> So we wondered what we might find in the students' practice, their ideas and values in action, some of which might be part of an intergenerational flow of ideas that somewhat mysteriously work to inform environmental education on the ground.

Like Cynthia, Neeson sees environmental education as:

> [D]irect experience in the environment...whatever landscape we inhabit is where to begin. We first need to study and participate....

Whinnie is teaching elementary school where, she says:

> It is important for my students to have concrete experience of exploring the environment they live in...then they have a reference point to extend out to other parts of the country and the world.
>
> My students do opinion pieces every week. First they read...then write.... They're learning to develop their own voice...to explain their thoughts and I'm trying to get them to think about why they think the way they do...where ideas come from.

Eunice, as education coordinator for a local non-profit organisation, believes that...

> Educators need to recognise that mental health and human humility are essential to a healthy and spiritual way of being. The school garden is one way of making that connection.

Her students already cared about the environment, already wanted to make a difference in the world when they came into Cynthia's classroom. However, under her guidance, they expanded their awareness and increased understanding of their own potential – a support Cynthia attributes to her professors (at the University of Windsor). The importance of reflexive work is echoed in both Cynthia's and her students' work, just as she was asked by her philosophy professor to examine her opinions and values and to attend to critical thinking. These are moral positions, it seems to us, operating as discursively produced identities that are integral to action and participation as environmentally responsible citizens. Summarising their own reflections on investigating learning as a culturally shaped activity, Sfard and Prusak (2005) endorse an increasingly argued conviction that the narrative turn in identity work will increase the researchability of educational phenomena. Questions about agency would become questions about discursive practices, as identity-building is equated with storytelling.

12.6 Stories as Precursors to Participation/Action in Environmental Education

Throughout our work, it has been clear to each of us that central issues in constructing those narratives for participants in our study include environmental influences through the life stages, particular sources of influence, as well as personal ways of knowing the environment. As Bonnett (2003) suggests, any thorough exploration of the idea of environmental education leads us to the metaphysics of education, that is, those motives that are working themselves out in our existence and thus participate actively in shaping our lives in fundamental ways. So whilst we mention certain influences at various stages of life, there seems to be something more complex about one's sense of the teaching of environmental appreciation than can be explained away by interpersonal, academic, and situational influences. As one of our research team expressed this rather common finding, some people have something within them that ties them to the natural world. This inner will cannot be explained easily, in terms of factors or influences.

In our intergenerational tracings, we think that we were able to see many ways that teachers participated in experiencing, perceiving, and knowing the world, involving intellect, the senses, emotions, and spirituality. Some people created experiences that were enabling, or emotionally infectious. For others, experiences in the environment occurred more spontaneously, for example, during outdoor excursions, or as part of planned educational experiences. And for others still, books and various media challenged teachers to expand their thinking or reflect on their emotions. Whether nature or sense of place itself fosters awareness, and perhaps, appreciation, the key move in our inquiry was to begin to see teacher's practices in terms of their identities rather than treat their stories as windows to another entity that stays unchanged.

If we could trace actual images of ideas that seem to cross people boundaries, we might want to disrupt some conventional notions of learning as a process of human cognition, and invite discussion about alternative conceptions – constructive, situative, and culturally critical. We might want to modify such notions to include more on the role of emotion and desire in learning. In all of this though, it seems to us that action/participation remains the most crucial part of learning in environmental education. While the origins and flows of ideas have proven elusive, there seems little doubt that some of our deeply held values are grounded in active participation. Some of what happens in classrooms is intellectual, but some is interpersonal, emotional, and meaningful in other ways. We believe that we are just beginning to understand how personal/social identities are implicated in participatory educational practice. Our methodologies seem to reveal the importance of locating ourselves in the world and understanding our values and beliefs and the ideas and actions that we can identify in ourselves. Yet, they remain inadequate and, in a sense, discomforting, precisely because they trouble meaning and motive rather than assumptions about curriculum and implementation. Unless we begin to address certain issues of methodology, such as the role of narrative as a starting point for collaboration, and genealogy and (auto)biography as a process of deconstructing research 'on' participation, we may continue to rely on more 'measured' approaches to participation as the route to change within environmental education. Perhaps good mentors simply provide interference that helps us truly attend to our intentions and conscious perceptions of direct experience. Where could we find more of these people who promote interference rather than solve our problems? It seems that mentors like Cynthia and her teachers are those who work on the self and on one's ability to participate meaningfully in practices of one's choosing, but how? We think that we have just begun to learn to approach complex narratives of inquiry as concepts such as participation demand. We have encountered narratives' incessant co-moulding (see Sfard and Prusak 2005) as an intergenerational flow of stories that are collective products, as pivot points between individual and social aspects of learning, yet we do not yet understand much about why individual actions reveal familiar resemblances.

References

Abram, D. (1996) *The Spell of the Sensuous*. New York: Vintage Books.

Bakhtin, M. (1981) *The Dialogical Imagination: Four essays*. Austin, TX: University of Austin Press.

Bakhtin, M. (1999) The problem of speech genres. In: Jaworski, A. and Coupland, N. (eds), *The Discourse Reader*. London: Routledge, pp. 121–132.

Ball, D. L. (1997) Developing mathematics reform: what *don't* we know about teacher learning – but would make good working hypotheses? In: Friel, S. and Bright, G. (eds), *Reflecting on Our Work: NSF Teacher Enhancement in K-6 Mathematics*. Lanham, MD: University Press of America, pp. 77–111.

Bauman, Z. (1996) From pilgrim to tourist – Or a short history of identity. In: Hall, S. and du Guy, P. (eds), *Questions of Cultural Identity*. London: Sage, pp. 18–36.

Berman, M. (1981) *The Reenchantment of the World*. Ithaca, NY: Cornell University Press.

Berman, S. (1997) *Children's Social Consciousness and the Development of Social Responsibility*. Albany, NY: SUNY Press.

Beyer, L. (1988) *Knowing and Action: Inquiry, Ideology, and Educational Studies*. London: Falmer.

Bonnett, M. (2003) Retrieving nature: education for a post-humanist age [special issue]. *Journal of Philosophy of Education*, 37, 4. Whole issue.

Bourdieu, P. (1984) *Distinction: A Social Critique of the Judgement of Taste*. Trans.Nice, R. London: Routledge & Kegan Paul.

Bourdieu, P. and Wacquant, L. (1992) *An Invitation to Reflexive Sociology*. Cambridge: Polity Press.

Bransford, J., Brown, A., and Cocking, R. (eds) (2000) *How People Learn: Brain, Mind, Experience and School*. Washington, DC: National Academy Press.

Bruner, J., and Haste, H. (eds) (1987) *Making Sense: The Child's Construction of the World*. London: Methuen.

Carr, W., and Kemmis, S. (1986) *Becoming Critical*. London: Falmer.

Chawla, L. (1994) *In the First Country of Places: Nature, Poetry and Childhood Memory*. Albany, NY: SUNY Press.

Clandinin, J., and Connelly, M. (2000) *Narrative Inquiry: Experience and Story in Qualitative Research*. San Francisco, CA: Jossey-Bass.

Cole, M. (1985) The zone of proximal development: where culture and cognition create each other. In: Wertsch, J. (ed), *Culture Communication, and Cognition: Vygotskian Perspectives*. New York: Cambridge University Press, pp. 120–161.

Cole, M. (1996) *Cultural Psychology*. Cambridge, MA: Belknap Press of Harvard University Press.

Day, C., Calderhead, J., and Denicolo, P. (eds) (1993) *Research on Teacher Thinking: Understanding Professional Development*. London: Falmer.

Elbaz, F. (1983) *Teacher Thinking: A Study of Practical Knowledge*. New York: Croom Helm.

Elbaz, F. (1991) Research on teacher's knowledge: the evolution of a discourse. *Journal of Curriculum Studies*, 23, 1: 1–19.

Foucault, M. (1981) *History of Sexuality, Volume One: An Introduction*. Harmondsworth: Pelican.

Gee, J. (2001) Identity as an analytic lens for research in education. *Review of Research in Education*, 25: 99–125.
Gilmore, L. (2001) *The Limits of Autobiography: Trauma and Testimony*. Ithaca, NY: Cornell University Press.
Goffman, E. (1959) *The Presentation of Self in Everyday Life*. New York: Doubleday.
Goodson, I. (ed) (1992) *Studying Teachers' Lives*. London: Routledge.
Grumet, M. (1987) The politics of personal knowledge. *Curriculum Inquiry*, 17: 319–328.
Gutiérrez, C., and Rogoff, B. (2003) Cultural ways of learning: individual traits and repertoire of practice. *Educational Researcher*, 32, 5: 19–25.
Hart, P. (2003) *Teachers' Thinking In Environmental Education: Consciousness and Responsibility*. New York: Peter Lang.
Heron, J. (1996) *Co-operative Inquiry: Research into the Human Condition*. Thousand Oaks, CA: Sage.
Heshusius, L. (1994) Freeing ourselves from objectivity: managing subjectivity, or turning toward a participatory mode of consciousness. *Educational Researcher*, 23, 3: 15–22.
Holland, D., and Lave, J. (eds) (2001) *History in Person: Enduring Struggles, Contentious Practice, Intimate Identities*. Santa Fe, NM: School of American Research Press; Oxford: James Currey.
Hutchison, D. (1998) *Growing up Green: Education for Ecological Renewal*. New York: Teachers College Press.
Kagan, D. (1992) Professional growth among preservice and beginning teachers. *Review of Educational Research*, 62: 129–169.
Kemmis, S. (1998) System and lifeworld, and the conditions of learning in late modernity. *Curriculum Studies*, 6, 3: 269–305.
Kemmis, S., and McTaggart, R. (2000) Participatory action research. In: Denzin, N. and Lincoln, Y. (eds), *Handbook of Qualitative Research*, 2nd edn. Thousand Oaks, CA: Sage, pp.567–605.
Krechevsky, M., and Stork, J. (2000) Challenging educational assumptions: lessons from an Italian-American collaboration. *Cambridge Journal of Education*, 30: 57–74.
Lave, J., and Wenger, E. (1991) *Situated Learning: Legitimate Peripheral Participation*. Cambridge: Cambridge University Press.
Louden, W. (1991) *Understanding Teaching: Continuity and Change in Teachers' Knowledge Empowerment*. New York: Cassell.
MacIntyre, A. (1981) *After Virtue*. New York: Macmillan.
Moore, S.W., Lare, J., and Wagner, K. (1985) *The Child's Political World: A Longitudinal Perspective*. New York: Praeger.
Noblit, G., and Dempsey, V. (1996) *The Social Construction of Virtue: The moral Life of Schools*. Albany, NY: SUNY Press.
Olson, J. (1992) *Understanding Teaching: Beyond Expertise*. Buckingham: Open University Press.
Orr, D. (1994) *Earth in Mind: On Education, Environment, and the Human Prospect*. Washington, DC: Prospect Island Press.
Poulson, L. (2001) Paradigm lost? Subject knowledge, primary teachers and education policy. *British Journal of Education Studies*, 49, 1: 40–55.
Reason, P. (ed) (1988) *Human Inquiry in Action*. London: Sage.
Richardson, V. (1996) Teacher education: research policy and practice. In: Leu, D. Kinzer, C., and Hinchman, K. (eds), *Forty-Fifth Annual Reading Conference Yearbook*. Chicago, IL: National Reading Conference, pp. 47–58.
Rogoff, B., Paradise, R., Arauz, R., Correa-Chávez, M., and Angelillo, C. (2003) First-hand learning through intent participation. *Annual Review of Psychology*, 54: 175–203.
Roth, W. M. (2004) Identity as dialectic: re/making self in urban school. *Mind, Culture, and Activity*, 11, 1: 48–69.
Russell, D. (1997) Rethinking genre in school and society: an activity theory analysis. *Written Communication*, 14, 4: 504–554.
Sarason, S. (1983) *Schooling in America: Scapegoat and Salvation*. New York: Free Press.

Sfard, A., and Prusak, A. (2005) Telling identities: in search of an analytic tool for investigating learning as a culturally shaped activity. *Educational Researcher*, 34, 4: 14–23.

Shepard, P. (1982) *Nature and Madness*. San Francisco, CA: Sierra Club Books.

Smagorinsky, P. (1995) The social construction of data: methodological problems of investigating learning in the zone of proximal development. *Review of Educational Research*, 65: 191–212.

Solomon, D., Schaps, E., Watson, M., and Battistich, V. (1992) Creating caring school and classroom communities for all students. In: Villa, R., Thousand, J., Stainback, W., and Stainback, S. (eds), *Restructuring for Caring and Effective Education: An Administrative Guide to Creating Heterogeneous Schools*. Baltimore, MD: Paul H. Brookes, pp. 41–60.

Sterling, S, (2003) Whole Systems Thinking as a Basis for Paradigm Change in Education: Explorations in the Context of Sustainability. Unpublished doctoral thesis, University of Bath.

Stevenson, R. (1987) Schooling and environmental education: contradictions in purpose and practice. In: Robottom, I. (ed), *Environmental Education: Practice and Possibility*. Geelong: Deakin University Press, pp. 69–82.

Taylor, C. (1989) *Sources of the Self: The Making of the Modern Identity*. Cambridge, MA: Blackwell.

Thomashow, M. (1995) *Ecological Identity: Becoming a Reflective Environmentalist*. Cambridge, MA: MIT Press.

Vygotsky, L. (1978) *Mind in Society: The Development of Higher Psychological Processes*. Cambridge, MA: Harvard University Press.

Wenger, E. (1998) *Communities of Practice: Learning, Meaning, and Identity*. Cambridge: Cambridge University Press.

Wertsch, J. (1991) *Voices of the Mind: A Sociocultural Approach to Mediated Action*. Cambridge, MA: Harvard University Press.

Wertsch, J., del Rio, P., and Alvarez, A. (1995) Sociocultural studies: history, action, and mediation. In: Wertsch, J., del Rio, P., and Alvarez, A. (eds), *Sociocultural Studies of Mind*. Cambridge: Cambridge University Press, pp. 1–34.

Wilson, T. (2002) Excavation and relocation: landscapes of learning in a teacher's autobiography. *Journal of Curriculum Theorizing*, 18, 3: 75–88.

Woods, P. and Jeffrey, B. (1996) *Teachable Moments: The Art of Teaching in Primary Schools*. Buckingham: Open University Press.

Chapter 13
Participatory Pedagogy in Environmental Education: Reproduction or Disruption?

Mary J. Barrett

Keywords environmental education, post-structuralist feminist analysis, classroom interaction, subject positioning, agency, participation

13.1 Introduction

Despite good intentions, participatory research and pedagogy can masquerade as an open process but still impose agendas that support particular versions of what constitutes appropriate thought, behaviour, and action. This chapter draws together two studies with some explication of feminist post-structural notions of the self in order to suggest ways in which feminist post-structural analyses might produce different readings of student responses to participatory pedagogies that have environmental change as their aim. Conceiving of the subject as discursively constituted opens up the self to both an interrogation into its construction and its possible reconstitution. It also opens up new ways of understanding student agency and helps bring to light how the discursive production of the self can limit students' ability to challenge dominant discourses and take up counter-hegemonic ones. Furthermore, a feminist post-structural analysis of power resists the tendency to 'blame the victim' when teachers or students do not 'get it right'.

13.2 Background

Although often extremely useful in initiating positive environmental change and providing space for marginalised voices to be heard, participatory approaches to research and pedagogy can be quite problematic. Despite good intentions, they can create the illusion of open processes but still impose agendas that support

---

College of Education, University of Saskatchewan, Saskatchewan, S7N 0X0, Canada, mj.barrett@usask.ca

A. Reid et al. (eds.), *Participation and Learning*, 212–224.

particular versions of what is appropriate thought, behaviour, and action (see Ellsworth 1989, 1997; Britzman 1995; Boler 1999, 2004; Kumashiro 2002a, 2004). In participatory research, full and equitable participation is elusive Boler (1999:193) and in education, (Fine *et al.* 2000; Robottom and Sauvé 2003), argues that an uncritical adoption of participatory processes can act as 'poisonous pedagogy', an example of 'the systemic ways we teach young people *not to notice* the cruelties and injustices inflicted upon them'. While in participatory development, Kothari (2001:141) suggests[1]:

> Participatory approaches can unearth who gets what, when and where, but not necessarily the processes by which this happens or the ways in which the knowledge produced through participatory techniques is a normalized one that reflects and articulates wider power relations in society.

In the context of these concerns, it is important to ask what is being disrupted and what is being reinscribed when engaging with the various discourses of participation.

Discourses produce and circulate values, beliefs, and notions of what is possible, doable, and acceptable. While people are often viewed as both the objects and subjects of participation, assuming that subjects are discursively constituted and any practice or perspective is oppressive in some ways and anti-oppressive in others (Kumashiro 2002b), we create a space for those 'post'-informed perspectives (postmodern, post-structural, postcolonial …) that can help interrogate the assumptions and effects of participatory pedagogies and research designs. For example, Boler (1999) and Pillow (2003) argue that although thoughtful critical reflection can often diminish their normalising and oppressive aspects, self-reflection is always partial and can still enable educators and researchers to remain comfortable in their blindness to their own complicity in reproducing oppressions. Feminist post-structural analyses can assist those involved in participatory education and research processes to identify their own complicity, but without the guilt, shame, and paralysis that often comes when discursively produced subjectivities are not foregrounded (Ellsworth 1989; Boler 1999; Davies 2000).

Because feminist post-structural analyses are able to interrogate ways in which power, politics, and subjectivities shape research and pedagogy, they can also help make some of the norms, foundations, and assumptions (i.e. normative discourses) that limit reflective vision visible and thus potentially revisable (Davies 2000). These analyses can enable researchers, educators, *and* students to identify effects of particular practices and perspectives (Kumashiro 2002b:17–18). In addition, they can be particularly useful in interrogating structures that espouse liberation and empowerment but may actually limit participants' 'free' engagement (see Kothari 2001).

For instance, within the assumptions of feminist post-structuralism, agency cannot exist outside the discursive (Butler 1993; Davies 2000; St. Pierre 2000). Even ideas such as independent agency and the self as an autonomous decision-maker are understood to be discursive productions. Basically, the 'choices' one

[1] While Kothari is referring to participatory development, similar claims can be made about many participatory processes in research and education.

makes are based on those available in discourse, and some carry more power, or cultural capital, than others. Thus, agency cannot exist outside discourse, but instead exists in appropriating or disrupting dominant discourses and taking up or rejecting unfamiliar ones (Davies 2000). In order to give open access to non-dominant and counter-hegemonic discourses, one of the main aims of feminist post-structural theorising is to make visible discursive structures and their effects, as well as their processes of production and regulation.

This work is particularly feminist (rather than simply post-structural) because of its activist agenda and emphasis on exposing power relations, particularly as they relate to gender, race, class, able-bodiedness, and sexual orientation. Post-structural theorists often aim to deconstruct discursive processes without necessarily having any social agenda beyond the academic task of deconstruction, earning them the reputation of being nihilistic. In contrast, feminist post-structural researchers most often engage in deconstruction with the goal of exposing inequities, oppressions, and exclusions in order to initiate change (Davies 2000; St. Pierre and Pillow 2000). There has also been some feminist post-structural theorising that examines ways in which humans remain the unmarked normal and thus exclude the voices of non-human Others (e.g. Armbruster 1998; Bell and Russell 2000; Russell 2005).

To date, however, theorising on how post-structural perspectives fit with 'the activist[2] objectives of much socio-ecological education' has been limited (McKenzie 2004b:187), although increasing. Most approaches to participatory educational processes and associated research assume a modernist notion of the self – a unified subject with independent consciousness and agency. Since post-structuralist notions of the subject assume that one is constituted through discourses, always in flux and often contradictory, they have significant implications for notions of agency, and thus, for participatory processes.

It is my hope that by drawing together two studies with some explication of feminist post-structural notions of the subject and self, readers might conceive of ways these analyses might open up useful new insights for both pedagogies that encourage student participation in environmental action and environmental education research. In the first study, Marcia McKenzie (2004a, 2006) highlights ways in which subject positions accessible to the students from within the school and community may restrict students from taking up the very positions that the socioecological educational programmes aim towards. In the second, Deirdre Barron (1995) illustrates ways in which gender and class discourses can make it difficult for students to take up conservationist or activist discourses.

It is also my hope that in reading this work, readers might engage the following questions. What does conceiving of the subject as discursively constituted make possible or impossible? (Kumashiro 2004) What does this assumption about the

[2] In the context of this chapter, I am interpreting activist, or action-oriented education to be education that works on developing students' desire and ability to engage in actions. To take action (as opposed to simply perform a behaviour) assumes making a decision to do something, rather than following prescribed behavioural objectives (Jensen 2004; Jensen and Schnack 1997). This action-oriented educational process is highly participatory in nature.

subject suggest about how students and teachers might negotiate changes that often require taking up different, unpopular, and/or counter-hegemonic discourses? (Whitehouse 2001; Boler 2004) What might feminist post-structural perspectives offer participatory pedagogy and research? In highlighting these studies, I am not claiming that feminist post-structuralism provides the 'right' or 'best' analysis, but do suggest that feminist post-structural approaches to research and pedagogy provoke different questions and provide insights that may not be acquired using other theoretical perspectives.

13.3 The Discursively Constituted Self

In her recent study of three different secondary school classes in British Columbia, Canada, Marcia McKenzie (2004a, 2006)[3] examined the contradictory discourses and resulting subject positions available to high school students in socioecologically focused courses. Schools included Hillview, a public high school in a rural working class community of 5,000; Kirkwood, a Montessori programme within an urban public high school; and Lawson College, a non-profit two-year International Baccalaureate school in a remote residential setting.[4] All three sites placed high priority on actively engaging students in social and environmental issues and included volunteer service and socioecologically based action projects as mandatory parts of the students' educational experience. Using discourse analysis to interrogate her observations of and interviews with teachers and students, McKenzie makes visible and provides a representation of how students attempted to correctly constitute themselves within discourses of neutrality, critique, achievement, diversity, consistency, knowing, activism, constitution of self, authenticity, and agency.

As the comments of students and teachers at all three sites seemed to indicate, students' notions of knowledge and themselves as knowers had significant implications for the ways in which they could engage in socioecological education. For instance, at Hillview, where discourses of educational neutrality and objective knowing tended to dominate, students seemed to understand information presented to them as true. In addition, while they emphasised how they were learning more *about* what is going on in the world, Hillview students expressed a sense of limited control over their environment and had strong doubts that one could even begin to change the world.

Discourses of educational neutrality and objective knowing were also apparent in the Kirkwood Montesorri programmes even though students there were involved in

[3] The description of McKenzie's study comes from these two references, and is necessarily a partial representation. Only in the case of page numbers for quotations will specific articles be referenced.

[4] While the schools vary in terms of dominant social class and depth of focus on social and ecological issues, all three sites were chosen for their exemplary pedagogy. School names are pseudonyms.

many change-oriented projects. At Kirkwood, teachers saw themselves as exposing students to different cultures and ways of understanding the world, yet claimed they were neither imposing any particular values on the students nor 'bumping up against' North American values. Teachers simultaneously saw themselves as neutral, and the programme as one that challenged the consumerism dominating much of Western society. In taking this approach, the teachers seemed to be positioning students as objective knowers – a position which appeared to constrain the students' abilities to actively care about, or engage in, social or environmental action. McKenzie suggests that this 'discursive tension' between educational neutrality and social critique 'is perhaps also not surprising' given that the Montesorri advocacy programme occurs within a public school 'where education in general is purported to be neutral' (2004a:65).

Students at the International Baccalaureate school, Lawson College, seemed to have access to different discourses around knowledge and being a knower. They saw their education as biased rather than neutral, and were explicitly taught to question sources of knowledge, how knowledge is gained, and the extent to which personal or ideological bias influences knowledge claims.[5] This more critical approach to knowledge seemed to both position Lawson students, and enable them to position themselves, as contingent knowers. As they suggested in their comments, the Lawson students found themselves becoming open to more diverse types of knowledge and engaging more deeply in environmental action than they had previously.

Students' understandings of ways in which they are constituted also appeared to have some influence on their engagement with participatory or action-oriented pedagogies. Several Lawson students seemed to see themselves as socially constructed as well as having some individual agency. McKenzie refers to these intersecting discourses of constitution and agency as 'contingent agency' (McKenzie 2004a:160) – a kind of agency that hinges upon both knowledge and identity being indeterminate and shifting. For instance, several Lawson students talked about exposure to new ideas, people, places, or experiences as significantly influencing their sense of themselves, while others spoke about how families and previous experiences produced who they were. Some also spoke of ways they both engage with, and push away, from those influences and how they are susceptible to falling back into old patterns when they go home.

At Lawson, students acknowledged some of the ways they were constituted by other discourses, yet still expressed some sense of self-determination. McKenzie (2006:201) suggests these intersecting discourses of constitution and contingent agency 'contrast to traditional understandings of agency as the capacity for choice and self-determination'. Students recognised themselves as multifaceted and shifting in relationship to the power held by the discourses that were constituting them.

[5] All students at Lawson College take a mandatory International Baccalaureate diploma programme course, *Theory of Knowledge*, which raises many of these questions. For a description, see: http://www.ibo.org/diploma/curriculum/core/knowledge/

In acknowledging their contingent agency, Lawson students appeared to be open to 'possibilities of a deeper reflexivity and more selective resistance to normative discourses' (p. 217), instead of feeling helpless in the face of larger social structures or blaming themselves for inability to make change.

Whereas the comments from Lawson students seemed to suggest some recognition of ways in which their actions and decisions were not completely their own doing, Kirkwood and Hillview students tended to attribute both their successes and failures to their independent abilities. McKenzie suggests that Hillview student conceptions of themselves as having independent agency appeared to limit them from moving beyond inactive caring to actively challenging many of the injustices about which they were learning. She suggests that at Hillview, intersecting discourses of awareness, limited agency, and inactive caring kept the student resistance within mainstream cultural narratives, and limited opportunities for student reflexivity and activism. She also suggests Hillview students' limited sense of agency appeared to be linked to their socio-economic class.

Like students at Hillview, McKenzie suggests many Kirkwood students had limited conceptions of their choices and abilities to make change, even though they were often active change agents in their school and community. At Kirkwood, many students exhibited a strong sense of 'individual power' coupled with some notion of themselves as discursively produced. Yet given that dominant discourses of individual power, educational neutrality, and economic achievement were not critiqued, the senses of student selves as change agents seemed limited at Kirkwood and often manifested in the form of lifestyle activism rather than any challenge to larger social structures. While claiming to support active participation and engaged citizenship, the discourses available to the students in both Hillview and Kirkwood programmes, both selected for their exemplary pedagogy, often appeared to contravene the programme and the teacher's own goals. McKenzie (2004a:iii) concludes that 'each group of students is to some extent "parroting" discourses common to their context', discourses which, in some cases, make it difficult for them to fully participate in their educational programmes.

13.4 The Power of Discourses Deirdre Barron

Deirdre Barron's (1995) study also engages a post-structural notion of discourse and the subject, and goes on to imply that given the power associated with dominant and normative discourses, change is much more complex than we might previously have thought. No matter how innovative their pedagogy, teachers cannot simply teach students to act or think differently. As students attempt to take up one discourse, they are simultaneously being produced by other (often more powerful) ones, making it difficult for them to take up counter-hegemonic storylines as their own.

Barron explores ways in which common dominant discourses of masculinity and femininity can make it difficult for elementary school students to take up counter-hegemonic discourses of environmentalism. She examines elementary school students' responses to the Dr. Seuss story, *The Lorax* (Seuss 1972), that involves the Once-ler, who aims to chop down all the trees to create 'thneeds' which 'everyone needs'; the Lorax, an androgynous creature who sets out to speak for and protect the trees; and a young child. Through a feminist post-structural analysis of students' discussion, Barron suggests that the students responded to the Once-ler's actions depend differently upon the discourses in which they were positioned, or positioned themselves.

Barron posits that when speaking from within a technological discourse, the children appeared to support the act of cutting down the trees, whereas when positioned within conservation discourses, they were hesitant to support tree-cutting. When it came to who could stop the Once-ler from cutting trees, gender discourses seemed to exert a powerful influence on students' responses. Most of the boys assumed an ability to act, while the girls seemed to question the possibility of their own agency. To the boys, the idea that they had a right to cut down trees appeared automatic; for the girls, the dilemma seemed not to centre around 'whether humans have the right to cut down trees', but rather, 'which humans had the right to stop the trees being cut' (Barron 1995:111). According to Barron, dominant discourses of femininity made the girls those who care for the trees and defer agency to appropriate authorities, while dominant discourses of masculinity made the boys the ones with the ability to take action.

Barron's study suggests how students' responses to a moral dilemma, such as the right to cut trees, can shift depending on their positioning within different discourses. She makes visible the ways in which positioning within discourses of femininity and masculinity, technocentrism or conservation, makes different responses available and more or less acceptable. Her study also highlights how the binaries that associate masculinity with culture and femininity with nature, are taken up and reproduced by students exploring the possibility of acting in response to an environmental dilemma. As long as these discourses constituting subject positions available to people and 'nature' remain invisible, Barron argues, they will continue to constitute students and nature in ways that reinforce rather than challenge hegemonic norms (see Plumwood 1993; Davies and Whitehouse 1997; Whitehouse 2002).

Just as significantly as making the gender, environmental and other discourses through which the student subjectivities are constituted visible, Barron's study highlights how 'the social requirement to construct oneself as a unitary being' (Barron 1995:115) can trigger internal struggle. For instance, if the girls were not pressed to take up humanist notions of the self (St. Pierre 2000) and construct themselves as a unitary 'girl', they may have been able to take up both discourses of femininity (caring) and masculinity (acting). Likewise, if students were allowed to embody contradictory discourses simultaneously and without condemnation, boys like Robert[6] could choose to protect the trees without the risk of failing to constitute himself as appropriately male.

---

[6] When challenged by the interviewer with the apparent contradiction between his views on the use of the machine and his earlier suggestion that the trees be saved, Robert positioned himself within the interventionist, rather than conservationist discourse, stating that he would cut down the trees if he needed money (Barron, 1995).

The desire to produce themselves as appropriately masculine or feminine made it difficult for the girls and boys to take up discourses such as independent agency and caring, respectively. Challenging dominating cultural narratives such as unfettered development and consumerism would also mean transgressing appropriate subject positions. Similarly, the humanist demands that one produces oneself as a coherent, essentialised, and non-contradictory subject, seemed to suggest that the students could not simultaneously take up contradictory discourses (see St. Pierre 2000). Instead, they responded to Barron's questions from within the available discourse that carried most cultural capital.

13.5 Implications of the Discursively Produced Subject

When advocates of environmental education, sustainability (e.g. Government of Canada 2002; UNCED 1993), or education in general (e.g. the UK's Department for Education and Skills 2003) call for student empowerment, participation, and action, the 'student' they allude to is most often assumed to be a fixed, rational, coherent individual capable of independent choice and action. By challenging this notion of the self, mismatches between programme goals and student learning like that encountered at Kirkwood and Hillview, as well as contradictions like those experienced by Robert in his responses to *The Lorax*, can be interpreted differently. For instance, rather than being seen as instances of student apathy, teacher incompetence, or personal hypocrisy, these tensions can be read as the effects of competing discourses.

Acknowledging that gaps between pedagogical aims and 'results' are often linked to available and dominant discourses rather than to individual failures or contradictory essential selves can open up new ways to understand and interpret student experiences in participatory pedagogies that encourage student activism. Continuing to assume a stable subject with independent agency can maintain limited notions of what is normal and appropriate (Butler 1992). By asking how meanings have been acquired and changed, how some meanings have 'emerged as normative and others have been eclipsed or disappeared', and what these shifts in meanings 'reveal about how power is constituted and operates' (Scott 1988:35, cited in St. Pierre 2000:484), feminist post-structural theorising can make the ways in which language, discourse, and desire work to enable and constrain student engagement visible (Davies 2000).

As McKenzie points out, when positioned as objective knowers and independent autonomous selves, students had difficulty fully engaging in issues that were rife with partial and politically charged knowledge. Similarly, when positioned within dominant discourses of femininity and the subject as unitary, the girls in Barron's study could not simultaneously care about and have the agency to speak for the trees. These two studies suggest that instead of condemning students for not caring, being apathetic, or simply unskilled in taking action, teachers, students, and educational researchers would benefit from examining discursive barriers to engagement, including those that may inadvertently be reproduced by teachers' actions within their own classroom. This does not mean ceasing to help students develop appropriate action skills, but it does, however, suggest assisting them in deconstructing ways in which they have been discursively produced – ways they have been constituted, and constituted themselves as, for example, a good student (McKenzie 2004a; Sammel 2004).

Conceiving of themselves as discursively constituted releases individuals from the pressure to produce themselves as unitary (Barron 1995; Bloom and Munro 1995) and opens the subject up to both interrogation into its construction and possible reconstitution (Butler 1993; Kelly 1997; Davies 2000). Rather than seeing themselves as fixed subjects or victims of a system beyond their control, students and teachers have more options for change, resistance, and reflexivity (Boler 1999, 2004; Kumashiro 2002b, 2004; McKenzie 2004a, 2006).

13.6 Implications of Attention to Subject Positions

Both studies show how the cultural narratives, or discourses to which students have access, make certain subject positions available and attractive to students and others much less accessible or desirable. Subject positions are the discursively produced storylines and corresponding 'conceptual repertoire and…location' (Davies 2000:89) from which one views and makes sense of the world. From a given subject position, only certain understandings of the world make sense. Depending upon what subject position(s) students have access to and take up, or are positioned in, their experiences of schooling, and what they see as relevant, will vary (Davies 2000).

The subject positions available to students reflect discourses in the schools and the broader community. In McKenzie's study, this was most evident in the contrast between the limited aspirations of the Hillview students for the future and the 'big plans for effecting change that many (mostly upper class) Lawson students express' (McKenzie 2004a:154). In Barron's (1995) study, it appeared as though for primary school girls, available and dominant discourses of femininity clashed with those of agency, and for the boys, discourses of masculinity eclipsed the option of caring.

The studies also suggest the need to pay attention to ways teachers and researchers (re)position students through speech and action (Davies 2000). Students in McKenzie's study made this positioning visible as they talked of their conceptions of knowledge, whereas Barron's spoke of how her interview questions pushed Robert to position himself as unitary.

The important point here is that educators who wish to engage students in action-oriented approaches to education may be asking students to take up subject positions to which they do not have access. As both McKenzie and Barron suggest, being locked into particular notions of who they could or should be, limits possible ways students can engage with issues raised in their classes. If teachers conceive of the subject as discursively constituted and recognise how they constantly reposition their students, they may then be able to work towards 'exposing' dominant discourses and giving students access to alternative ones. In addition, they may be more able to position students differently (Laws and Davies 2000). It is important to note, however, that at the same time teachers may be working to open up subject positions for students, they are simultaneously, and powerfully, being positioned themselves – often in ways that make it difficult for them to challenge normative notions of teaching and learning (Kumashiro 2002b).

13.7 Further Discussion

Given that some discourses hold more power than others, some subject positions are more likely to be taken up than others. As student positioning within dominant discourses shifts, the power a student holds in relation to teachers, parents, and peers also changes. In asking students to formulate and articulate their own beliefs or speak up against hegemonic norms, teachers may be asking students to contradict notions of what it means to be a proper person, teen or student in their local context – in other words, to position themselves as the illegitimate Other (see Kumashiro 2004; Whitehouse 2001). An analysis of power relations perpetuated through discourse highlights how some responses may be much more possible than others (Kumashiro 2002b, 2004). Scrutinising cherished beliefs and assumptions may provoke strong emotions (Boler 2004) and it takes skilful negotiation of what are often competing subject positions to successfully and simultaneously position oneself as both teenager and environmental activist (Whitehouse 2001).

What all this suggests for educators is that rather than assuming that most teenagers lack the interest, ability, or courage to take up unpopular subject positions (see Sammel 2004), and thus student engagement in action-oriented education is blocked from the outset, students could be taught to use words and actions to resist those very structures that may inscribe them (Davies 2000) as incapable or unconcerned. Furthermore, through their talk and action, teachers may make different subject positions more accessible to students, shift power relations in the classroom (Boler 2004; Kumashiro 2004), and open up possibilities for what Barron refers to as 'radical environmental reform' (Barron 1995:117).

The power invested in maintaining particular subject positions as normal (Davies 2000) means that making these changes can be an uphill battle. As these two studies illustrate, basing one's analysis on notions of the subject as discursively constituted, and assuming that teachers' talk and action continuously (re)position students, highlights how reproduction of particular notions of normal can constrain student engagement in socio-political action. It is not enough to introduce students (or teachers) to counter-hegemonic discourses and assume they can adopt them, even if they sincerely wish to do so.

Unless dominant cultural narratives of participation are made visible and in some cases, disrupted, participatory approaches to research and education risk reinscribing the status quo and reproducing familiar subject positions as the unmarked normal. Helping students understand how they are constituted by discourse and constituting discourses simultaneously, can help ameliorate this risk (see Boler 2004; Kumashiro 2004; McKenzie 2004a, 2006), as can helping students understand ways in which they can use language and everyday practices to resist dominant inscriptions and negotiate multiple subject positions. And as Davies (2000:71) claims, locating 'sources of the contradiction' in available discourses can make it 'possible to examine the contradictory elements of one's subjectivity without guilt or anxiety and yet with a sense of moral responsibility'.

13.8 Limitations of Critique

Like participatory processes where 'tools provided can limit the performance' (Kothari 2001:149), feminist post-structural analysis also has limits and must interrogate what it might itself be (re)inscribing. For instance, Barron and McKenzie constitute identity and agency in specific ways. Barron's study seems to essentialise particular notions of what it means to be a girl or boy by focusing on single discourses (e.g. that of girls as caring and boys as active agents). Her analysis also reproduces the subject as unitary by failing to account for multiple, often overlapping and conflicting subjectivities available to the primary school students. Similarly, in naming discourses such as 'achievement', 'knowing', and 'agency', McKenzie tends to (re)produce them as fixed entities rather than constantly shifting conduits of power.

As St. Pierre and Pillow (2000:6) suggest, it is important to turn a feminist post-structural analysis back upon itself to examine 'the functions and effects of any structure or grid of regularity that [theorists] put into place, including those poststructuralism itself might create'. While feminist post-structural analyses provide no 'sure ground' (Butler 1995:131) from which to learn or research, I suggest the kind of expansive critique it enables can reduce the risk of a participatory educational or research project (re)inscribing oppressive agendas and assumptions, even as it disrupts others. As the two highlighted studies illustrate, this kind of critique can be particularly useful given the change-oriented aims of much environmental education and the challenges students and teachers encounter when negotiating dominant and counter-hegemonic discourses.

Acknowledgments My thanks to Marcia McKenzie for her feedback on an earlier version of this chapter. I gratefully acknowledge the financial support of the Social Sciences and Humanities Research Council of Canada.

References

Armbruster, K. (1998) 'Buffalo gals, won't you come out tonight': a call for boundary-crossing in ecofeminist literary criticism. In: Gaard, G. and Murphy, P. (eds), *Ecofeminist Literary Criticism: Theory, Interpretation, Pedagogy*. Chicago, IL: University of Illinois Press, pp. 97–122.

Barron, D. (1995) Gendering environmental education reform: identifying the constitutive power of environmental discourses. *Australian Journal of Environmental Education*, 11: 107–120.

Bell, A. and Russell, C. (2000) Beyond human, beyond words: anthropocentrism, critical pedagogy, and the poststructuralist turn. *Canadian Journal of Environmental Education*, 25, 3: 188–203.

Bloom, L.R. and Munro, P. (1995) Conflicts of selves: nonunitary subjectivity in women administrator's life history narratives. In: Hatch, J.A. and Wisnewski, R. (eds), *Life History and Narrative*. London: Falmer, pp. 99–112.

Boler, M. (1999) *Emotion and Education*. New York: Routledge.

Boler, M. (ed) (2004) *Democratic Dialogue in Education: Troubling Speech, Disturbing Silence*. New York: Peter Lang.

Britzman, D. (1995) Is there a queer pedagogy? Or stop reading straight. *Educational Theory*, 45, 2: 151–165.

Butler, J. (1992) Contingent foundations: feminism and the question of 'postmodernism'. In: Butler, J. and Scott, J. (eds), *Feminist Theorize the Political*. New York: Routledge, pp. 3–21.

Butler, J. (1993) Bodies that matter. New York: Routledge.

Butler, J. (1995) For a careful reading. In: Benhabib, S., Butler, J., Cornell, D., and Fraser, N. (eds) *Feminist contentions: A philoscophical exchange*. New York: Routledge, pp. 127–143.

Davies, B. and Whitehouse, H. (1997) Men on the boundaries: landscapes and seascapes. *Journal of gender Studies* 6, 3: 237–254.

Davies, B. (2000) *A Body of Writing*. Oxford: Rowan & Littlefield.

Department for Education and Skills (2003) *Working Together: Giving Young People a Say*. London: The Stationery Office/DfES.

Ellsworth, E. (1989) Why doesn't this feel empowering? Working through the repressive myths of critical pedagogy. *Harvard Educational Review*, 50, 3: 297–324.

Ellsworth, E. (1997) *Teaching Positions: Difference, Pedagogy, and the Power of Address*. New York: Teachers College Press.

Fine, M., Weis, L., Wessen, S., and Wong, L. (2000) For whom? Qualitative research, representations and social responsibilities. In: Denzin, N. and Lincoln, Y. (eds), *Handbook of Qualitative Inquiry*. New York: Sage, pp. 107–131.

Fraser, N. (ed) (1994) *Feminist Contentions: A Philosophical Exchange*. New York: Routledge, pp. 127–143.

Government of Canada. (2002) *A Framework for Environmental Learning and Sustainability in Canada*. Ottawa: Queen's Printer.

Jensen, B. and Schnack, K. (1997) The action competence approach in environmental education. *Environmental Education Research*, 3, 2: 163–177.

Jensen, B. (2004) Environmental and health education viewed from an action-oriented perspective: a case from Denmark. *Journal of Curriculum Studies*, 36, 4: 405–425.

Kelly, U. (1997) *Schooling Desire: Literacy, Cultural Politics, and Pedagogy*. New York: Routledge.

Kothari, U. (2001) Participation: the new tyranny? In: Cooke, B. and Kothari, U. (eds), *Participation: The New Tyranny?* New York: Zed Books, pp. 139–152.

Kumashiro, K. (2002a) Three readings of D. Butin's Commentary. *Educational Researcher*, 31, 3: 17–18.

Kumashiro, K. (2002b) *Troubling Education: Queer Activism and Antioppressive Pedagogy*. New York: RoutledgeFalmer.

Kumashiro, K. (2004) *Against Common Sense: Teaching and Learning Toward Social Justice*. New York: RoutledgeFalmer.

Laws, C. and Davies, B. (2000) Poststructuralist theory in practice: working with 'behaviourally disturbed' children. *Qualitative Studies in Education*, 13, 3: 205–221.

McKenzie, M. (2004a) *Parrots and Butterflies: Students as the Subjects of Socio-ecological Education*. Unpublished doctoral dissertation. Burnaby, BC: Simon Fraser University.

McKenzie, M. (2004b) The 'willful contradiction' of poststructural socio-ecological education. *Canadian Journal of Environmental Education*, 9: 177–190.

McKenzie, M.D. (2006) Three portraits of resistance: the (un)making of Canadian students. *Canadian Journal of Education*, 29, 1: 199–222.

Pillow, W. (2003) Confession, catharsis, or cure? Rethinking the uses of reflexivity as methodological power in qualitative research. *Qualitative Studies in Education*, 16, 2: 175–196.

Plumwood, V. (1993) *Feminism and the Mastery of Nature*. New York: Routledge.

Robottom, I. and Sauvé, L. (2003) Reflecting on participatory research in environmental education: some issues for methodology. *Canadian Journal of Environmental Education*, 8: 111–128.

Russell, C. (2005) 'Whoever does not write is written': the role of 'nature' in post-post approaches to environmental education research. *Environmental Education Research*, 11, 4: 433–443.

Sammel, A. (2004) *Teachers' Understandings and Enactments of Social and Environmental Justice Issues in the Classroom: What's 'critical' in the Manufacturing of Road Smart Squirrels?* Unpublished doctoral dissertation. London/Ontario: University of Western Ontario.

Scott, J. (1988) Deconstructing equality-versus-difference: or, the uses of poststructuralist theory for feminism. *Feminist Studies*, 14, 1: 33–50.

Dr. Seuss (1972) *The Lorax*. London: Collins.

St. Pierre, E. (2000) Poststructural feminism in education: an overview. *Qualitative Studies in Education*, 13, 5: 477–515.

St. Pierre, E. and Pillow, W. (eds) (2000) Introduction: inquiry among the ruins. *Working the Ruins: Feminist Poststructural Theory and Methods in Education*. New York: Routledge, pp. 1–24.

UNCED (1993, 3–14 June 1992) *Agenda 21: Programme of Action for Sustainable Development. Rio declaration on environment and development.* Paper presented at the United Nations conference on environment and development, Rio de Janeiro.

Whitehouse, H. (2001) 'Not greenies' at school: investigating the discourses of environmental activism in regional Australia. *Australian Journal of Environmental Education*, 17: 71–76.

Whitehouse, H. (2002) Landshaping: a concept for exploring the construction of environmental meanings within tropical Australia. *Australian Journal of Environmental Education*, 18: 57–62.

Chapter 14
Elusive Participation: Methodological Challenges in Researching Teaching and Participatory Learning in Environmental Education

Paul Hart

Keywords environmental education, research, methodology, participation, learning

14.1 Introduction

The purpose of this methodology-focused chapter is to consider the issues encountered in researching participation from the perspective of those subjectivities necessary for its occurrence in education and environmental education. It is one thing to discuss dilemmas of participatory education itself as a potential dwelling point for the conceptualisation of work in environmental education. It is quite another to consider where human predispositions for such work have come from. The chapter outlines a collaborative attempt to explore how a focused search for origins of meaning in environment-related practices of teachers has become constructed and disrupted as an inquiry process. My methodological interest is in the process of the inquiry as it emerged at the intersection of narrative interpretive methods to capture and represent evolving ways of knowing, and some obvious constraints and tensions concerning some fundamental conditions through which social processes are both generated and shaped. The chapter begins with an account of the basics of research processes that have spanned more than a decade of change in our thinking about such inquiry, in order to illustrate how initial assumptions about social reality are constructed (e.g. that environmental education, almost by definition, is a participatory process), and have re-emerged as questions concerning broadened contexts of meaning-making. The chapter concludes with a view of environmental education as a social practice, centred both on how teachers construct their school experiences and what this may mean for practical realities such as participation or participatory action within the institutions of education.

Faculty of Education, University of Regina, Regina, Saskatchewan, S4S 0A2, Canada, hartpa@uregina.ca

A. Reid et al. (eds.), *Participation and Learning*, 225–240.

14.2 Background and Context

> [I]n many ways, I am striving for something that I will never gain; a level of understanding and knowledge that enables … thinking about the question that guides my experience …. This reflects the fortuitous and overwhelming challenge of the postmodern age: how to find meaning in the world, given the proliferation of ambiguity and uncertainty. ... Here is the challenge: How can I construct an ethical and moral foundation for my actions, if I must also accept the temporality of interpretive meaning? What makes any world view more or less acceptable than any other? (Thomashow 1995: 202)

If the purposes of inquiry include some understanding of precursors to participation in environmental education, then in terms of 'sense making' and 'community building' amongst teachers, according to Thomashow (1995) a place to begin is through the willingness of people: to look deeply within themselves, to understand their motivations and aspirations, to articulate (as best they can) their environmental values and where they come from, and to describe the application of these values in their personal/professional decisions.

Educational researchers with an interest in ecological identity and social agency can themselves begin this process by reflecting on their own learning, their own literary autobiographies, as a basis for understanding themselves as researchers. Thomashow's (1995) emphasis on both discursive and practical knowing with research is evident in his work on teachers' interpretations of classroom experience as processes of personal sense-making within a complex maze of social connections. This emphasis matters for those of us with an interest in tracing the origins of something as elusive as participation, embedded as it is within layers of human thought. Of course, as researchers, we are also aware of the constitutive power of cultural and institutional details of discourse, to the extent that beginning where teachers are also means locating them within the larger landscapes of everyday life.

The study from which the methodological issues discussed here were derived began as a narrative study of the philosophical underpinnings of teachers' experiences in environmental education constructed autobiographically (see also Chapter 12 by Paul Hart, this volume). As a group of qualitative-oriented researchers, we[1] were convinced by arguments about the potential of narrative inquiry to help us and our participants make sense of teachers' practices in terms of their intentions, desires, beliefs, and values (Polkinghorne 1988, 1995; Connelly and Clandinin 1990; Witherell and Noddings 1991; Clandinin and Connelly 1996; Conle 1999, 2000). Without rehearsing the case for narrative as a valid way of knowing (see, e.g. Bruner 1986, 1996; Greene 1994; Jalongo and Isenberg 1995), we accepted the notion that teachers' practical knowledge is primarily narrative in form. It seemed to us that unless we could understand their perspectives, our knowledge of teaching and learning would remain fragmented and disconnected. In acknowledging the power relationships inherent in such a statement, we also recognised that telling

[1] I use the words 'we' and 'our' to refer to our team of research associates that included Ann Camozzi, Susan Gesner, Christine Robertson, Loraine Thompson, and Judith McPhie at various stages of the research funded by the Social Sciences and Humanities Research Council for Canada over a period of approximately ten years.

stories is not enough. The potential for teachers' personal narratives becomes evident only in reflection – as teller and listener, writer and reader delve beneath the surface of anecdote to examine motives, implications, and connections. This potential is, in fact, the socially interactive power of such inquiry (see Craig 2000).

As we expanded this autobiographical emphasis in the hope of bending back ours minds on stories of the past that help form teacher identities (Coffey 1999), we entered more deeply into personal experience methods. In our conversations with teachers, we encourage reflection about things such as self-concept, identity, and agency, that is, on guiding assumptions behind pedagogical repertoires that framed their educational journeys historically. According to Coffey (1999), it may be through such methods that researchers themselves are forced to confront how fieldwork research and textual practice work to construct, reproduce, and implicate selves, relationships and personal identities. It is these aspects that constitute the substance of various dilemmas that we now confront as ways of addressing the integrity of our research work (see Goodson and Sikes 2001). These dilemmas have to do with power relationships between individual lives and social settings, locating the story, and frameworks for interpreting it, within history and with representation. Epistemologically, our dilemma concerned the adequacy that could be given to teacher accounts as constitutive of their experiences versus actively constituting these experiences. Methodologically, this translated into problems of documenting how the roots of teacher actions, systems of motivation, and the ties between the two, are evidenced in their practical reasoning. In other words, how did teachers themselves orient to school systems, as well as to shared meanings (as part of their views of teaching), to account for environmental education in their practices?

In our more recent inquiries, we were well aware that while there was no way to deepen understandings of teachers' professional work in environmental education other than by working and conversing with them about their sense of meaning, something more was needed to enable critical reflection on their processes of learning and personal/social change over time. In this new work, we widened our interpretive frame beyond narrative and life history methods that, although well suited to engage the highly personalised and uniquely contextualised nature of teaching, lack social reflexivity concerning the constitutive quality of systems of discourse. Our interest has shifted from an almost exclusive focus on agency, to one that is more the ways in which teachers are constituted, not causally, but through stories of experiences where their practical ideas could be traced as lines of descent focused on activity in environmental education. We view this emerging interest as genealogical in that it goes beyond descriptive analysis to engage elements of discursive practices that made participation in environmental education possible. This genealogical interest is focused on exploring conditions of possibility and the emergences of environmental sensitivities as part of teachers' subjective knowledge; ideas useful in searching for clues to teachers' thoughts, desires, and intents by moving between past and present subjectivities (see Meadmore *et al.* 2000).

While genealogical work offers analysis of the singularity of events, for example, how particular instances of environmental education came to be (constituted) in a certain way, this methodology recognises a multiplicity of causes and conditions that constitute environmental education practice (Scheurich and McKenzie 2005). What is important is that lines of descent can be traced for the emergence of ideas that guide those practices, but also for discontinuities and those events closest to present thought, as represented geographically (within Canada, the UK, the USA, Australia, or South Africa, for example). Each of these geographical locations seemed to provide particular contextual expressions of environmental education to guide inquiry (rather than the arbitrary use of generalised historical events, such as UNESCO environmental (educational) conferences, as time frames). Genealogical method, in rejecting the notion of an idealised march of progress, as well as a tendency toward accounts of environmental education based on historical high points such as international conferences and final meanings/definitions/principles, provides a methodological means for a theoretical shift that permits individuals to look at themselves as agents, but also at how they were formed as subjects (see Meadmore *et al.* 2000).

Environmental education sensibilities, as geographically constituted within local contextual practices, were thus traced, not as a notion of progress (as an arbitrary construction), but as a series of pragmatic interpretations or useful fictions (see Gough 1999), of human desires brought about by sociocultural practices where learning, knowing, and power structures cross. We are currently in the process of constructing 'small stories' of taken-for-granted practices and tactics that have special geographical historical significance. We are also scrutinising these stories in anticipation of the possibility that our inquiry may serve to problematise current local thought and practice in environmental education, risking that 'special resentment' which Rorty (1979) recognises might occur when beliefs so central to our desires are challenged.

Refusing a final vocabulary by treating the category 'environmental education' as a mere moment in the journey of thinking differently, we attempted to work across some sociocultural boundaries (Meadmore *et al.* 2000). As we worked with teachers (sometimes as students) and their mentors across educational institutions, our far from Foucauldian genealogical inquiry created what several critical and post-structural commentators describe as 'rhizomatic pathways', multiple tracings searching for the flow of ideas across people and time. We tried to provide openings for teachers, mentors, and researchers to (re)imagine environmental education in the present, not in terms of the past, and challenged ourselves to provide those spaces to think differently. By refusing closure on authority or authenticity in what counts as environmental education, we risked exposure to forms of critique in our reflexive accounts of processes, choices, and the limits of our experience (see Ball 1990; Paechter 1996).

Case reports constructed from the study consisted of writings that attempted to capture faculty/teacher/student professional relationships, or at least strong connections that individuals used to make sense of their discourses-practices as environmental education commitments (often deep, complex, and extending far beyond these connections into personal philosophies). Each of the case studies had a conceptual structure based in teacher/student professional relationships, set within teacher's education experiences. Although our conversations were with individuals, rather than groups, and at times far removed from actual experiences in teacher's education, we could sense a passion and certain genuineness in the recounting of stories that may have credibility across institutions. The ethos of our interpretive work in seeking out 'emic' meanings, i.e. those recognisably held by the people in each case, remained strong (see Stake 2000). Even so, we justify our intensive analysis of a small body of cases on the grounds that readers have opportunities to learn from each case as an attempted authentic representation of experience rather than as abstract generalisation. Of course, critical appraisal of the authenticity, beyond participant scrutiny, is crucial to the credibility of our cases.

Excerpts from one of the cases in our study are presented in Chapter 12 of this book. The cases provide small evidence of the vastness of the accumulated data. My intent here is to focus on methodological dilemmas and issues that arose in researching peoples' intentions, desires, and actual practices as an instance of the difficulty other researchers might encounter in pursuing something as elusive as the roots of participation as part of personal/social commitments to environmental education. Barone's (1997) detailed account of a teacher/student relationship is used to illustrate several of these methodological issues from the researcher/writer's point of view, because such accounts cannot yet be found in environmental education literature.

14.3 Methodology Issues and Challenges: Relationships Between Self, Field, and Text

In a collaborative educational (auto)biography, Barone (1997) shaped an (auto)biographical account of the subtle traces of a teacher's impact on a former student's life story. It represents the joint efforts of a researcher, a teacher, and a former student of the teacher to craft a narrative, as life story, but cast within the student's reconstruction of that life story. Within this composition, he raises the issue of how students believe teachers can affect their lives over time. First, Barone cautions readers not to expect the (re)storying to be a perfectly rounded seamless structure. Rather, it is characterised by erratics, meanderings, and interruptions. Second, it is partial and incomplete, although 'vouched for' by the student as an honest version of her meaning as extracted from certain life experiences. Finally, it is recognised for its fictional quality, as a product justified only as employed in the service of legitimate educational purposes.

I was drawn into this story because it resonated well with my own similar attempts to extend autobiographical work to genealogical study. I could relate instantly to the 'manner in which good stories rattle commonplace assumptions and disturb taken-for-granted beliefs' as Barone (1997:223) put it, highlighting the heuristic functions of a work of literature as social research. Teachers within our study of participation in environmental education were constantly referring to the influences of people, books, and/or places in their philosophical accounts of how they came to value the environment in their educational experiences. Following a brief quotation, adapted from Barone's attempt to develop a research account of the teacher's experience (Barone 1997:226–233), I discuss issues of representation and legitimisation in qualitative/narrative-based research as they relate to my own struggle to understand participation within stories of environmental education.

> For so long the source of these feelings was a mystery, lying asleep beneath the surface of her understanding. ... But she saw now ... that they arose from her need for recognition as a person ... she saw now ... generations ... things passed on from then to now ... it had taken her years to realize ... [he] had taught her to look inward ... he had taught her this through his actions. ...
>
> The words now ricocheted through time ... old words now, but still performing new magic, still illuminating her world. ... She had tried to become to her students the kind of person [her teacher] had been to her. Humane, concerned, compassionate, demanding ... and, as she had thanked [him], her students had thanked her. ... Generations, she thought again, things passed on from here to now.
>
> The qualities of purity and innocence were rooted deep in her family, were of the medium – the culture – that had nourished her as a child, and therefore of the green mountains that so subtly shaped the culture. It was why she loved the land ... thanks to [her teacher] she would always be a part of these hills ... well, that was part of the picture anyway ... there was much more to her than that.

Looking at Barone's (1997) text illustrates how working through a genealogical method demands from the researcher a strong grasp of the epistemological and theoretical tensions involved in asking how educational practices function as they do. The positioning of environmental education within the politics of schooling, as a kind of embodiment of a different participatory world view (see Reason 1988), constituted for us as researchers conditions for the genealogical examination of how environmental education came to be a part, even a small part, of school culture. In order to find clues about the conditions of possibility for environmental education in schools, we traced the discourses-practices beyond the limits of autobiography, not to show some sort of linear progression of ideas as much as the possible lines of descent to past practices. Genealogy concerns itself with the productive, rather than the negative or inhibitive forms of power – as an extended archaeology of how teachers are gradually, progressively, and materially constituted through a multiplicity of connections, desires, ideas, materials, and thoughts (see Meadmore *et al.* 2000). It does not seek in-depth investigation or an uncovering, but the elaboration of small stories, subtle shifts in thinking, details as clues (see Dreyfus and Rabinow 1982). As Barone (1997) suggests, this search may be characterised as a series of irregular movements along multiple paths (as with rhizomes) between the past and present for possibilities/emergences of particular discourse-practices. It is not an excuse for sloppy history, and does not imply that any arbitrary construction will do. Rather, it calls for pragmatically oriented historical interpretation of those cultural practices where knowledge and power cross as 'dialectic of discourse and the everyday' (Smith 1990:202).

For Convery (1999), whether autobiographical or genealogical, such narrative-based accounts are a privileged resource heavily influenced by identity considerations. It may be more useful to consider identity as multiple when the retrospect narrative offers the attraction of fixing meaning to one's actions in the reconstruction of a desirable or a preferred identity. This normal process of recounting self-descriptions that compensate for perceived inadequacies, helps create a sense of security with self. Genealogical work may help to disrupt certainty in such interpretations by providing multiple stories/perspectives in which the creation of identities through theory passes across people intersubjectively. Barone (1997) maximises the non-mimetic capacities of autobiography through his emphasis on autobiographical extension, as a self-representational practice knowable through metonymy.

Although such work may assist researchers in differentiating between teachers' narrative intentions and learners' experience, it may also help researchers to avoid tendencies to uncritically celebrate implied values in their eagerness to sponsor the teacher's voice. As Shields (2002) puts it, there is no epistemological, hermeneutic, phenomenological, or psychoanalytic dance that will allow us to flawlessly inscribe the marks of our true self. We are left with concrete, ethical dilemmas in which self-expression becomes suspect and demands reflexivity. It seems important as researchers to distinguish between metonymy and metaphor in extending our autobiographical work through genealogy. Metonymy emerges through patterns over time, unlike metaphor which depends only on a meaningful relationship grasped in the present. Thus, whereas autobiography uses metaphor as a mirroring of an essential self in text, genealogy recognises the continual production of identity as a kind of patterning, sustained through time by the modes of production that create it (Gilmore 2001:101).

It seems important in our zeal as researchers to acknowledge the danger of assuming moral self-assurance in our interpretive practices that may foreclose possibilities for alternative interpretations, reflexivity, and self-criticism. We should be attempting, through our genealogical accounts, to destabilise teachers' single interpretations of their experiences to discover what they inevitably conceal as much as what they tend to reveal (Convery 1999). The problem remains that there is neither a rational procedure nor an empirical method for this discovery process. This is not a reason to abandon the narrative sensibility underlying such methods, but to reconceptualise it as a rhetorical technique with all its uses and dangers.

Because narrative and genealogical work involve knowing our stories as embedded in the myths, folklore, and histories of culture (nature), and as frames for our identity both as participants and researchers, we can use the method to become more conscious and critical in exploring our perspectives beyond narrative. In this context, self-expression becomes suspect rather than self-evidently necessary and emancipatory. In our attempts as researchers to represent real people in texts that transcend their materiality as an emblem for a person's striving, our autobiographical work *appears* as a metaphor for identity, beyond proof of what simply is. However, such representation of the real is so crucial to autobiographical study that genealogical extension warrants coming at it from a new direction, through metonymic claims of contiguity wherein the person writing extends to self through time and place. This conception of the extension of the self gets at the limits of text representation itself. Instead of respecting the sufficiency of text, the genealogy integrates each text within another as an extension of self or identity (and its agency).

Story interpretation is inevitably hermeneutic and always open to question: Why is this story being told now, under these circumstances, by this narrator? Although we accept a certain indeterminacy and the essential incontestability of stories, we remain interested in the troublesome shape of narrative as cultural and historical fictions. Our problem, as researchers/writers, is in achieving consciousness of what we are actually doing in narrative construction and insisting on critical, intersubjective, and reflexive analysis of our plausible fictions. Whilst there is less need now than a few years ago to rehearse these arguments, some elements are discussed here because they provide a basis for our interpretations of the historical extensibility of narrative into (auto)biography and genealogical work.

14.4 Applying Research Theory to Research Practice: Moving Beyond the Armchair

In our work with teachers, we believed that we could not track their environmental sensitivity and participatory tendencies without some form of retrospective accounting of the meaning of various experiences, hence our interest in tracing the links between *discourse-in-practice* (as constitutive and meaningfully descriptive within varied institutional or cultural sites) and *discursive practices* (as subjectivities already embedded and embodied in the discursive conventions of social interaction). Our interviewing process was conversational, intended to encourage meaning-making through the narrative recounting of significant experiences in memory. Memory work, as we viewed it, was a form of self-invention. However, a turn towards others (as mentors) via memory seems a turn towards the self as producer of counter images – as an occasion for self-representation with the discourses of schooling and environmental education. Within this interactive process, we explain our interest in how teachers come to participate in environmental education, not hesitating to relate our own experiences as they seemed appropriate within the give-and-take of conversation.

We were not interested in whether teachers' accounts represented the truth as much as what each participant thought about how certain experiences seemed to be implicated in their constructed longitudinal versions of their lives, that is, what they thought they did in what circumstances and for what reasons (in their own words). We were listening to how each individual shaped her/his own story of development (of thought *and* practice), constituted of numerous smaller stories and events and constrained by the structures and events of cultural life as well as the demands of the drafting of a story. Many 'turning points' emerged in teachers' remembered histories, seemingly (re)produced by access to some new experience as an awakening of deep-seated beliefs and values, often culturally recognisable.

Our rough attempt to extend autobiographical methodology made it clear that personal accounts of lived experience should be taken as open-ended, time- and context-bound and indeterminate (see Bohman 1991). Although our methods were situational and specific, (auto)biography seemed useful in penetrating personal reality, giving details about character, agency, and experience. Within crucial *interactive* episodes, we were able to trace lines of individual and collective activity and perhaps even see aspects of self brought into being (see Becker 1971; Hones 1998), aware of questions about ethical and moral responsibility (see Grumet 1987, 1990; Measor and Sykes 1992; *Environmental Education Research*, 5(4), 1999). We were particularly sensitive to the tendency in autobiography to underplay social forces

(Schilling 1992; Holstein and Gubrium 2005) and used the genealogical method (e.g. relationships across time) to compensate (Meadmore *et al.* 2000).

Critical and postmodern commentators have assisted researchers in recognising a tendency within some methodological genres to ignore the ideological processes that inform teachers' (and researchers') autobiographical accounts (e.g. individualist versus social conceptions of agency). Recognising memory as a product of discourse originating in the world as well as one's mind is to find a base in conversation, culture, and personal relationships that helps to expose how ideology might operate to establish a consensual view of past personal experiences as significant and memorable. Identity, it seems, is a function of representation; that is, the revised Real emerges through a reconceptualisation of what the author claims as evidence. Critical work on embodied knowing is needed here as a way to facilitate genealogical work on discursive and practical knowing (Weedon 1987). Genealogical accounts permit the writer to take multiple runs at self-representation, as a way to explore the possibilities present within autobiography rather than produce a unity to the problem of representing identity. Convery (1999) suggests that if researchers were to confront teacher story-tellers sensitively with evidence of how identity claims may be constructed through narrative strategies, the resulting collaborative deconstruction of these narratives might enable exploration into how and why teachers' selves are 'theorised in discourse' (MacLure 1993:337). While we believe this kind of interaction would assist in understanding how notions such as participation are nuanced into environmental education practice, we have yet to pursue our inquiries to this level of exploration.

An autobiography is a fragment of a theory (Gilmore 2001:12). A genealogy is an assembly of theories of selves interacting, as multiple selves and multiple representations of personal identity in relation to significant others – a politic of representations. Each participant in our study was encouraged to engage in situating themselves within the flow of environmental education ideas and ideal. We needed a place where such methods of interaction could begin – in other words, a methodological predisposition based on some understandings of both methods and methodologies, as well as their groundings. We opted to remain within the narrative genre, albeit overlain with the excavation and relocation of ideas in relation to each other. Thus, rather than unearthing a composition through self-reflective questions (as in, for example, an action-research process), the idea was to scrutinise the questions.

Rather than autobiographic work that draws back on experience in order to see more clearly (see Grumet 1991), we focused (genealogically) on geographical (space) and temporal shifts (both theoretical and practical) as movement or desire to move 'elsewhere from somewhere' (see Kamler 2001). Wilson (2002) explains that this movement does not lie in the confessing 'I' of much personal narrative, but in the response of one person to another (interactively). Dimensions of learning as sociocultural and interactive (i.e. participative) seem implicated in such tracings. In this context, we could say that the cultural work performed in genealogy profoundly concerns the representations of environmental educators. Although this idea may appear to taint the process ideologically, it seems important as a way to represent oneself as a speaking subject. We see the lives of our research participants as composed of a variety of social networks with others who may understand lives differently, but it is the very existence of these differences that seems crucial to understanding our inquiries about participation in environmental education. In fact, the grounds of our interest in participation lie within the epistemological problematics of how we understand other people's lives *and* how we, as researchers, justify our knowledge claims about what we know about them.

14.5 Rethinking and (Re)Locating Narrative Work in Research on Participation in Environmental Education

In our attempts to search for meaning in teachers' desire to practice forms of environmental education, it seemed to us that our participants tended to talk about remembered experiences that had led them to environmental education as a more participatory form of pedagogy, almost by definition. In response to our questions about where these ideas had come from, they described how our process of 'bending back their minds' on their own stories helped them to 'see' inside some of their ideas and interactions, that is, to reconnect with their motives and to think more deeply about where they had come from. Their thoughts of early-life experiences were often overlain by readings or coursework in teacher education, where teaching 'mentors' had caused them to question their taken-for-granted assumptions and values, much as Barone's (1997) account described earlier in this chapter. As their interpretations of their remembered experiences grew more complex, our methodological challenge broadened. In retrospect, as inquirers, both participants and researchers had moved without much effort beyond descriptive autobiographical kinds of narrative to intergenerational interactions with 'ideas'. For example, participation as a concept was no longer simply assumed to be a part of environmental education – it was examined more explicitly as a fundamental concept, so much taken for granted in our assumptions about what environmental education practice really means. Although great potential derives when methodology compliments research interest, the process is a dynamic that has its limits. Discursive and practical ways of knowing and learning, while useful in generating stories, images, metaphor and exploring the relationship between belief and action, involve more creative metonymic strategies that accommodate tacit self-understandings from our past but as (re)produced within more recent interpersonal relationships and social/cultural political structures. Huberman (1993) describes these strategies as post-structural notions of sustained interactivity which remain elusive even when approached through such methods as (auto) biography and life history (Tierney 2000), autoethnography (Ellis and Bochner 2000), and genealogy (Meadmore *et al.* 2000).

The challenge is to move beyond the transformative illusion that such narrative strategies retrospectively effect. As Denzin (1989) cautions, it is necessary to destabilise the illusion that we understand when we do not, or that we have found meaningful coherent lives where none exist. Such a sceptical interpretation of research on teacher narratives troubles the notion that researchers' academic texts can result from inviting teachers to construct preferred identities. If learning is to occur in such situations, we must get beyond assigning a special role to agency and relocate learning in the journey – perhaps across the same landscape as narrative could describe – but with more focus on how our readings as teachers and researchers are a product of our intertextual and cultural formations. In other words, we use the narratives from oral/life history as genesis narratives that struggle against their location in all too familiar terrain. We engage interactively in the meaning of environmental education through our close examination of seemingly general concepts such as the participatory focus in environmental education rhetoric.

Good narrative constitutes an invitation to participate, to live the narrative (as a whole) vicariously. The teachers in our study used several personal experience methods to story and restory their experiences – methods such as oral history, personal writing, letter writing, and stimulated recall using photos and educational materials to draw attention to what they viewed as being important to their provision of environmental education (see Clandinin and Connelly 2000). We regarded such ideas as 'participations', with their notions of environmental education as positionings in their thinking about the nature and meaning of phenomena. We engaged participants in troubling their stories as well as their positions in terms of authority, plausibility, and meaning. We resisted the tendency to become immersed in what Bruner (1990) calls the 'swarm of participations' that act to distribute the self across occasions in use such as those inherent in the familiar characterisations of the 'caring' professor or the 'activist' father. Well aware of the epistemological burdens of the narrative genre, our recurrent narrative forms were produced from wider spatial/temporal interactions because we believed intergenerational connecting could help us look at ways people learn socio-culturally, as they create meaning from their lives through their available discourses-practices.

The possibility of tracing ideas, as precursors to participatory activity in environmental education and as lines of descent that might reveal emergences of particular discourses-practices, created openings for scrutinising both narrative (as life history/autobiography) and genealogical methodologies in an active search for clues to teachers' thoughts, desires, and assumptions (see Scheurich and McKenzie 2005). Genealogical work seemed to provide some added potential of opening richer contextual background dimensions. However, in moving from narration to genealogy, we were conscious about crossing epistemological boundaries – shifting focus from simple discursive and practical knowing to an examination of the sociological roots of production. While this shift was fundamental to our inquiry because of the focus on self, identity, and agency, we were uncertain how profound this emphasis could become in terms of how we came to judge our own knowledge claims (see Gilmore 2001). Scepticism about autobiographical work notwithstanding, we had chosen to compound such accounts by two- or threefold.

In producing several case-type instances of local contextual teachers' discourses-practices, constructed through tracings of underlying ideas of pragmatic interpretations of human motivations across temporal cultural spaces, we wonder if it was a worthy endeavour. The resulting 'small stories' across generations provide only fleeting marginalised instances in which knowledge and power seem to merge. Useful, perhaps beyond mere fiction, we believe we could see lines of descent as small emergences (with discontinuities) of ideas closest to the hearts of these people. Participation as an unexamined assumption of environmental education was one of these ideas. Our work provides evidence, we think, for the notion that contextually driven frames, rather than notions of an idealised march of progress of historical high points (e.g. the UNESCO environmental education conferences, declarations, and agendas), are what really drive local school practices/final meanings. Perhaps we have missed some things, but by refusing a final reading on what counts as environmental education, or as participation within environmental education, our pursuit of 'limited experiences' has begun. By refusing closure on our own authority and what counts (or in this case what leads to teacher's participation in environmental education), at least we opened some spaces for reflexive critique – spaces to think differently about environmental education.

14.6 Conc(ol)lusions[2]

In this chapter, I have attempted to trouble what it means to inquire into human qualities as precursors to participation in environmental education. Our research team set out to find ways to open up the possibility of fluidity between the idea categories of teachers/mentors and students across generations. We explored ways of interacting with educators as ways of speaking and writing that traced an assumed flow of ideas tied to participatory aspects of learning. We wanted to know if teacher's stories, framed autobiographically and genealogically, could make sense of teaching environmental sensibilities within institutional contexts. With teachers, we explored the possibilities of multiple sources of their agency, at the same time exploring the constitutive force of discursive practices through which both education and environmental education are carried out. As much as this work was revealing of the power of learning through participation across generations of people, many methodological issues resulted from our privileging of both discourses-in-practice and discursive practices. We can only see these issues after having lived through the inquiry process in reflexive interpretation (Alvesson and Sköldberg 2000).

Influenced by approaches to narrative inquiry that have the potential for tracing ideas genealogically, our exploration of various cultural/discursive threads (revealed in teachers' stories) through which they became 'environmental' educators, helped us to rethink what it means to be an 'environmental' educator, that is, what it means to educate 'as a participatory practice' in the context of current schooling practices. Inevitably, this rethinking involved questions of (self)consciousness, identity, and

[2] I am indebted to Kathy Nolan for this idea as distinct from 'conclusions'.

agency that centre methodological debates (see Archer 2000). Problems arose as our work struggled to get beyond a view of discourse separate from and setting limits to teacher's agency. This struggle caused us to think more broadly about our inquiry process as an ontological issue of post-structural and critical realist positionings. While we can assume that discourses-in-practice may operate to set boundaries to teachers' actions, we cannot overlook the power of people to act differently. As Bhaskar (1979) argues, beneath any empirically demonstrable pattern of events, there must be a mechanism conceivable in ontological terms that operates to constitute actual occurrences in the world.

In some ways, tracing environmental meaning over generations seemed to open up possibilities for the exploration of relationships, as intersubjective connections between teachers and students, as well as those between the researchers telling stories and the actual people themselves. However, as researchers, we soon find that, despite best intentions, we imprint our personal stories onto teachers' writing. Post-structural theory calls into question the authority of the author and breaks divisions between the teller and the told. Personally, because I found myself strangely distant from data that others had collected, and because intimate first-hand observations/interactions were not always available, I worked through some of my own stories as a way of getting inside the experience of being in environmental education (see Davies 2003). Rather than denying that stories had to be constructed in order to believe them, I live with(in) the contradiction (Bhaskar's (1993) performative contradiction) by attempting to remain true enough to participants (i.e. necessary fictions) and still carry the mark of the inquirer. According to Price (2004), this implies an epistemology which allows for both the real and constructed aspects of truth – a pragmatic position which perhaps straddles post-structural values of decentring subjects but without completely linguistic senses for what it means to be (a critical realist position?). Using both approaches implies a methodology which allows for both real and linguistic aspects of truth (Price 2004).

What seems crucial for environmental education researchers in this story of methodological tension is to consider a theory that neither elides the referent nor neglects the socially produced character of judgments about it. Bhaskar (1993) describes such a critical realist position in terms that there are both transitive and intransitive aspects of truth. Research methods that are intransitive have no implied relationship between researcher and subjects – those that are transitive require the existence of relationships. For genealogical work that attempts to look at the transitive aspects of reality (such as participation as a product of meaning within environmental education work), learning occurs through relationships with others. If environmental education implies a participatory dimension, and if environmental education research attempts to avoid performative contradiction or what Bhaskar (1993) calls theory–practice inconsistency, the possibility of ontological and epistemological fallacies arises. That is, if the transitive aspects of reality (i.e. participation as a notion formed in relationships) is the component of reality through which we must know intransitive aspects, then learning itself must have a participatory component. In research, the extreme post-structuralist position where everything is a construct, an illusion of linguistic activity, is incomplete without acknowledging that although, as researchers, we invented our participants' stories, we also recognise that in some ways they exist independently of us. Price (2004) describes the need to bring constructed *and* real aspects of human ideas and relationships together is not only intellectually necessary, but morally imperative. She invokes Latour (2004) and Haraway (2003) in the search for research methodologies that look for spaces or relational understandings of the social world. The implications for participation in such methodological spaces are unavoidable.

Rather than using our efforts to search for knowing as something that exists inside individual heads, as intransitive facts/ideas, we should be searching for relational knowing from several points of view (as well as our own), and for changing/arranging networking relationships for different forms of participation, as something that requires our actions and natures in dynamic relational spaces. Agency, identity and self-consciousness can then be construed in relational terms as learning through participation.

References

Alvesson, M. and Sköldberg, K. (2000) *Reflexive Methodology: New Vistas for Qualitative Research*. London: Sage.

Archer, M. (2000) *Being Human: The Problem of Agency*. Cambridge: Cambridge University Press.

Ball, S. (1990) Self-doubt and soft data: social and technical trajectories in ethnographic fieldwork. *Qualitative Studies in Education*, 3, 2: 157–171.

Barone, T. (1997) Among the chosen: a collaborative educational (auto)biography. *Qualitative Inquiry*, 3, 2: 222–236.

Becker, H. (1971) *Sociological Work: Method and Substance*. London: Allen Lane.

Bhaskar, R. (1979) *The Possibility of Naturalism*. Brighton: Harvester Press.

Bhaskar, R. (1993) *Dialectic: The Pulse of Freedom*. London: Verso.

Bohman, J. (1991) *New Philosophy of Social Science*. Cambridge, MA: MIT Press.

Bruner, J. (1986) *Actual Minds, Possible Worlds*. Cambridge, MA: Harvard University Press.

Bruner, J. (1990) *Acts of Meaning*. Cambridge, MA: Harvard University Press.

Bruner, J. (1996) *The Culture of Education*. Cambridge, MA: Harvard University Press.

Clandinin, J. and Connelly, M. (1996) Teachers' professional knowledge landscapes: teacher stories – stories of teachers – school stories – stories of schools. *Educational Researcher*, 25, 3: 24–30.

Clandinin, J. and Connelly, M. (2000) *Narrative Inquiry: Experience in Story and Qualitative Research*. San Francisco, CA: Jossey-Bass.

Coffey, A. (1999) *The Ethnographic Self: Fieldwork and the Representation of Identity*. London: Sage.

Conle, C. (1999) Why narrative? Which narrative? Struggling with time and place in life and research. *Curriculum Inquiry*, 29, 1: 7–32.

Conle, C. (2000) Thesis as narrative or "what is the inquiry in narrative inquiry?" *Curriculum Inquiry*, 30, 2: 189–214.

Connelly, F. and Clandinin, J. (1990) Stories of experience and narrative inquiry. *Educational Researcher*, 19, 4: 2–13.

Convery, A. (1999) Listening to teachers' stories. Are we sitting too comfortably? *Qualitative Studies in Education*, 12, 2: 131–146.

Craig, C. (2000) Stories of schools/teacher stories: a two-part invention on the walls theme. *Curriculum Inquiry*, 30, 1: 11–41.

Davies, B. (2003) *Shards of Glass: Children Reading and Writing Beyond Gendered Identities*. Cresskill, NJ: Hampton Press.
Denzin, N. (1989) *Interpretative Biography (Qualitative Research Methods)*. London: Sage.
Dreyfus, H. and Rabinow, P. (1982) *Michael Foucault: Beyond Structuralism and Hermeneutics*. New York: Harvester Wheatsheaf.
Ellis, C. and Bochner, A. (2000) Autoethnography, personal narrative, reflexivity: researcher as subject. In: Denzin, N. and Lincoln, Y. (eds), *Handbook of Qualitative Research*, 2nd edn. Thousand Oaks, CA: Sage, pp. 733–768.
Gilmore, L. (2001) *The Limits of Autobiography: Trauma and Testimony*. Ithaca, NY: Cornell University Press.
Goodson, I. and Sikes, P. (2001) *Life History Research in Educational Settings: Learning from Lives*. Buckingham, UK: Oxford University Press.
Gough, N. (1999) Rethinking the subject: (de)constructing human agency in environmental education research. *Environmental Education Research*, 5, 1: 35–48.
Greene, M. (1994) Epistemology and educational research: the influence of recent approaches to knowledge. *Review of Research in Education*, 20: 423–464.
Grumet, M. (1987) The politics of personal knowledge. *Curriculum Inquiry*, 17, 3: 319–328.
Grumet, M. (1990) Retrospective: autobiography and the analysis of educational experience. *Cambridge Journal of Education*, 20, 3: 321–326.
Grumet, M. (1991) Curriculum and the art of daily life. In: Willis, G. and Schubert, W. (eds), *Reflections from the Heart of Educational Inquiry: Understanding Curriculum and Teaching Through the Arts*. Albany, NY: SUNY Press, pp. 74–89.
Haraway, D. (2003) Online interview video clip. (http://www.egs.edu/video/donna-harraway.ram) (retrieved 30 January 2004).
Holstein, J. and Gubrium, J. (2005) Interpretive practice and social action. In: Denzin, N. and Lincoln, Y. (eds), *The Sage Handbook of Qualitative Research*, 3rd edn. Thousand Oaks, CA: Sage Publications, pp. 483–505.
Hones, D. (1998) Known in part: the transformational power of narrative inquiry. *Qualitative Inquiry*, 4, 2: 225–248.
Huberman, M. (1993) *The Lives of Teachers*. (Trans. Neufeld, J.) New York: Teachers College Press.
Jalongo, R. and Isenberg, J. (1995) *Teachers' Stories: From Personal Narrative to Professional Insight*. San Francisco, CA: Jossey-Bass.
Kamler, B. (2001) *Relocating the Personal: A Critical Writing Pedagogy*. Albany, NY: SUNY Press.
Latour, B. (2004) Why has critique run out of steam? From matters of fact to matters of concern. *Critical Inquiry*, 30, 2: 225–248.
MacLure, M. (1993) Mundane autobiography: some thoughts on self-talk in research contexts. *British Journal of Sociology of Education*, 14, 4: 311–322.
Meadmore, D., Hatcher, C., and McWilliam, E. (2000) Getting tense about genealogy. *Qualitative Studies in Education*, 13, 5: 463–476.
Measor, L. and Sykes, P. (1992) Visiting lives: ethics and methodology in life history. In: Goodson, I. (ed), *Studying Teachers' Lives*. London: Routledge, pp. 209–233.
Paechter, C. (1996) Power, knowledge and the confessional in qualitative research. *Discourse: Studies in the Cultural Politics of Education*, 17, 1: 75–84.
Polkinghorne, D. (1988) *Narrative Psychology*. Albany, NY: SUNY Press.
Polkinghorne, D. (1995) Narrative configuration in qualitative analysis. *Qualitative Studies in Education*, 8, 1: 5–23.
Price, L. (2004) Participatory curriculum development: lessons drawn from teaching environmental education to industry in Zimbabwe. *Environmental Education Research*, 10, 3: 401–408.
Reason, P. (ed) (1988) *Human Inquiry in Action*. London: Sage.
Rorty, R. (1979) *Philosophy and the Mirror of Nature*. Princeton, NJ: Princeton University Press.
Scheurich, J.J. and McKenzie, K.B. (2005) Foucault's methodologies: archaeology and genealogy. In: Denzin, N. and Lincoln, Y. (eds), *The Sage Handbook of Qualitative Research*, 3rd edn. Thousand Oaks, CA: Sage Publications, pp. 841–869.

Schilling, C. (1992) Reconceptualising structure and agency in the sociology of education: structuration theory and schooling. *British Journal of Sociology of Education*, 13, 1:69–87.

Shields, F. (2002) Who do we think we are? The ethics and politics of autobiographical curriculum. *Journal of Curriculum Theorizing*, 18, 4: 15–30.

Smith, D. (1990) *Texts, Facts, and Femininity*. London: Routledge.

Stake, R. (2000) Case studies. In: Denzin, N. and Lincoln, Y. (eds), *Handbook of Qualitative Research*. Second edition. Thousand Oaks, CA: Sage, pp. 435–454.

Tierney, W. (2000) Undaunted courage: life history and the postmodern challenge. In: Denzin, N. and Lincoln, Y. (eds), *Handbook of Qualitative Research*, 2nd edn. Thousand Oaks, CA: Sage, pp. 537–553.

Thomashow, M. (1995) *Ecological Identity: Becoming a Reflective Environmentalist*. Cambridge, MA: Teachers College Press.

Weedon, C. (1987) *Feminist Practice and Poststructuralist Theory*. Oxford: Basil Blackwell.

Wilson, T. (2002) Excavation and relocation: landscapes of learning in a teacher's autobiography. *Journal of Curriculum Theorizing*, 18, 3: 75–88.

Witherell, C. and Noddings, N. (1991) *Stories Lives Tell: Narrative and Dialogue in Education*. New York: Teachers College Press.

Chapter 15
Student Participation in School Ground Greening Initiatives in Canada: Reflections on Research Design Decisions and Key Findings

Janet E. Dyment

Keywords school ground greening, student participation, research methods, participatory research, research design decisions

15.1 Introduction

This edited collection brings together voices and experiences from around the world, all charged with a similar task: to address participation in the context of environmental, health and sustainability education and to do so within a critical framework. Participation viewed from a critical perspective adds to a growing body of literature that seeks to inform and strengthen understandings of this frequently used phrase while urging resistance to 'jargonisation' and encouraging sophisticated consideration of 'the good, the bad and the ugly' of participation.

In this chapter, I add to this evolving discourse of participation by reflecting on the dilemmas I faced during my doctoral dissertation research on participation within the context of a school ground greening initiative in Canada. In this project, I explored the nature of children's participation, and several aspects of both the research findings and research process are worthy of exploration and comment. Through the reflections in this chapter, I hope to illustrate the kinds of information that can and cannot be gathered about the nature and context of participation through research approaches like mine. I also present some of the key findings that emerged vis-à-vis the limits and possibilities of youth participation in school ground greening initiatives. The chapter is of particular interest to those who have ever struggled or grappled with research design decisions when trying to understand issues related to participation.

Centre for Human Movement, University of Tasmania, Locked Bag 1330, Launceston, Tasmania 7250, Australia, Janet.Dyment@utas.edu.au

A. Reid et al. (eds.), *Participation and Learning*, 241–255.

15.2 Participation in School Ground Greening

Evidence of the wide-ranging benefits of green school grounds is mounting. These benefits extend, potentially, to student learning, environmental awareness, teacher motivation, social behaviour and relationships, and health and safety (Dyment 2005a). School grounds can thus be considered and developed as sites to enhance these various dimensions of school life.

While the majority of these benefits influence students *after* the greening process is completed, a modest body of research has also explored the benefits that emerge for young people during the *process* of greening (Titman 1994; Kenny 1996; Hart 1997; Moore and Wong 1997; Hunter, Layzell, and Rogers 1998; Rickinson, Sanders, Chillman, Doyle, and Jameson 2003; Dyment 2004, 2005a). Researchers have developed many terms (e.g. 'action competence', Jensen and Schnack 1997; 'genuine participation', Simovska 2000; 'environmental praxis', Fawcett, Bell, and Russell 2002; 'participatory democracy', Wals and Jickling 2000) and frameworks (e.g. 'Ladder of Participation', Hart 1992, 1997) to discuss the importance of meaningful youth participation in initiatives such as greening projects.

While these researchers differ somewhat in their interpretation and uptake of notions of participation, they generally agree that a critical component of school ground greening (and other such initiatives) is to ensure that young people's voices and concerns are considered during the greening process. They problematise and resist 'token' approaches to including students in greening initiatives, for example, where they are included in tree planting events only. Some researchers place particular emphasis on the notion that students should be involved in the problem identification phase of greening projects (e.g. Hart 1997). Many argue that young people have a right to participate in decisions that relate to their quality of life and contend that students, when given the opportunity, will be able to critically evaluate their spaces, identify alternatives, and evaluate the outcomes (see, for example, Chapter 18 by Barratt and Barratt Hacking, this volume). All these researchers agree that one of the biggest, if not the biggest outcomes of school ground greening is allowing young people to acquire skills related to democracy, participation, and citizenship during the process of greening. And finally, they assert that young people will carry these skills into adulthood, allowing them to become political, engaged, and reflexive adults who know their rights and responsibilities as members of a community.

In looking for additional support for child and youth participation in greening initiatives, many point to international documents that recognise and support the notion that young people can and should be involved in decisions related to civic life. For example, the 1989 United Nations Convention of the Rights of the Child (CRC) is a set of universal standards for the protection and development of children (UNICEF 1990). The CRC contains a series of 'participation articles' that articulate that young people are independent, thinking individuals who are capable of being involved in decisions that affect them (e.g. Articles 12–15, 17).[1] Other international conferences have built on the foundations laid in the CRC and have sought to clar-

[1] Interestingly, the CRC also contains an article closely related to the issue of school ground greening. Article 31 addresses children's right to play, rest, and leisure.

ify and strengthen the nature of children's participation (e.g. Agenda 21, Local Agenda –21, see UNCED 1992; WCED 1993).

15.3 Reflections on Researching Participation: My Methodological Leanings

There is a range of approaches to researching participation, from 'insiders' working together to determine both research content and process, to 'outsiders' maintaining considerable distance from research subjects and using pre–post designs, questionnaires, attitude scales, and tests of factual knowledge. There are challenges inherent to all approaches. Participatory action research can be limited in generalisability with its focus on one or a few sites, and tensions can arise between researchers and other participants who have different conceptions of the problem and process of research. Outside experts, on the other hand, even in spite of their theoretical and methodological sophistication, can lack the personal engagement and contextual sophistication necessary for rich understanding.

While there is a general recognition in the field of education that *all* approaches have inherent strengths and limitations, it seems to me that those who currently research participation in environmental and health education are now valorising more participatory approaches. Indeed, some argue that this is the only way to research participation. Similarly, during the meetings in Copenhagen where this book had its genesis, there was strong support for participatory research and significant critique of 'sanitised apolitical empiricism ... used to look at the political act of participation' (Dyment, journal notes, June 2004). I wonder whether more traditional strategies risk being sidelined in the wave of growing interest in participatory approaches? If so, it troubles me as I believe that they can still offer insight into participation.

To clarify, I believe that participatory approaches are more congruent with the topic at hand and can contribute rich, deeply meaningful understandings. Nevertheless, I must admit that I have made strategic research design decisions that resulted in my methods not being fully commensurate with my ideological and methodological leanings.

I was not initially inclined to use participatory approaches in my research. Early in my career, as a student working within scientific or pseudoscientific communities, I had a firm grasp of positivist takes on rationality, objectivity, and truth. My early research experiences were conducted within this paradigm and I felt forced to detach myself from my research projects in order to maintain an 'unbiased' perspective. Entering the fields of environmental studies and education, I was delighted to encounter writers like Patti Lather (1991) who argued that researchers could choose to investigate issues that they believe in, that research could contribute to social justice, that it could be emancipatory and a form of praxis. I heeded her advice and began to investigate subjects of greater personal and political interest to me.

In making the shift towards postpositivist and post-postpositivist genres of inquiry, I have sought to interrogate my role as researcher (Harding 1986; Kirby and McKenna 1989; Lather 1991; Heshusius 1994). Eschewing the idea that a researcher should or could be an impartial observer of events, I have come to acknowledge the influence I have had upon the design, analysis, and dissemination of my studies. I was forced to examine how various aspects of my identity (e.g. related to race, class, gender, ability, sexuality, species, etc.) influenced all phases of my research projects. I wanted to consider how identity had influenced *what* I had chosen to study (e.g. school ground greening projects), the *design* of my study (e.g. qualitative and quantitative), and *where* I chose to study (e.g. Ontario vs. other provinces; Canada vs. USA), as well as *whom* I chose to work with (e.g. teachers vs. students vs. community members vs. more-than-human entities).

In shifting perspectives, I also had to come to terms with the limits of objectivity, recognising multiple constructions of reality (Lather 1991; Guba and Lincoln 1994; Denzin and Lincoln 2000). Donna Haraway's (1988) ideas about 'situated knowledges' and 'partial perspectives' relieved my feelings of pressure to produce generalisable results and find *the* (singular) answer to complex problems. I tried not to feel discouraged by the impossibility of objectivity and the limits of generalisability, but instead I attempted to revel in the notion that my work could be woven together with other research contributions in a multi-hued tapestry of understanding.

Such an approach is more conducive to those of us who identify ourselves as both environmentalists *and* researchers. In her article 'Environmental Education Researchers as Environmental Activists', Karen Malone (1999) reflects upon her own journey as a doctoral student required to confront this issue. Drawing on critical feminist writing, Malone describes the similarities between feminist education research and environmental education research and argues for an activist approach to environmental education research. She contends that:

> as researchers in environmental education we are engaging in a political act…[I]f environmental education emerges from environmentalism and if environmentalism is a social movement, we are in essence generating knowledge to advance a social movement. (p. 175)

Critical of the valuing of traditional apolitical research, Malone issues 'a challenge to environmental education researchers and all critical researchers to move outside the "academy" and develop partnerships with schools and communities and become directly involved in environmental activism' (p. 176). Connie Russell (my doctoral supervisor) has noted elsewhere (Russell 2003) that this can be quite a challenge for any researcher, let alone a new one. Indeed, it can easily lead to a sort of paralysis and therefore throughout my dissertation, she repeatedly encouraged me to try not to 'get stuck' in one role or the other, but to seek balance in my dual roles of researcher and environmentalist.

Given this brief description of my methodological journeys, a reader might well expect that I now always align myself with postpositivist or post-postpositivist methodologies and choose corresponding strategies. While this has been so for some of my endeavours, I am not nearly that consistent. Other factors besides methodological leanings come into play, including my research partners and the audience(s) I desire for my research. In the next section, I will illustrate these tensions through presenting some of my dissertation research. I hope it will also provide an example of how traditional methods still have the potential to inform understandings of participation.

15.4 Case Study: My School Ground Greening Research

I investigated school ground greening in an entire school board in Canada's largest city, Toronto. One aspect of the project was to examine if and how various stakeholders (students, parents, teachers, principals, school board officials) were participating in the school ground greening movement. I was certainly not the first to explore the issue of participation in school ground greening projects, other researchers have noted the importance of student participation (Kenny 1996; Hunter *et al.* 1998; Mannion 2003) and described student participation at particular schools (Moore and Wong 1997). While those research initiatives outlined benefits, concerns, drawbacks, challenges, and enablers of participation in greening projects at individual schools, it was not yet known whether these factors were broadly representative of a large number of schools. My study thus helped fill this knowledge gap by exploring stakeholder participation in school ground greening across a large number of schools.

The Toronto District School Board (TDSB) was an ideal study site since so many of its schools have begun the process of greening. Approximately 20% of the more than 500 schools in the board have initiated greening projects. The schools in my study differ in a variety of ways. Some schools have only 200 students, while others have more than 1,500; some schools are located in Toronto's wealthiest neighbourhoods, while others are located in its poorest; some of the greening projects are new, while others are more than ten-years old; some have budgets of only a few hundred dollars, while others have had access to hundreds of thousands of dollars; and some have very complex designs including ponds, murals, outdoor classrooms, vegetable gardens, mazes, and butterfly gardens, while others are much more modest and consist of only a few trees.

In order to understand the nature and extent of stakeholder participation in the greening projects across the TDSB, I had to make a series of research design decisions. How would I best get a handle on stakeholder participation in the greening initiatives across the entire board? While I was initially drawn towards a more participatory approach (e.g. spending extensive amounts of time in a single school, working on projects with stakeholders, having stakeholders define research questions, etc.), I realised that adoption of this approach would prevent understanding the nature of participation across a large number of schools. As well as this, the pragmatic considerations that plague most doctoral research students, such as limited amounts of time, also influenced my research design decisions. Further, the TDSB and Evergreen, a non-governmental organisation (NGO) that facilitates greening initiatives in Canada, had offered financial and logistical support for the project and were interested in 'hard numbers' from my research. They, in fact, were hoping for a *Closing the Achievement Gap* (Lieberman and Hoody 1998) type of report that could be used to influence policy.

Given these competing interests, making my research design decisions was a laborious process. I was faced with balancing my own methodological leanings, logistical constraints, a desire to influence policy, and the realities of partnering with a school board and an NGO. How best to proceed then? In the end, I decided to use a mixed-methods approach of questionnaires and interviews.

Methods

My questionnaires were designed to understand trends across a large number of schools. They were piloted on two occasions and refereed by a jury of six professionals and academics who were familiar with survey design and/or school ground greening research and practice. Four questionnaires were distributed to 100 schools with green school grounds and were completed by a principal, a teacher involved in the greening project, a teacher *not* involved, and an involved parent. It was anticipated that those invited to complete the questionnaires would have varying levels of experience in greening efforts and offer a variety of perspectives.

The follow-up interviews allowed for a deeper exploration of the results at five schools, selected to ascertain more fully if and how socio-economic status influenced greening projects. One school from each 'category' of socio-economic status (i.e. very high, high, medium, low, very low) was selected. I visited the five schools and spoke to the principals, teachers, and parents who were involved in the greening project. A total of 22 individuals were interviewed (five principals, seven teachers, and ten parents). During the interviews, participants were asked how various stakeholders were involved in the greening initiatives.

The questionnaires were statistically analysed to understand basic trends in participants' responses and to explore if and how responses differed as a function of individual characteristics (e.g. age, gender, experience), school characteristics (e.g. number of students, socio-economic status of the school catchment area) and greening project characteristics (e.g. history, amount of funding). Data from the interviews were fully transcribed and thematically coded.

Illustrative results

Through analysis of the questionnaires and interviews, I was able to identify ways in which various stakeholders (students, teachers, parents, principals, school board) were or were not involved in the process of school ground greening. In this chapter, I present only a few key results related to student involvement in the greening initiatives. While students were not themselves participants in this study, adult interpretations of their roles were telling.

In the questionnaires, respondents were asked a series of questions about stakeholder involvement. One of the questions asked, *Who provided the initial involvement?* Analysis of the questionnaires revealed that while a wide array of individuals had been involved in initiating greening projects, students were rarely (1.4%) reported as being instrumental in providing the initial motivation (Table 15.1).

Table 15.1 Individuals who provided the initial motivation for school ground greening projects

Individuals/Groups	%
Teachers	41.8
Individual parents	22.6
Principal	18.5
Parent/teacher committee	8.2
Other	3.4
Community members	2.0
School board staff or trustees	2.0
Students	1.4

Note: N = 146 respondents.

Rather, the most commonly cited initiators were teachers (41.8%), followed by individual parents (22.6%) and principals (18.5%).

At the five case study schools, the interviews confirmed that students rarely provided the initial motivation or were involved in problem identification. While the questionnaires could only illustrate a quantitative representation of student involvement (e.g. Table 15.1), the interviews helped me to explore *why* the students were not involved in this phase. Many reasons were offered. Several interviewees suggested that students were too young to realise that 'things could be different' (Parent, School A). Another parent contended that 'at that age, you just accept what you have ... you never question things' (Parent, School B). One principal who was concerned about time and curriculum demands indicated 'it would take too long to get them to do that' (Principal, School D). Funding application deadlines also seemed to constrict the amount of student involvement in the early phases. An involved teacher indicated that her students were not involved in the early phases of the process of greening because 'we found out about a proposal at the beginning of February and we had to have it in by February 19th. So we just rushed through the visioning process ... I thought that we'd just apply for this, we'll ask for this and this and go from there ... get the students involved once we have more money' (Teacher, School C). In this instance, interview data fleshed out questionnaire data.

On some occasions, however, the data from the questionnaires and interviews served to highlight contradictions. For example, on the questionnaire, study participants were asked to indicate the levels of involvement of a variety of individuals on a 4-point Likert scale (1 = not at all involved, 4 = very involved). Examples were provided to illustrate 'initial phases' (e.g. problem identification, visioning, designing, fundraising, initial planting) and 'on-going maintenance' (e.g. weeding, watering in summer, harvesting). Questionnaire respondents indicated that students, as well as most other stakeholders (except school board staff and trustees), were involved, to some degree, in the *initial* phases of greening projects (Table 15.2). The individuals receiving the highest ranking for involvement included principals ($M = 3.39$), teachers ($M = 3.37$), and students ($M = 3.27$), as well as individual parents ($M = 3.20$). Respondents were then asked to indicate the level of involvement of individuals during *ongoing* maintenance. Respondents indicated that students ($M = 3.04$),

Table 15.2 Mean level of involvement of individuals during initial and ongoing maintenance phases of school ground greening initiatives

		Initial		Ongoing	
Types of stakeholders	*N*	*M*	SD	*M*	SD
Students	97	3.27	0.822	3.04	0.910
Teachers	98	3.37	0.831	2.96	0.924
Individual parents	96	3.20	1.094	2.75	1.217
Parent/teacher committee	97	2.84	1.098	2.25	1.058
Community members	97	2.48	1.034	1.93	0.992
Principals	97	3.39	0.755	2.72	1.067
Custodial staff	97	2.87	0.896	2.62	0.973
School board ground maintenance staff	95	2.67	1.006	1.99	0.965
Other school board staff or trustees	83	1.98	1.023	1.39	0.682

Note: Level of involvement was based on a 4-point scale (1 = not at all involved, 2 = not very involved, 3 = somewhat involved, 4 = very involved).

teachers (M = 2.96), individual parents (M = 2.75), principals (M = 2.72), and cleaning staff (M = 2.62) were the most involved. The questionnaires, then, indicated that students were very involved in both the initial and ongoing phases of greening projects.

The follow-up case study interviews at the five schools allowed me to gather a much more complex understanding of how students were involved in the initial and ongoing phases of the school ground greening projects. When asked to describe student involvement, all interviewees reported that students were involved in contributing design ideas. Schools used a number of different techniques for gathering these. At one school, students were asked to complete surveys with their parents. At another school, student representatives elected to a Student Council sought design ideas from their own classes and worked as a Council to generate a vision for a new school ground. At three of the schools, students were asked to create maps and drawings of their ideal school ground. It appears that student ideas were thus sought out and seriously considered during the design phase of the projects. This finding is consistent with the questionnaire data (Table 15.3). Once the design ideas were solicited, however, adult committees always made the final design decisions and no students were involved.

Student involvement tapered off considerably in fund-raising, with adults doing the large majority of fund-raising and planning for the projects in the case studies. Many of the interviewees shared this principal's perspective on children's interest and ability to be involved in such activities: 'When you consider that they are young children and they don't really care to know about a lot that goes into writing grants, organisation or ideas ..., that for them might be a little boring' (Principal, School B). One involved parent, who had very little experience in organising such an initiative, also indicated that when 'adults are in over their heads' with fund-raising and planning, 'it is difficult to know how to include students' (Parent, School C). A notable exception was a grade-5 class who were actively involved in raising funds through a partnership programme with a financial institution's 'Business Entrepreneur

Table 15.3 Student participation in various phases of greening initiatives. (Adapted from Jensen 2000.)

	Involved or not	Problem identification	Visioning and designing	Fund-raising	Planting	Maintenance
Students suggest/common decisions						
Students suggest/students decide						
Teacher suggests/common decisions			X		X	X
Teacher informs/students accept or reject	X			X	X	X
Teacher decides/no room for student influence		X				

Note: Some of the phases of greening contain more than one 'X', to reflect the range of responses across the schools profiled.

Programme'. This class made presentations about the greening project to approximately 25 corporations in the city and raised CDN $30,000.

Students were involved in the actual plantings at all five of the case study schools. At some schools, the planting occurred over a very short period and the entire student body was involved. One parent reminisced about how the plantings occurred over the course of one week: 'The students planted all the trees throughout the week…they went out and took each class at a time and picked the trees that they wanted and planted them and then watered them for the rest of the year.' At other schools, the planting occurred over a longer time period and small groups of students were involved throughout the process. Despite the reported active student involvement in planting, it appears as though student involvement in the planting was sometimes actually quite regulated and overshadowed by adults' interests. During the interviews, a parent recalled: 'Sometimes, I get particularly bad-tempered with all these children around, because they won't do what I wanted them to do' (Parent, School B). Another teacher recalled her frustrations in having the students involved in the plantings: 'I've gone back on the weekends a few times, and moved their plants around … I just didn't have the heart to tell them at the time, but I really couldn't just leave them that way' (Teacher, School C).

Student involvement in maintenance of the greening projects varied. At two of the schools, students could join clubs to help to maintain the garden (e.g. Garden Club, Environmental Club). At two other schools, students were less involved with maintenance and adults assumed the responsibility (e.g. teachers, parents, principal). At these schools, students could help the adults if they liked but there was no formal way of facilitating such involvement. At another school a principal expressed concern that the maintenance of the garden could get in the way of formal teaching, suggesting that 'There's too much for children to weed … [I]f one were to devote the time in the curriculum to doing what needs to be done, it's a saw-off from other things that are needed to be done' (Principal, School C). The interviews thus contradicted the questionnaire data that indicated that students were quite involved in maintenance (Table 15.3).

15.5 Reflections on Findings

Based on these results across a range of schools, it appears that while students in this school board provide a considerable amount of time and labour to the actual project, they are not overly involved in problem identification, visioning, or planning phases of greening. I would thus argue, as others have, that much room exists for students to become more involved in earlier phases of the greening projects. While it is laudable and important to include them in the designs, actual plantings, and maintenance, much is being lost when they are not involved in earlier phases, particularly the problem identification phase (Hart 1997).

In Roger Hart's 'Ladder of Participation' (1997, see also Chapter 2 by Roger Hart, this volume), he sets out eight 'levels' of participation, ranging from 'Manipulation or Deception' (the lowest 'Rung #1' on the ladder), to 'Child Initiated, Shared Decisions with Adults' (the highest 'Rung #8' on the ladder). In reflecting upon the questionnaires and interviews, it seems as though there are a range of 'rungs' represented within the present study. At some schools, for example, I found elements of 'Manipulation or Deception' (Rung #1) whereby, as Hart (1997:40) describes, 'an adult designs a garden, has children carry out simple planting, and then tells journalists and photographers that the children designed and built the garden'. At other schools, I found elements of 'Tokenism' (Rung #3), whereby adults are interested in giving children a voice, but have not 'begun to think carefully and self-critically about doing so' (p. 41). The highest level of participation that I understood to be occurring in the five schools profiled in the follow up case studies was 'Consulted and Informed' (Rung #5), whereby students' understood the nature of the project, their opinions were seriously considered and they were kept abreast of developments.

It thus appears that while students in this selected board are involved, to a degree, in school ground greening initiatives, room exists to broaden the scope of their involvement. Additional 'rungs' on Hart's Ladder include 'Adult Initiated, Shared Decisions with Children', 'Child-Initiated and Child-Directed', as well as 'Child-Initiated, Shared Decisions with Adults' (Rungs #6–#8 respectively). The work of Hart (1997), Hunter *et al.* (1998), Moore and Wong (1997), and Titman (1994) provide compelling evidence that increased student involvement in all aspects of a greening initiative, will generate numerous benefits for the students and the school community, as well as the greening project.

Student participation in greening initiatives in this school board can also be analysed and explored by using frameworks other than the Ladder of Participation. For example, emerging out of Jensen and Schnack's (1997) action competence work, Jensen (2000) has developed a matrix of participation that is slightly more sophisticated than the

Ladder framework, in that it highlights how the level of student participation may vary at different stages of a project. It is also a helpful tool for discussion and evaluation that makes many implicit assumptions about participation more visible. Jensen acknowledges, for example, that it may not be appropriate or possible for students to be involved in some phases of initiatives, while in other phases, they may be the primary drivers and decision-makers. When the student involvement reported in this study is analysed in light of Jensen's matrix, it once again becomes clear that students in this school board are not involved nearly as much as they could or should be throughout the greening initiatives (Table 15.3).

There are numerous reasons why students are not involved in more authentic ways in the greening projects profiled in this study. Chawla (2002) summarised some of the challenges that limit young people's involvement in authentic planning processes, noting, amongst others, the following barriers: time constraints; a poor understanding of child's capabilities; a belief that adults can adequately represent the perspectives of children; a belief that children are unskilled and unreliable; a lack of understanding as to how to facilitate participation; and a fear of politicising children, etc. It appears that many of these barriers were present for the questionnaire respondents and interviewees in this research.

Additionally, one needs to look at the culture of schooling. In discussing the potential implications of the deschooling movement for environmental education, Weston (1996) notes how schools have little room for anything beyond preordained, teacher-directed activities (see also Hargreaves 2003). It should be no surprise then that more participatory approaches face challenges. As Robottom and Sauvé (2003) assert, a common problem for participatory research, and environmental education generally, is the instrumentalist framework of schooling – how does one foster greater participation within the context of the status quo of technocratic rationality? Lousley's (1999) study of Environment Clubs in the TDSB demonstrated this well. Schools were wary of controversy and teachers and administrators in her study desired to 'depoliticise' the activities of any of clubs which dared stray too far from banal efforts such as recycling paper in the school. (See also Whitehouse 2001 for a description of the ways in which a group of students, too, were wary of being seen as overly political.)

While school ground greening, at first glance, hardly seems radical, its potential to disrupt anthropocentrism (Bell and Russell 1999) and to encourage greater participation of students suggests that, as with Lousley's Environment Clubs, there are likely to be efforts to depoliticise the movement. Planting trees in a preordained location is one thing; students identifying for themselves their unhappiness with various aspects of schooling, such as being imprisoned indoors for large periods of time, is quite another. Feminist post-structuralism may offer interesting insights here (see Chapter 13 by Barrett, this volume and McKenzie 2004). Rather than envisioning hordes of conservative teachers zealously squelching any sign of 'real' participation by students, or conspiratorially allowing students 'token' participation whilst maintaining control, it is perhaps more helpful to look at the way the very discourses of schooling constrain certain possibilities (Dyment and Reid 2005). How might understanding of what constitutes 'authentic' and 'democratic' participation, 'agency' and 'empowerment' differ in the discourses of schools and the discourses of participatory forms of environmental education?

15.6 Reflections on Research Design Decisions

There are three areas of the research process that deserve further consideration. First, the findings offered in this chapter illustrate the obvious point that data from different sources can shed different light on the same issues. The responses I heard from case study interviewees did not always support the findings in the questionnaires. Sometimes, my data sets contradicted each other. I acknowledge that I felt unprepared for and continuously struck by these differences; I found solace realising that others have found these contradictions as well in environmental education research (see Hart and Nolan 1999:28). It is interesting to ponder now how the questionnaire design itself influenced the stories I heard. For example, closed-answer questions like asking respondents to rank involvement of various individuals gave me insight into some aspects of who was involved, when they were involved, and how they were involved, but my understanding was still quite limited given that the questionnaire construction forced respondents to answer within certain confines. Like others who have used questionnaires, I felt, at times, 'restricted by both questions and methods (that were) incapable of understanding the complexity' (Hart and Nolan 1999:25). I was relieved to complement my findings from the questionnaires with case study interviews because they allowed me to ask further questions about student involvement in the greening initiatives. The apparent contradictions between the questionnaires and the case studies certainly reinforced the value of having a mixed-methods approach. My understanding of student involvement in school ground greening would have been even more limited had I chosen to only collect one form of data (and, of course, had I added another strategy such as ethnographic field work, I would have had another perspective on school ground greening that might well have contradicted the questionnaires and case study interviews).

Second, while the questionnaires revealed trends in participation across a large number of schools, the interviews added some depth of understanding of these trends. Noting such trends across many schools, rather than focusing on the experiences in a single school, emphasised to me the widespread lack of student participation. Identification of these trends adds weight to recommendations to the School Board and Evergreen that they need to focus more on student involvement.

Third, my research was considered very useful by both research partners, the TDSB and Evergreen. A summary report (Dyment 2005a) and peer reviewed publications (Dyment 2004, 2005b, 2005c; Dyment and Reid 2005) have already been published and upper-level administrators within the TDSB are using them to promote discussion of how they might better realise the potential of greening initiatives. A key factor in explaining its immediate popularity within policy circles is the presence of numbers. Other reports in environmental education circles that have adopted a quantitative approach (e.g. Learning Through Landscapes 2003; Lieberman and

Hoody 1998) are often seen to be 'convincing', 'striking, and 'comprehensive' (Evergreen 2000:5–6, in discussing the Lieberman and Hood report). While I, as a researcher, may decry the hegemony of numbers and need to work on educating policy makers, administrators, and the general public about the value of other approaches, at this moment in time it seems strategic to make use of research strategies that speak to decision-makers. Nevertheless, I worried about the ways in which my close collaboration with the TDSB and Evergreen influenced, consciously or not, my research decisions. Furthermore, I worried about falling into the trap identified by Audre Lorde (1984) of trying to use 'the Master's tools to dismantle the Master's house'.

15.7 Conclusion

In this chapter I have highlighted the ways in which students are and are not involved in greening efforts in the TDSB. In light of these findings, I believe efforts must be directed towards changing and enhancing the role of students in school ground greening projects. Research must continue to explore the factors that limit and enable student participation, examine the skills needed by educators who want to adopt participatory approaches, and appraise the role that could be assumed by a school board in creating a board-wide culture that endorses truly authentic student involvement.

I have also shared some of the research design dilemmas that emerged for me while performing this research. I wish to reiterate that, as in any programme of research, different research methods will shed light on different things. I was taken by the metaphor that making research design decisions could be compared to choosing a room in an observatory for viewing the night sky (Reid, Anchorage, Alaska, 2003, personal communication).[2] Depending on which room a researcher chooses, one will be able to see different, albeit sometimes overlapping parts of the night sky. If you are in Room A, you will be able to see stars in one part of the night sky, if you are in Room B, you will see stars in another part of the sky, and so on. There is no single room that will allow a researcher to see all of the night sky. Of course, one can leave the building altogether for stargazing purposes, but there will still be limits to our knowing. Like Haraway (1988), I do not believe that any of us are capable of the 'god trick' of seeing everywhere from nowhere. I thus wish to caution readers of this book to not dismiss *any* approaches too quickly. While I understand the reasons for the pendulum swing towards more participatory research approaches, there remains value, for now, in traditional approaches as well.

Acknowledgments I gratefully acknowledge support from the Social Science and Humanities Research Council of Canada, the Toronto District School Board and Evergreen.

[2] This metaphor owes its roots to an early conversation between Alan Reid and his colleague, Stephen Gough, at the University of Bath.

References

Bell, A.C. and Russell, C.L. (1999) Life ties: disrupting anthropocentrism in language arts education. In: Robertson, J.P. (ed), *Teaching for a Tolerant World, Grades K-6: Essays and Resources*. Urbana, IL: National Council of Teachers of English, pp. 68–89.

Chawla, L. (2002) Toward better cities for children and youth. In: Chawla, L. (ed), *Growing up in an Urbanised World*. London: UNESCO/Earthscan, pp. 15–34.

Cooke, B. and Kothari, U. (eds) (2001) *Participation: The New Tyranny*. London/New York: Zed Books.

Denzin, N.K. and Lincoln, Y.S. (2000) Introduction: the discipline and practice of qualitative research. In: Denzin, N. and Lincoln, Y. (eds), *Handbook of Qualitative Research*, 2nd edn. Thousand Oaks, CA: Sage, pp. 1–28.

Dyment, J.E. (2004) 'At that age, you just accept what you have…You never question things': a case study of student participation in school ground greening projects. *Children, Youth and Environments*, 14, 1: 130–152 (see also http://cye.colorado.edu).

Dyment, J.E. (2005a) *Gaining Ground: The Power and Potential of Green School Grounds in the Toronto District School Board*. Toronto, Ontario: Evergreen. Available at http://www.evergreen.ca/en/lg/lg-resources.html.

Dyment, J.E. (2005b) Green school grounds as sites for outdoor learning: barriers and opportunities. *International Research in Geographical and Environmental Education*, 14, 1: 24–41.

Dyment, J.E. (2005c) 'There's only so much money hot dog sales can bring in': the intersection of green school grounds and socio-economic status. *Children's Geographies*, 3, 3: 307–323.

Dyment, J.E. and Reid, A. (2005) Breaking new ground? Reflections on green(ing) school grounds as sites of ecological, pedagogical and social transformation. *Canadian Journal of Environmental Education*, 10: 286–301.

Evergreen (2000) *Nature Nurtures: Investigating the Potential of School Grounds*. Toronto, Ontario: Evergreen Foundation.

Fawcett, L., Bell, A.C., and Russell, C.L. (2002) Guiding our environmental praxis: teaching and learning for social and environmental justice. In: Filho, W.L. (ed), *Teaching Sustainability at Universities: Towards Curriculum Greening: Environmental Education, Communication, and Sustainability*. New York: Peter Lang Scientific Publishers, pp. 223–228.

Guba, E. and Lincoln, Y.S. (1994) Competing paradigms in qualitative research. In: Denzin, N.K. and Lincoln, Y.S. (eds), *Handbook of Qualitative Research*. Thousand Oaks, CA: Sage, pp. 105–117.

Haraway, D. (1988) Situated knowledge: the science question in feminism and the privilege of the partial perspective. *Feminist Studies*, 14, 3: 575–599.

Harding, S. (1986) *The Science Question in Feminism*. Ithaca, NY: Cornell University Press.

Hargreaves, A. (2003) *Teaching in the Knowledge Society: Education in the Age of Insecurity*. New York: Teachers' College Press.

Hart, P. and Nolan, K. (1999) A critical analysis of research in environmental education. *Studies in Science Education*, 34: 1–69.

Hart, R. (1992) *Children's Participation: From Tokenism to Citizenship – Innocenti Essays #4*. New York: UNICEF Publication.

Hart, R. (1997) *Children's Participation: The Theory and Practice of Involving Young Citizens in Community Development and Environmental Care*. London: Earthscan.

Heshusius, L. (1994) Freeing ourselves from objectivity: managing subjectivity or turning toward a participatory mode of consciousness. *Educational Researcher*, 23, 3: 15–22.

Hunter, J., Layzell, J., and Rogers, N. (1998) *School Landscapes: A Participative Approach to Design*. Winchester, UK: Hampshire County Council.

Jensen, B.B. (2000) Participation, commitment and knowledge as components of pupils' action competence. In: Jensen, B.B., Schnack, K., and Simovska, V. (eds), *Critical Environmental and Health Education Research Issues and Challenges*. Copenhagen NV: Research Centre for Environmental and Health Education, The Danish University of Education, pp. 219–238.

Jensen, B.B. and Schnack, K. (1997) The action competence approach in environmental education. *Environmental Education Research*, 3, 2: 163–178.

Kenny, K. (1996) *Grounds for Learning: A Celebration of School Site Development in Scotland*. Winchester, UK: Learning Through Landscapes.

Kirby, S.L. and McKenna, K. (1989) *Experience Research Social Change: Methods from the Margins*. Toronto, ON: Garamond Press.

Lather, P. (1991) *Getting Smart: Feminist Research Pedagogy with/in the Postmodern*. New York: Routledge.

Learning Through Landscapes (2003) *National School Grounds Survey*. Winchester, UK: Learning Through Landscapes.

Lieberman, G.A. and Hoody, L.L. (1998) *Closing the Achievement Gap: Using the Environment as an Integrated Context for Learning*. Ponway, CA: Science Wizards.

Lorde, A. (1984) *Sister Outsider: Essays and Speeches*. Trumansberg, NY: Crossing Press.

Lousley, C. (1999) (De)Politicizing the environmental club: environmental discourses and the culture of schooling. *Environmental Education Research*, 5, 3: 293–304.

Malone, K. (1999) Environmental education researchers as environmental activists. *Environmental Education Research*, 5, 2: 163–177.

Mannion, G. (2003) Children's participation in school grounds developments: creating a place for education that promotes social inclusion. *International Journal of Inclusive Education*, 7, 2: 175–192.

McKenzie, M. (2004) The 'willful contradiction' of poststructural socio-ecological education. *Canadian Journal of Environmental Education*, 8: 177–190.

Moore, R.C. and Wong, H.H. (1997) *Natural Learning: The Life History of an Environmental Schoolyard*. Berkeley, CA: MIG Communications.

Rickinson, M., Sanders, D., Chillman, B., Doyle, P., and Jameson, N. (2003) *Grounds for Improvement Secondary Action Research Programme: Year 2 Report*. Hampshire, UK: Learning Through Landscapes.

Robottom, I. and Sauve, L. (2003) Reflecting on participatory research in environmental education: some issues for methodology. *Canadian Journal of Environmental Education*, 8: 111–128.

Russell, C.L. (2003) Minding the gap between methodological desires and practices. In: Hodson, D. (ed), *OISE Papers in STSE Education*, Volume 4. Toronto, ON: Imperial Oil Centre for Studies in Science, Mathematics, and Technology Education, pp. 485–504.

Simovska, V. (2000) Exploring student participation within health education and health promoting schools. In: Jensen, B.B., Schnack, K., and Simovska, V. (eds), *Critical Environmental and Health Education Research Issues and Challenges*. Copenhagen NV: Research Centre for Environmental and Health Education, The Danish University of Education, pp. 29–43.

Titman, W. (1994) *Special Places, Special People: The Hidden Curriculum of Schoolgrounds*. Surrey, UK: World Wildlife Fund, UK.

UNCED (1992) In *Agenda 21, the Planetary Vision Green Plan*, Chapter 36. New York: United Nations.

UNICEF (1990) *Convention on the Rights of the Child*. New York: UNICEF.

Wals, A.E.J. and Jickling, B. (2000) Process-based environmental education: seeking standards without standardizing. In: Jensen, B.B., Schnack, K., and Simovska, V. (eds), *Critical Environmental and Health Education Research Issues and Challenges*. Copenhagen NV: Research Centre for Environmental and Health Education, The Danish University of Education, pp. 127–149.

WCED (1993) *Agenda 21-Sustainable Development Action Programme: Rio Declaration on the Environment and Development*. New York: United Nations.

Weston, A. (1996) Deschooling environmental education. *Canadian Journal of Environmental Education*, 1: 35–46.

Whitehouse, H. (2001) Not greenies at school: discourses of environmental activism. *Australian Journal of Environmental Education*, 17: 71–76.

Chapter 16
Researching Participation using Critical Discourse Analysis

Debbie Heck

Keywords critical discourse analysis, citizenship education materials, discourses of citizenship, action competence, research reflexivity

16.1 Introduction

This chapter discusses my reflections as a researcher on the use of critical discourse analysis as a research approach for understanding notions of participation within education curriculum materials. It identifies my journey through the process of identifying a suitable approach to explore my research problem, and the chapter begins by presenting the research problem that I sought to explore. The remainder of the chapter has three parts. The first part charts the conceptual and theoretical frame for a research design that meets the dual challenge of being located within a critical paradigm and developing a rigorous research approach to convince bureaucrats who might ordinarily favour quantitative forms of research. The second part of the chapter explores critical discourse analysis as a research approach that met this dual challenge. The design and outcomes of the research process provide a contribution towards making the theoretical practical in terms of critical research approaches. The final part details recommendations for researchers and practitioners interested in using critical discourse analysis to explore participation, identifying possible future directions for environmental and health education research.

16.2 Background and Research Problem

In my capacity as an academic in Australia, part of my work is with professional associations such as the Australian Association for Environmental Education and the Australian Federation of Societies for the Study of Society and Environment. It

Griffith University, Gold Coast Campus, G06 3.56 PMB 50, Gold Coast Mail Centre, Queensland 9726, Australia, d.heck@griffith.edu.au

A. Reid et al. (eds.), *Participation and Learning*, 256–267.

was through that work that I became involved in a process of commenting on citizenship education materials developed by the Federal level government in Australia. As it is the States and Territories in Australia that are responsible for education and curriculum, this creates a challenge because any nationally developed material needs to be acceptable to all States and Territories, and each can have different and sometimes competing political persuasions to the Federal government funding the materials development. Given this situation, curriculum materials developed by the Federal government will tend to be adopted voluntarily by teachers in schools, that is, they are not mandatory.

'Discovering Democracy' was the largest ever curriculum materials development project established in Australia when it was announced in 1997. At the launch of the school curriculum programme by Dr. David Kemp, Federal Minister responsible for schools, the Minister highlighted the need for young people to understand the 'history and operations of Australia's system of government and institutions, and the principles that support Australian democracy' (Kemp 1997:3). It was at this meeting that I first questioned the contribution these materials might be able to make towards citizen participation. I began to reflect on the relative level of power invested with the Minister, the role of politics in terms of shaping and driving the creation of national curriculum materials, and the mechanisms that might be used to promote voluntary use or adoption of the materials in schools. It was clear to me that all of these aspects might influence the type of citizenship education that would be advocated in the materials developed and hence the approach to young citizen's participation.

Given my background in environmental education, I also hold that the environment is a central part of our everyday life and issues related to the environment should therefore be part of citizenship (Buckingham-Hatfield 1997:31). Hence, the development of national citizenship education materials should include references to the environment in its broadest sense, including the social, economic, political, and ecological. At the time I was cynical of the process for developing these materials, yet also hopeful that the type of citizenship education could more closely resemble the notion suggested by Corcoran and Pennock (1995:6):

> If we are to have an ecologically literate citizenry which can exercise effective stewardship of the planet's precious natural resources, if we are to be able, through democracy, to reduce consumption and make the changes in lifestyle which are necessary to sustain the earth's vast populations, then we need citizens who are both knowledgeable about environmental issues and skilled in participatory governance.

Indeed, increasing levels of concern that citizens have for the environment has been argued to be the spark in the development of a new social movement for environmentalism (Christoff 1996:159; Lipschutz and Mayer 1996:50). In Australia, evidence for these new social movements comes from citizens joining groups such as 'Greenpeace' to participate in protests and other forms of activism outside formal government processes (Gilbert 1997:79). Thus I felt that this curriculum materials project could offer opportunities for the development of forms of participation that would support the development of young people as environmental citizens within this context.

My concerns about the directions for the development of the curriculum were justified even further by what emerged at a series of focus group discussions held with teachers commenting on the plans for the materials. Immediately it was clear that in fact the materials were going to be very focused on a narrow definition of citizenship and citizenship education. While some of the comments from the focus groups were considered, only minor changes were evident in the plans for the materials. Thus at a very early stage in the development of the materials, the different levels of power and influence could be seen to be playing out throughout the processes of materials production.

The research problem emerged as a concern for how to identify the type of participation advocated within the citizenship education materials in a way that would challenge the current processes of this production by government agencies. Making issues of power more overt was going to be a real challenge, as I now show.

16.3 Conceptual and Theoretical Framework

One of the dilemmas faced by a researcher is to match the research with the theoretical framework it seeks to contribute towards (Gough and Reid 2000:54). This notion is a very real issue when identifying research approaches that explore participation within environmental education and health education. The theory most often associated with research related to participation is critical theory (Hart and Nolan 1999:37). It is the emancipatory and social transformation aspects of critical forms of inquiry that align so well with the aims of research into participation. Critical theory and critical research provide a mechanism to explore issues of power- and knowledge-production processes (Robottom and Hart 1993; Kincheloe and McLaren 2000). They challenges the status quo (see Hart and Nolan 1999:32) and make clear the political nature of research (Gergen and Gergen 2000). Critical research seeks to be emancipatory to forge a better understanding of the link between knowledge and actions. Sauvé and Berryman (2003:175) suggest that critical research seeks to achieve three aims: deconstruction, reconstruction, and change. However, it also offers the researcher a degree of reflexivity to identify the context within which the research has been undertaken by whom, under what conditions, and for what purpose (p. 174). It was for all of the above reasons that I felt this research needed to be based within the critical research paradigm.

The dilemma for the researcher in matching a critical research paradigm to this research was the need to deconstruct texts; identify how to reconstruct the process of the text development, dissemination, and consumption; and identify the views on participation and the power relationships that influenced the types of participation evident in the materials. The most pressing issue was to identify a form of research or inquiry that had a critical frame and that would also be considered legitimate by the other stakeholders in the research – the bureaucrats, policy makers, curriculum developers and teachers – all of whom have different levels of power and influence in the process of materials development and who value different types of research approaches and styles.

It is for the above reasons that research designs typically associated with participation in education, including cooperative inquiry, participatory action research, and action learning, were explored but were deemed not relevant in this context. Peter Reason (1994) suggests that these approaches to research have been developed from a more participative world view. However, for the purpose of addressing the research question here, these approaches to research were not going to be considered rigorous by bureaucrats who do not appreciate nor readily accept qualitative research from an interpretive paradigm, much lest a critical paradigm.

The emphasis on the analysis of the texts to identify the types of participation evident in the forms of citizenship and citizenship education promoted in the materials led to an exploration of the notion of discourse. A reading of *Discourse*, by Sara Mills (1997), provided an overview of the various approaches to discourse analysis. The outcome of this reading was the identification of critical approaches to discourse analysis that are concerned with uncovering the social effects of discourse practices and the way social practices shape discourses and create different power structures (Fairclough 1995:23). It was this contribution towards social change through a critical research paradigm that convinced me that critical discourse analysis could be the most appropriate for the exploration of notions of participation in education.

Writers such as van Dijk (1985:1) ask the question: What is the social purpose for undertaking discourse analysis? In this case it was to ensure that teachers, teacher educators, curriculum developers, government departments, and agencies could consider the hegemonic influences embedded within the curriculum materials they are involved in developing or using in classrooms. This knowledge will allow these groups to consider how this issue might be redressed in classrooms and through future curriculum development processes. Hence, the research sought to consider what discourses of citizenship were within the text and what purposes these particular discourses serve within the text.

Fairclough's (1992) approach to critical discourse analysis provides the researcher with the ability to describe and critique discourses. This is achieved through examination of texts according to three distinct phases. Phase one is a description of the text in terms of the vocabulary, grammar, and textual features evident (Fairclough 1989:109). Phase two is interpretation or analysis. This includes the examination by the researcher of the way participants develop, distribute, and use the texts (p. 141). The third and final phase is explanation. This phase involves the examination of how social practices are reproduced within society as a product of the process of text construction and use in the particular circumstance under investigation (p. 162). The three phase approach to discourse analysis translates into a three-dimensional conception of discourse (Fairclough 1992). The conception of discourse and the three phases of discourse analysis are diagrammatically represented in Figure 16.1. In summary, Fairclough (1992:73) asserts that:

> Every discoursal instance has three dimensions: it is spoken or written language *text*; it is an *interaction* between people, involving processes of producing and interpreting the text; and it is part of a piece of *social action* – and is some cases virtually the whole of it. [emphases in original]

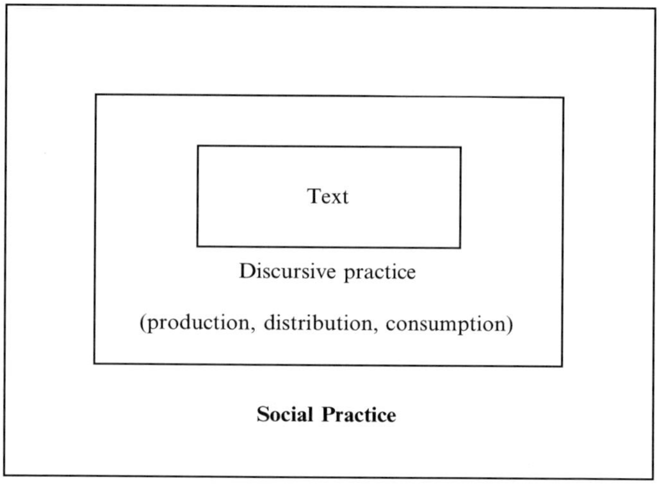

Figure 16.1 Three-dimensional conception of discourse and discourse analysis (Based on: Fairclough 1995:98)

One of the chief concerns with this approach to research was whether it would be considered credible or rigorous. The main work in the area of validation in terms of discourse analysis has been undertaken within social psychology. Potter and Wetherell (1987) have identified four techniques that can be employed within discourse analytic studies to validate the findings of such research. Three of these were relevant in this study including: coherence, participant orientation, and fruitfulness. (The fourth is 'creating new problems'.)

Coherence relates to whether the discourse analysis process selected, coherently deals with the discourse samples selected for the study (Fairclough 1992:238). Potter and Wetherell (1987) indicate that it is essential that the discourse analysis can account for patterns within the text as well as exceptions to those patterns. A researcher needs to indicate that the discourse analysis process used to examine a text provides a clear and consistent approach to analysing the text. The development of a comprehensive and well-documented text detailing the way Fairclough's critical discourse analysis approach would be actioned, was an important part of the development of coherence and credibility in research.

Participants' orientation – or what participants or those involved in the process of text, production, consumption, and distribution *see* – is an important part of validating a study of this type. Potter and Wetherell (1987:170) suggest that if research indicates through a discourse analysis that two competing discourses are at work in a setting and this is not acknowledged by the participants in this setting, to some extent the research would be considered suspicious. An interview process with participants in my research provided an opportunity for validating findings through discussions with participants who have been engaged in the process of development, distribution, and consumption of the sample texts. This is an essential part of the process to ensure rigour within critical discourse analysis.

Fruitfulness related to whether the discourse analysis process provides a useful strategy to describe, interpret, and explain the text and the discourses operating within the text. Potter and Wetherell (1987:171) ask whether the research makes 'sense of new kinds of discourse' and/or has it been able to 'generate novel explanations'. If the answers to these questions is yes then the research would be considered fruitful and it should be well respected.

In summary, the use of critical discourse analysis can generate many dilemmas for the researcher especially in terms of how to develop credible research within the critical frame. It also presents issues of how to practically achieve these outcomes. The story of the research design detailed in the next section provides an explanation of how these were addressed in my research.

16.4 Critical Discourse Analysis – the Research Design

My goal in this chapter is to illustrate and reflect on the use of critical discourse analysis as a research method for exploring those notions of participation and citizenship education evident in the federal curriculum materials, and the social processes and power relationships that allowed those particular discourses to remain dominant within the texts. Sauvé and Berryman (2003) highlight a plethora of theoretical papers about critical research and the scarcity of research reports that actually use the advocated approaches. Critical discourse analysis as a research method is very similar in that the theoretical approach is widely discussed while examples of the approach in action are very difficult to locate. It is for this reason that I draw on an example of critical discourse analysis research as part of the reflective process within this chapter.

The first phase of the critical discourse analysis made use of semiotic analysis of the texts to allow for the development of a description of ways that 'participation' was evidenced in the text. This process identified two distinct views of participation within the citizenship education materials.

Participation as legal status

The first view of participation was based on a 'legal status' view of citizenship (see Gilbert 1996; Sears and Hughes 1996). In this view citizens participate in the process of informed voting. The curriculum materials promoted the view that students needed to know how the processes of government operated, provided them with the skills and abilities to gather information about candidates, and encouraged them to make informed choices. In addition students needed to know how the process of voting operated to ensure that they were able to lodge a valid vote.

Two examples from the materials of the 'legal status' view of citizenship participation were evidenced in texts on the 'Federal System' and texts on 'Justice'.

Activities within the text on the Federal System illustrate the focus within the materials on content knowledge. These texts engaged students in drawing a diagram of how the two houses of federal parliament operate (Curriculum Corporation 1998). Students also explored Parliament House through an interactive CD-ROM 'Parliament at Work'. The CD-ROM provided information about the jobs of various people within Parliament House including elected members and employees (Media 1999). In this view of citizenship participation means that students need to be knowledgeable about how the system works so that they can vote and representatives take on the various parliamentary roles. Voting in elections is a legal requirement in Australia and this curriculum supports the view of citizens meeting this legal requirement to participate. A second example illustrates the focus on the legal consequences of law within the texts related to the topic of 'Justice'. The teaching notes suggest a range of questions the teacher can use with their class to engage students in the process of exploring a story about justice. The main purpose of this exploration is to reinforce the notion that if you do not follow the law you will be punished. This is supported by questions within the text such as (Curriculum Corporation 1998:14):

> What was the law that was broken?
> How did the punishment relate to the law that was broken?

Participation as public practice

The second view of participation advocated in the materials was a 'public practice' view (Gilbert 1996; Sears and Hughes 1996). Citizen participation in this view is acknowledged as a voluntary activity. In this view to participate provides the option of going beyond the legal requirement to vote and asserts that students need to know how to participate in government and to let those in power know about the issues and concerns of citizens. This view of participation, as a voluntary activity that allows citizens to operate within political and legal structures, is the most predominant within the school materials.

An example of evidence of the 'public practice' approach to citizen participation is found in a text group titled 'People Power' from the upper primary materials (Curriculum Corporation 1998:127–128). This text group consisted of four activities. The first is a teacher-led discussion followed by a matching activity, where students define the meanings of different citizen-action strategies. In the second activity students in small groups identify the strengths and weaknesses of using different citizen-action strategies within specific scenarios. The third activity is a largely teacher-led discussion about the consequences of citizen-led action followed by individual student report writing about a particular people-power movement they have already studied and the consequences of that movement. The final teaching and learning activity is a teacher-led discussion on the notion that not all popular movements are positive, and cites the example of Nazi Germany. Two extension activities are suggested: a class forum on people-power strategies in the present day, and the opportunity for students to explore significant individuals involved in citizen action.

The predominance of teacher-led discussions within this text denotes a focus on the development of content knowledge about how citizens can participate. This is evident in the following examples of sentence themes: 'Display the final focus question …', 'Deepen students' understanding', 'Invite students', and 'Discuss the issues that might arise …'. The cognitive processes identified within the text are frequently requested of students and relate to them learning information and content. The students are to engage in processes such as: 'identify', 'suggest', and 'think'. The focus within these texts is thinking about how other people have participated as examples of knowledge about what is possible. This emphasises the voluntary and optional nature of citizen participation within the current government structures.

The second phase of the critical discourse analysis process involved interviews with participants involved in the development of the materials. The outcome of this process was clear confirmation from participants that the two approaches to citizen participation identified in the grammatical and semiotic analysis were the ones advocated in the materials (see below for alternatives).

The final phase of the critical discourse analysis was the identification of the social processes that influenced the development of these two citizen participation discourses within the materials. As an example, one of the key social processes that influenced the development of this project was that it was funded by the government. In addition, there was bipartisan support (i.e. across cross-party political lines) for the development of an education project to engage Australians in democratic processes. These social processes strongly influenced the development of the materials and the type of citizen participation advocated in the materials. One of the participants explained the government's concern regarding the lack of engagement in democracy in Australia generally as:

> Concern over political extremism that arises out of ignorance and the awareness that many people in Australia felt alienated from their political institutions and political processes, that there was quite a sense of powerlessness amongst many citizens who felt that they didn't really know how to influence the political institutions that they have. (Interview 1:1)

The outcome of these concerns meant that the 'official knowledge' of the texts reflected the British heritage of Australia's democratic history and the formal structures and functions of government. One social process used to achieve this end was the use of a group of experts. The Civics Education Group was established and appointed by the Minister responsible for education to oversee the development of the materials while the final approval for publication rested solely with the Minister. The Civics Education Group consisted of senior academics from history and law while education was represented by senior bureaucrats from state and Catholic education systems. The view of citizen participation developed in the materials was the type those producing the materials felt the Minister and the Civics Education Group would approve. The layered approach to the approval of materials for consideration by the Civics Education Group was part of the social process that limited citizen participation to the 'legal status' and the 'public practice' approaches. The power the Civics Education Group had over the development of these materials and their content is reflected in the following example from the data:

> We, when I say we, professional associations and teachers only were on the first lot of reference groups giving feedback and found it quite difficult. ... We felt that the Civics Education Group has decided what would be in the kit and we might have some input to do a bit of the tweaking of the activities of the resources.... (Interview 8:7)

The teachers and professional associations were given a clear message in response to some of their comments and suggestions. These included, for example

> In no way are you going to chop out this whole section on charters because it is a critical part of the understanding of history of democracy. So I admire the group in that, for their passion, they were passionate about what they wanted in there.... (Interview 8:7)

This participant went on in the interview to question the role one group has in 'dictating' what should or should not be included in this particular set of materials. The example demonstrates the types of social processes that operated during the development of these materials and some of the potential reasons why the 'legal status' and 'public practice' approaches to citizen participation were so consistently represented within the materials. The message within the materials was a focus on students being informed about how to participate rather than providing opportunities for active participation. That is, the official knowledge of citizenship in Australia focused on responsibility of a 'good citizens' to be informed and the *prerogative* to decide whether or not to participate further.

16.5 Recommendations for Researchers and Future Challenges

As researchers we are faced with many competing challenges as we strive to develop a better understanding of participation. In the case of this research, critical discourse analysis proved a useful tool for uncovering and identifying the discourses evident within the text, validating the identification of those discourses and exploring the social processes that provided the context for those discourses to develop.

The challenge of this research for me as an academic was how to ensure the critical intent of the work. I can write academic journal articles to inform my colleagues; however, this does not target the groups as a researcher I began being most concerned about. One of the mechanisms I have identified for achieving this aim is to explore ways of developing the skills and abilities of teachers as practitioners who make use of curriculum materials. Teachers need to have skills that allow them to critically reflect on the process of development of curriculum materials. There is a need to be able to evaluate the materials in terms of who has developed them and how they are going to be used. I intend to develop my teaching in this way through my involvement in graduate teacher education and master's degree work in education.

In regard to the research process I suggest three main implications for further consideration by other researchers with an interest in education and participation. The first is how to identify the different types of citizen participation promoted in educational settings. In this research, critical discourse analysis was a useful research approach that provided both a broad definition of the terms in a text and a clear process for unpacking the discourse of participation evident within those texts.

The second consideration is how to identify and explain the social cultural practices that influence the development of different views of participation in education. Through these explanations educators can be made aware of how these social processes operate and provide avenues to allow social processes to be challenged. This research has illustrated that socio-cultural practices have shaped educational discourses in relation to participation. This is significant as educators seek to identify why in the case of this particular study, the 'legal status' and 'public practice' discourses of participation were evident in school materials, while other more radical approaches to participation such as 'democratic participation' were not identified at all in the materials. Uncovering why these materials developed particular views of participation challenges educators and researchers generally to understand and reflect on the impact of enshrining dominant views and silencing others.

Understanding these processes within the community of educators can enable the community to challenge the taken-for-granted assumptions and pedagogical processes developed within prepared curriculum and materials. The educational materials analysed in this research lacked a focus on approaches to citizenship education and participation that reflected ideas and pedagogy that went beyond the nation state. For example, the materials did not provide opportunities for teachers and students to engage in creative ways of problem solving, the ability to explore the need for action competence, the notion of respect for different cultural views, different concepts of power, and the role of politics as both a public and a private arena for participation. As environmental and health educators we need to challenge such materials and the curriculum settings in which they are used to ensure they make a contribution towards the development of a sustainable future.

The third consideration is how the materials are used by practitioners. While a particular discourse of participation may well be identified in the materials and those involved in the process of development might well agree that it was the view of participation those in positions of power sought to portray, the question remains does this view of participation translate into the practice of teachers? Teachers are well known as practitioners who change and adapt materials often drawing on and hybridising a range of different sources. The question is whether teachers actively resist the reading of these texts in these two main ways, and develop in their classrooms different discourses of participation. Further research into participation in education could thus explore the notion of how curriculum and materials can be transformative rather than reproductive in nature.

The future for research in participation needs to continue to strive to find mechanisms for studying participation in education through a critical frame. This is a key challenge that all environmental and health-education researchers face. Further research undertaken with regard to participation in education could include:

- Exploring different views of participation and the social processes of construction of these views within curriculum materials for a range of different disciplines, interdisciplinary fields, and cultural contexts
- Exploring how dominant views of participation in educational intuitions are interpreted by educators in a range of different educational settings
- Researching approaches that explore participation at the level of the 'curriculum in use'
- Comparing and contrasting different discourses of participation evident within curriculum materials within and between different nations to examine similar or different social processes and understandings of participation within socio-ecological and sustainability processes

References

Buckingham-Hatfield, S. (1997) Changing lives, changing spaces: environmental education and citizenship. *Nordisck Arkitekturforskning*, 10, 1: 31–38.

Christoff, P. (1996) Ecological citizens and ecologically guided democracy. In: Doherty, B. and de Geus, M. (eds), *Democracy and Green Political Thought: Sustainability, Rights and Citizenship*. London: Routledge, pp. 151–169.

Corcoran, P. and Pennock, M. (1995) Democratic education for effective environmental stewardship. *The Green Teacher*, 44, 6–10.

Curriculum Corporation (1998) *Discovering Democracy: Upper Primary Units*. Carlton, Victoria: Curriculum Corporation.

Fairclough, N. (1989) *Language and Power*. London: Longman.

Fairclough, N. (1992) *Discourse and Social Change*. Cambridge: Polity Press.

Fairclough, N. (1995) *Critical Discourse Analysis: The Critical Study of Language*. London: Longman.

Gergen, M. and Gergen, K. (2000) Qualitative inquiry: tensions and transformations. In: Denzin, N. and Lincoln, Y. (eds), *Handbook of Qualitative Research*, 2nd edn. Thousand Oaks, CA: Sage, pp. 1025–1046.

Gilbert, R. (1996) Education for active and informed citizenship. In: Gilbert, R. (ed), *Studying Society and Environment: A Handbook for Teachers*. Melbourne: Macmillan, pp. 107–124.

Gilbert, R. (1997) Issues for citizenship in the postmodern world. In: Kennedy, K. (ed), *Citizenship Education and the Modern State*. London: Falmer Press, pp. 65–81.

Gough, S. and Reid, A. (2000) Environmental education research as profession, as science, as art and as craft: implications for guidelines in qualitative research. *Environmental Education Research*, 6, 1: 47–57.

Hart, P. and Nolan, K. (1999) A critical analysis of research in environmental education. *Studies in Science Education*, 34: 1–69.

Kemp, D. (1997) *Realising the Promise of Democracy*. Paper presented at the Curriculum Corporation Fourth Annual Conference: The Curriculum Agenda 1997 and Beyond, Landmark Parkroyal Sydney.

Kincheloe, J.L. and McLaren, P. (2000) Rethinking critical theory and qualitative research. In: Denzin, N.K. and Lincoln, Y.S. (eds), *Handbook of Qualitative Research*, 2nd edn. Thousand Oaks: CA, Sage, pp. 279–313.

Lipschutz, R. and Mayer, J. (1996) *Global Civil Society and Global Environmental Governance: The Politics of Nature from Place to Planet*. Albany, NY: State University of New York.

Media, I. (1999) *Parliament at Work [CD ROM]*. Carlton, Victoria: Curriculum Corporation.

Mills, S. (1997) *Discourse*. London: Routledge.

Potter, J. and Wetherell, M. (1987) *Discourse and Social Psychology: Beyond Attitudes and Behaviour*. London: Sage Publications.

Reason, P. (1994) Three approaches to participative inquiry. In: Denzin, N.K. and Lincoln, Y.S. (eds), *Handbook of Qualitative Research*. Thousand Oaks, CA: Sage, pp. 324–40.

Robottom, I. and Hart, P. (1993) *Research in Environmental Education: Engaging the Debate*. Geelong, Victoria: Deakin University Press.

Sauvé, L. and Berryman, T. (2003) Researchers and research in environmental education: a critical review essay on Mark Rickinson's report on learners and learning. *Environmental Education Research*, 9, 2: 167–180.

Sears, A. and Hughes, A. (1996) Citizenship education and current educational reform. *Canadian Journal of Education*, 21, 2: 123–142.

van Dijk, T.A. (1985) Introduction: the role of discourse analysis in society. In: van Dijk, T.A. (ed), *Handbook of Discourse Analysis. Volume 4: Discourse Analysis in Society*. London: Appleton Century Crofts, pp. 1–8.

Chapter 17
Youth Participation in Local Environmental Action: An Avenue for Science and Civic Learning?

Tania M. Schusler[1] **and Marianne E. Krasny**[2]

Keywords environmental action, youth, participation, science education, civic engagement

17.1 Introduction

This chapter considers environmental action as an avenue for developing young people's capabilities for democratic participation as scientifically literate citizens. From the literature, we describe parallels between civic education conceptualised as civic engagement and science education approached as inquiry-based learning. We suggest six guiding principles for youth participation in local environmental action: *youth as contributors, genuine participation, deliberate action, inquiry, critical reflection*, and *positive youth development*. We illustrate these principles by applying them to local and national environmental programmes in the USA. The engagement of individual youth and the depth of their learning may vary widely even when programmes incorporate the guiding principles. We suggest future research directions around youth experiences, guiding principles, educator practices, participant characteristics, educational setting, impacts on adults and community, and culture. Such research will enhance understanding of environmental action and its contribution to science and civic learning.

17.2 Background

Environmental education practice in the USA often focuses on promoting personal responsibility and environmentally conscious individual lifestyle choices. However, it does not always adequately address the economic and political structures that limit

[1]Department of Natural Resources, Cornell University, Fernow Hall, Ithaca, NY 14853–3001, USA, tms23@cornell.edu

[2]Department of Natural Resources, Cornell University, Ithaca, NY 14853, USA, mek2@cornell.edu

A. Reid et al. (eds.), *Participation and Learning*, 268–284.

the freedom of individuals to make those choices. For example, one's freedom to buy local products or eat organic foods is restricted by the globalisation and industrialisation of agriculture – particularly if one cannot afford the luxury of paying more for environmentally responsible products. Environmental education in the USA could benefit from a political economy approach (Youngman 2000) that better addresses the political and economic structures within which individual actions take place. In this chapter, we consider how such an approach involving participation in local environmentally focused actions can integrate science education and civic engagement.

Emmons (1997:35) describes environmental action as 'a deliberate strategy that involves decisions, planning, implementation, and reflection by an individual or group. The action is also intended to achieve a specific positive environmental outcome, either small or large.' To improve access to locally produced fresh foods, for example, people might create a farmers' market, farm-to-school lunch programme, or community-supported agriculture. Participation in such local environmental action, which occurs at the intersection of ecological, economic, social, and political systems (Dryzek 1997), will also provide opportunities for integrating science and civic education, particularly if such efforts are based on both citizens' interests and sound science (e.g. a farm-to-school lunch programme that incorporates scientifically based nutritional guidelines). Such an approach is consistent with environmental education guidelines that emphasise knowledge and skills in both science and citizenship (NAAEE 2004). Youth grappling with environmental issues may develop understandings of environmental science and political processes, and skills in scientific inquiry and civic engagement, all of which are crucial to participation in a democratic society.

In this chapter, we illustrate examples of the extent to which science and civic education can operate in concert rather than conflict. Drawing on literature on civic engagement, science education, and youth development, we set out six guiding principles for youth participation in local environmental action. Following this, we apply the principles in a commentary on current environmental action programmes in the USA. The chapter concludes with suggestions for further research, the results of which, we suggest, could aid educators in the practice of engaging youth in action to improve their local environment and provide opportunities for science and civic learning.

17.3 Integrating Science and Civic Education

Civic education is a complex enterprise involving a variety of cognitive, conceptual, and attitudinal strands (Torney-Purta *et al.* 2001). Approaches to civic education vary immensely, in part reflecting fundamental differences in their conceptualisation of what it means to be a 'good' citizen (see, for example, Gibson 2001; Battistoni 2002; Youniss *et al.* 2002; Kirlin 2003). We conceive of civic education in relation to environmental action along the lines of a growing body of literature on youth civic engagement. Camino and Zeldin (2002:214) define civic engagement as 'being able to influence choices in collective action' and they recognise that citizen engagement – long a bedrock of democracy – is the purview of every citizen, not simply officials and professionals. Skelton and colleagues (2002:9) offer a definition of *youth civic engagement* as 'young citizens developing civic skills and habits as they actively shape democratic society in collaboration with others'. Gibson (2001) suggests a broad conception of youth civic engagement that includes a wide range of indicators, such as: voting; knowledge and understanding of political processes; development of attitudes supporting democratic practices; critical thinking skills; ability to use information sources, including the news media; interaction and deliberation skills; and participation in civic activities like volunteering, service, or fund-raising for local causes (cf. Chapter 16 by Heck, this volume). Examples of pathways for youth civic engagement include public policy consultation on youth issues, community coalitions for youth development, youth infusion in organisational decision-making, youth organising, and school-based service learning (Camino and Zeldin 2002).

Youth participation in local environmental action reflects civic education based in the traditions of participatory democracy, public work, and social justice (Battistoni 2002) because it includes youth directly in democratic processes. It can also involve collective action towards some public purpose (e.g. creating a community garden, changing local policy to protect water quality), and can address the root causes of problems. Through civic engagement young people can develop understanding of civic concepts and gain civic skills, including those related to political knowledge, critical thinking, communication, public problem solving, civic judgment, civic imagination and creativity, community/coalition building, and organisational analysis (Battistoni 2002). This learning, in turn, can increase young people's ability to exert influence in public affairs (Newmann 1975) through enabling them to play an informed and active role in the political systems of power and decision-making (Fien 1993), to make choices rather than accept the prescriptions of others (Freire 1973), to hold experts to account, and to insert their own knowledge into the public discourse (Fischer 2000).

To exert influence in contemporary, science-laden public policy discussions, however, also requires some familiarity with science. In public debates ranging from regulation of genetically modified organisms to attempts to mitigate global climate change, crucial normative assumptions are often buried in technical analyses with little opportunity to question or examine the science itself (Fischer 2000). Scientifically literate citizens may have the ability to assess the value of knowledge in a particular context and to participate in the social negotiations that produce knowledge (Roth and Désautels 2004). They may also be capable of critically evaluating the scientific evidence touted by politicians, corporations, or environmental organisations, and other interest groups. Scientific literacy can provide individuals with greater control over their lives by enabling them to make better-informed personal decisions (e.g. health care, nutrition, risk acceptance); the capacity to participate in science-laden policy debates at local (e.g. drinking water quality), national (e.g. regulation of genetically modified organisms), and international (e.g. global warming) scales; and the desire to realise broader economic and personally rewarding opportunities, through often well-compensated and stimulating scientific and technological careers. Scientific literacy is one form of knowledge among many that increases the resources (i.e. knowledge, skills, strategies, understandings) upon which one can draw to participate in public life.

In examining how science education can contribute to civic literacy, it becomes obvious that the scientifically literate citizen must understand more than just scientific concepts and facts. The science education reform movement in the USA emphasises the importance of inquiry-based learning to developing scientific literacy. According to the US *National Science Education Standards* (NRC 1996:2):

> Inquiry is central to science learning. When engaging in inquiry, students describe objects and events, ask questions, construct explanations, test those explanations against current scientific knowledge, and communicate their ideas to others. They identify their assumptions, use critical and logical thinking, and consider alternative explanations.

This approach to science learning involves some of the same skills that are required for civic participation, including problem solving, planning, decision-making, and discussion with peers. While to participate in public debates on issues involving science, students also must have an understanding of the 'Nature of Science' – that science is tentative, empirically based, subjective, necessarily involves human inference, imagination, and creativity, and is socially and culturally embedded (Lederman 1998).

The extent to which science education is in concert or conflict with civic education depends to a large extent on how one conceptualises both. Conceiving science as a set body of facts to be assimilated and civic education as learning about governmental structure and the electoral process offers little opportunity for integration. In addition, science is often misleadingly portrayed as value-free and apolitical: characteristics counter to civic participation. Yet, scientific practice involves several characteristics of genuine civic engagement, including questioning assumptions, understanding cause and effect relationships, considering alternative explanations, and debating critically within a community. In short, both science and civic education emphasise critical thinking. Whether in discovering what factors affect the water quality of a stream, or analysing how racial discrimination affects educational and economic opportunities, the habit of asking critical questions is an essential dimension of both science and civic education.

In *Democracy and Education*, Dewey equated thinking with inquiry and explained, 'We sometimes talk as if "original research" were a peculiar prerogative of scientists or at least of advanced students. But all thinking is research, and all research is native, original, with him who carries it on' (Dewey 1916:148). Fortunately, the narrow conceptions of science as a body of facts and civics as fulfilling one's obligation to vote do not reflect the potential for science and civic education to work together. An area of overlap exists between science and civic education; thus, programmes can be designed to foster both science and civic learning among youth.

17.4 Principles of Youth Environmental Action

In this chapter, we suggest six principles that might guide thinking and future research about youth participation in local environmental action that are consistent with providing opportunities for science and civic learning. These principles are:

youth as contributors, genuine participation, deliberate action, inquiry, critical reflection, and positive youth development.

Youth as contributors

Young people have both the right and responsibility to participate in decisions affecting their environment and are capable of making valuable contributions to their communities and society (Hart 1997; de Winter 1997; Eames-Sheavly 1999). Skelton and colleagues (2002:6) explained that young people are best seen not as future citizens but 'as co-creators of a thriving democracy and of the healthy civic practices of the environments in which they live, work, play and learn'. Youth participation in environmental action reflects a fundamentally different relationship between young people and adults – one that requires a sharing of power, for example – than that typically prevalent in our schools, youth programmes, and communities. Youth participation may also influence adult perceptions of youth. Research conducted in the context of youth governance has demonstrated that youth voice in organisational decision-making can positively influence adults and organisations (Zeldin *et al.* 2000).

Genuine participation

Participation in environmental action provides opportunities to experience democracy in authentic situations where youth can contribute and influence outcomes. People learn to participate in a democracy through the *exercise* of democracy, for that knowledge, as Freire (1973) stated, can only be assimilated experientially. At the heart of democratic processes, participation occurs in many forms with varying degrees of influence exerted by participants. Some seemingly participatory processes are deceptive. Decoration, tokenism, and manipulation do not meaningfully involve youth but rather advance predetermined adult agendas (Hart 1997). Genuine forms of participation, such as consultation and shared decision-making, are distinguished by honesty and clarity about the extent of young people's power and the opportunity for youth to choose to participate to the maximum of their ability and interest (Hart 1997). Through participation, youth can learn civic concepts (such as decision-making structures) and skills (such as communicating and negotiating) that increase their ability to influence public affairs.

Deliberate action

Schnack (1994:190, in Simovska 2000:30) defines *action competence* as the 'capability – based on critical thinking and incomplete knowledge – to involve yourself as a person with other persons in responsible actions and counter-actions for a more humane world'. Two key distinctions between environmental action and activity are intentionality and targeting the root causes of a problem (Jensen and Schnack 1997). Of these, we find intentionality most important in distinguishing action from activity. For example, youth participation in an activity initiated and organised by adults, such as an environmental clean-up, while beneficial would not necessarily constitute action because it lacks deliberate choice or intent of the young people involved. Jensen and Schnack (1997) also argue that such a clean-up would not constitute action because it focuses on symptoms (e.g. removing trash and debris) rather than causes of environmental degradation. Actions that do not directly address root causes have the potential, however, to contribute indirectly to solving environmental problems (Bishop and Scott 1998). For example, a clean-up initiated by youth might draw public attention to the issue of littering or illegal dumping, which might lead a community to consider other actions to eliminate these sources of degradation. When youth take action to effect change, they can acquire skills related to planning, public speaking, fund-raising, and organising community support, as well as learn about civic-related concepts such as public purpose and power. Regardless of whether or not their efforts are successful, engaging in collective action enables youth to think critically about the kind of world they want to live in. It also can enhance their understanding of social, economic, and political systems as they identify opportunities for and obstacles to realising their vision.

Inquiry

Fusco and Barton (2001) and Roth and Lee (2004) view youth as potentially active producers and creators of scientific knowledge that contributes to community action in collaboration with peers, educators, and community members. Because it occurs at the interface of natural and social systems, young people's environmental research can involve a suite of quantitative and qualitative research methods ranging from water quality and soil analysis to interviews and participatory mapping (Doyle and Krasny 2003). Youth can engage in multiple aspects of the research process, including defining research questions, collecting and analysing data, interpreting results, and communicating conclusions. Thus, the research dimension can provide opportunities for youth to learn both science concepts (e.g. non-point source pollution, epidemiology of lead poisoning, changes in land use over time) and skills (e.g. aerial photo and map interpretation, Geographic Information Systems, interviewing, document analysis, synthesising, and communicating results) (Mordock and Krasny 2001).

Critical reflection

Reflection – thinking about what one is doing to more fully understand its meaning – is essential to both science and civic education. Lederman (1998) has demonstrated that students do not implicitly learn about the Nature of Science and scientific inquiry simply by doing science. Such understanding is better facilitated through an 'explicit reflective approach' in which the educator explicitly points out aspects of the Nature of Science and scientific inquiry highlighted by students' experiences, and encouraging students to reflect on the implications that such aspects have for the way they view scientists, scientific knowledge, and the practice of science (Lederman 1998). Similarly, political knowledge and civic skills do not automatically develop from the experience of civic engagement or community service (Battistoni 2002). In this context, one particular approach for encouraging critical reflection on civic engagement to consider is 'conceptual organising'. This involves the explicit introduction of political ideas (e.g. citizenship, democracy, freedom, public life, power, self-interest, leadership, diversity, accountability) to challenge youth to reflect on and draw meaning from their actions, consider the broader implications of their work, and situate it in a larger public sphere (Boyte *et al.* 1999).

Positive youth development

Following a comprehensive review of existing studies, the National Research Council (NRC) in the USA compiled the characteristics of settings that promote positive youth development. These include physical and psychological safety, appropriate structure, supportive relationships, opportunities to belong, positive social norms, support for efficacy and mattering, opportunities for skill building, and integration of family, school, and community efforts. The NRC also identified 28 personal and social assets that facilitate positive physical, intellectual, emotional, and social development for youth. Participation in youth environmental action can foster many of these, including critical thinking and reasoning skills, good decision-making skills, confidence in one's personal efficacy, optimism coupled with realism, connectedness or perceived good relationships and trust with peers and adults, and commitment to civic engagement (Eccles and Gootman 2002).

17.5 Applying Principles of Youth Environmental Action to Youth Programmes

A growing number of local projects and state or national programmes in the USA involve youth in action to enhance their local environment (Table 17.1). Many of these programmes also reflect the guiding principles described earlier. The programmes' goals are ambitious and their anecdotal success stories can be inspirational.

In the following section, the projects are considered in more detail in terms of how they relate to the six principles outlined above, starting with the *Earth Force* programme, and then the Seneca Falls Landfill Project and the Garden Mosaic Programme, whose approaches are compared in Table 17.2.

Table 17.1 Examples of environmental action programmes in the USA

Earth Force operates programmes in approximately 100 school and community-based organisations primarily in eight US metropolitan areas. It is a non-profit, national civic participation and service learning programme in environmental education, designed to teach middle-school-aged youth the knowledge, skills, and attitudes necessary to becoming active citizens influencing environmental decisions in their communities (www.earthforce.org).

Garden Mosaics is an informal science education and community action programme designed to connect youth and elders in investigating the mosaic of plants, people, and cultures in gardens, in learning about science concepts and practices, and in acting together to enhance their community. Operating in cities across the USA and several international sites, the programme involves youth aged 10–18 from diverse ethnic, cultural, and economic backgrounds in activities that take place in urban community gardens, as well as in home and school gardens. Participants conduct investigations and action projects and then share the results of their efforts on the programme web site (www.gardenmosaics.org). Garden Mosaics is funded by a grant from the National Science Foundation (NSF) Informal Science Education programme, to the Cornell University Department of Natural Resources in Ithaca, NY.

National Public Radio's weekly environmental news programme, *Living On Earth*, engages middle and high school students in science-based explorations of their local environment, and in production of original audio journalism about their findings that airs over local and national radio (www.loe.org/edu). Major funding for the Living On Earth Ecological Literacy Project is provided by the National Science Foundation and the Corporation for Public Broadcasting.

During the pilot phase of the *Project Wild Science and Civics: Sustaining Wildlife* programme, students planned and constructed school and community wildlife habitats, organised a cross-cultural water festival along the Rio Grande River, and studied the impact of airport expansion on wildlife (www.projectwild.org/ScienceandCivics.htm). Project WILD was established in 1983, and is administered by the Council for Environmental Education (CEE) and is co-sponsored by the Western Association of Fish and Wildlife Agencies (WAFWA).

High school students participating in *Shaw EcoVillage's Ecodesign Corps* in Washington, DC have developed a mural celebrating Shaw activists; protected and restored trees; identified grant opportunities for local businesses; educated workers about environmental safety; and created designs for neighbourhood development oriented around public transit (www.shawecovillage.org). Shaw EcoVillage, founded in 1998, is a 501(c)3 organisation whose mission is to develop youth leaders to be catalyst for sustainable change in our urban neighbourhoods.

Teenagers participating in *Urban Community Action Planning for Teenagers* in Worcester, MA, led planning, developed partnerships, and raised funds to renovate a neglected neighbourhood park. Another group engaged in research to document rubbish (from paper litter to sofas and car parts) in their neighbourhood, assessed people's perceptions of the problem, and proposed possible solutions (Ross and Coleman 2000) (www2.clarku.edu/departments/idce/cp/research/research.shtml). UCAP is a participatory, systematic community development approach, adapted from Participatory Rural Appraisal (PRA), developed by Ross and Coleman.

Earth Force

Evaluation results from the *Earth Force* programme (www.earthforce.org) lend evidence to the importance of several of the guiding principles. *Earth Force* incorporates a six-part problem-solving process that guides young people in assessing their local environment; selecting an issue for further study; analysing relevant public policy and community practices; identifying options for affecting change; and developing, implementing, and evaluating an action project. Pre- and post-surveys

Table 17.2 Examples of how principles of youth environmental action were applied in the Seneca Falls Landfill Project and Garden Mosaics programme

	Related student (or adult as noted) activities	
Principle	Seneca Falls Landfill Project	Garden Mosaics
Youth as contributors	Creating opportunities for community learning through panel discussion	Contributing to online databases used for educational, scientific, and community development purposes
	Contributing to research results on student attitudes through presentations to community groups	
Genuine participation	Debating and agreeing on a process for selecting a community-based research project	Working with adults in the garden and community to define and carry out an action project (e.g. reclaiming a vacant lot to create a new community garden)
	Collectively developing a mission statement and timeline for achieving it	
	Soliciting school board support for project and funding for field trips	
	Working in teams to plan and conduct project tasks, such as preparing a press release, inviting guest speakers, and developing informative posters for display at panel discussion	
Deliberate action	Initiating ideas for action (e.g. panel discussion of community experts, survey of student attitudes) and bringing them to fruition	Initiating ideas for action (e.g. enhancing an existing garden through an art project, building a compost bin) and bringing them to fruition
Inquiry	Conducting library and online research about landfills	Brainstorming research questions
	Asking own questions of community experts	Conducting interviews and observations and taking measurements in gardens and neighbourhoods
	Designing and implementing survey of students' attitudes	
	Debating interpretations of survey results	Compiling results into reports or on datasheets
	Presenting project to peers at a multi-school research congress	
	Communicating results to community groups	Discussing the implications of results

(continued)

Table 17.2 (continued)

	Related student (or adult as noted) activities	
Principle	Seneca Falls Landfill Project	Garden Mosaics
		Reporting results to online databases
		Participating in an international community of youth and adults conducting research about people and practices in community and urban gardens
Critical reflection	Participating in class discussions reflecting on what it means to be a community member and how social science affects people's lives	Participating in discussions of what assets community gardens provide to communities and where else in their neighbourhood such assets can be found
	Writing in response to questions posed in journal assignments	Reflecting on the processes and outcomes of action projects
Positive youth development	Educators provide:	Settings, programme materials, and educators, gardeners, and other adults provide:
	Physical and psychological safety	Physical and psychological safety
	Appropriate structure	Appropriate structure
	Supportive relationships	Supportive relationships
	Opportunities to belong	Opportunities to belong
	Positive social norms	Positive social norms
	Support for efficacy and mattering	
	Opportunities for skill building	Support for efficacy and mattering
	Integration of school and community efforts	Opportunities for skill building
		Integration of school and community efforts

used in the evaluation of the 2001–2002 programme found statistically significant increases in participants' civic skills, including knowing where to find information, how to contact adults for information, what it takes to change rules and laws, how to work with others, and acting in ways to protect the environment over the long term. The evaluation also found statistically significant declines on several attitudinal measures, including:

- Belief that each person should do what he or she can to protect the environment
- Commitment to working on environmental issues now and later in life
- Belief that the participant personally can make a difference
- Belief that people working together can solve community problems

- Belief that it is important to listen to people on all sides of an issue
- Belief in the importance of finding long-term solutions
- Attention paid to environmental issues (Melchior and Bailis 2003)

A further examination of these results reveals that impacts differ according to whether or not participants had actually conducted an action project. Youth at sites that involved them in such projects showed more positive civic attitudes than those that did not. Furthermore, programme duration had a strong effect on results, with participants in longer duration programmes (lasting more than 18 weeks) showing substantially more positive impacts and almost none of the negative impacts appearing in the overall analysis. In contrast, participants in shorter-duration programmes showed declines on a wide range of measures and almost no positive impacts (Melchior and Bailis 2003). Long-duration programmes might provide participants with more time to participate genuinely, engage in deliberative action that is successful in creating change (or understand why their actions were unsuccessful and learn from that experience in a way that is empowering rather than demoralising), contribute in meaningful ways, and critically reflect on their experiences.

Seneca Falls Landfill Project

The *Seneca Falls Landfill Project* provides a second example of how the guiding principles can be applied to designing a youth environmental action project (Table 17.2). This project involved 60 high school biology students, aged 14–16 and of mixed academic abilities, in an upstate New York school classroom (Tompkins 2005). The students worked collectively in defining the following mission: 'To learn and share how Seneca Meadows Landfill affects our community and others beyond it'. To realise their mission, students gathered information on environmental, economic, and social impacts of the landfill on their community, took a field trip to the landfill and to a National Wildlife Refuge downstream of the landfill, invited guest speakers into the classroom, and developed posters, fact sheets, and press releases to share with the community. The students also initiated and organised a panel discussion so that other students and community members could learn about the landfill. Panel members included a landfill representative, environmentalist, community educator, town supervisor, and engineer from the state regulatory agency. Finally, recognising that youth voices were largely absent from discussions around the future of the landfill, the students conducted a survey of their peers' knowledge and opinions about the landfill and shared their results in public presentations to community groups.

Results from the Landfill Project show that even when one applies the principles of youth environmental action, the engagement of individual students and depth of their learning may vary widely. A selection of student responses to the question: 'What did you learn by participating in this project?' is illustrative:

> In this project, I learned that young people *can* make a difference. We have really proved this in our project. I've also learned how poorly our society handles issues like garbage – more should be done to recycle.
>
> I learned a lot about our town as a whole and how little or much they care. I got to see group work finally pay off. And specifics about landfills – leachate, et cetera.
>
> During this project I learned many new things! I learned about how to research people's opinions and thoughts. Sometimes it was frustrating, but in the end I was very proud of what our class stuck through and accomplished.
>
> This year, I've learned about the real scientific method. It takes a lot of work to be educated, but it's worth it. Now, I'm glad I know more about the environment and people of our community. I also learned that science is everything: social, biochemical, et cetera. This year was a great learning experience!
>
> Not much. The reason why was because this project was boring.
>
> It really didn't interest me. If I'm not interested, I don't pay attention.
>
> I didn't learn anything.

What caused some students to participate enthusiastically and view the project as a valuable learning experience, while others chose to participate as little as possible and claimed to have learned little through the process? One explanation might lie in students' motivation for participation. Some students were genuinely interested in the project, while others participated reluctantly because it was a required part of their coursework. Thus, designing programmes that take into account student backgrounds and motivations can be key to programme success.

Garden Mosaics

Garden Mosaics (www.gardenmosaics.org) is a national (and increasingly international) programme that integrates civic action, multicultural understanding, and science learning (Table 17.2). Youth participating in *Garden Mosaics* conduct investigations focusing on the role of community gardens in their neighbourhood, the connection of gardeners' planting practices to their cultural heritage, and urban weed control. The youth then report the results of these investigations to online databases, which are used for a number of purposes. For example, results from the Community Garden Inventory are used by the American Community Gardening Association to build a case for the importance of these urban settings for community development and food security (Saldivar-Tanaka and Krasny 2004). Results from the Weed Watch investigation are used to help a Cornell University agronomist develop an environmentally sound urban weed control programme. Stories about gardeners' planting practices and cultures are posted online where they serve as an educational tool for web site visitors, including youth, gardeners, and others interested in plants and people. Through engaging youth in collecting data that are used for political, scientific, and educational purposes, *Garden Mosaics* incorporates inquiry, youth as contributors, and positive youth development principles.

In addition to posting the results of their investigations online, *Garden Mosaics* youth have also used their interviews and observations to help define an action project that benefits the garden and wider community. To illustrate, in the process of talking with gardeners, youth in Sacramento, California learned that the available plots in a neighbouring community garden did not meet the demand for gardening space among community members. The youth worked with a landscape architecture student to design a community garden space adjacent to their school garden. Youth participating in *Garden Mosaics* write a report on their action projects, including a section on their reflections about the work they accomplished, and post them on the *Garden Mosaics* web site. Thus, the *Garden Mosaics* action projects address issues related to genuine participation, deliberate action, and critical reflection.

It should be noted that even though a programme addresses the guiding principles, its success may still be variable. Similar to what was observed in the Seneca Falls Landfill Project, youth in some *Garden Mosaics* projects never developed interest in the activities, whereas youth in other groups became very engaged and experienced significant outcomes in terms of developing positive relationships with elders, learning about gardens, plants, and gardeners, and conducting a project to benefit their community. Although educator skills and enthusiasm may explain some of this variability, even the same educator may have very different results depending on the group of youth. In describing a programme in Allentown, Pennsylvania that went well, the programme evaluator wrote:

> The Master Gardener who led the programme was particularly enthusiastic about the youth-elder connections during the programme; he brought in 'expert' visitors on a weekly basis. The youth interviewed these visitors, documented what they learned through photographs and posters, and applied their knowledge to the care of the garden. Youth learned about planting practices, pest management, and water management, among other topics. … At the end of the season, youth demonstrated their newly gained expertise during a game of 'Garden Jeopardy' (in which every question was correctly answered!) and through the creation of a summary poster of their experiences; they also took an observer on a successful Garden Tour and conversed easily about the garden. The evaluation team feels that the Allentown site illustrated the potential of the Garden Mosaics programme. When the programme is thoughtfully implemented with an appropriate audience, enthusiastic youth learn about numerous aspects of garden science through interactions with elders; youth are then able to apply that knowledge in a variety of ways (S. Thompson, unpublished Garden Mosaics project report).

17.6 Future Research

The three programmes, all of which incorporate principles of youth environmental action, illustrate that participating in local environmental action can contribute to positive learning experiences for some youth but not for others. Research can illuminate the additional forces (e.g. educator practices, youth motivation, curriculum, funding, institutional support) that influence the impact of youth environmental action on participants. We propose several areas for future research:

Science and civic education

How does participating in local environmental action influence young people's views of citizenship and civic engagement, themselves in relation to their community, and science and its relevance to their lives? What civic and scientific skills do youth gain, and through what experiences do they develop those skills?

Principles of youth environmental action

How are the guiding principles exhibited in youth environmental action programmes occurring in diverse contexts? What additional principles apply? How does each of the guiding principles relate to programme success?

Educator practices

How do specific educator practices create opportunities for youth participation in local environmental action? Some educators seem to be 'naturals' in developing ownership, empowerment, and leadership skills among youth but little systematic analysis has addressed how they do so. Researchers can contribute to building and transmitting an accepted body of knowledge about practice by investigating what educators who successfully facilitate youth participation in environmental action actually do, and how they are trained and supported in doing it (P/PV 2000).

Participant characteristics

How do youth background, motivation, and demographics interact with programme design, educator characteristics, and other factors to determine impacts for individual youth?

Educational setting

How does the educational setting (e.g. formal classroom vs. non-formal community, computer resources available) in which youth environmental action occurs influence young people's learning experience?

Impacts on adults and community

How does youth environmental action benefit communities (e.g. through new green spaces or new policies)? How does it specifically impact adults and other members of the community (e.g. through adults changing their attitudes about youth)?

Culture

To what extent is youth participation in local environmental action relevant in different cultures? How are the guiding principles applied in programmes from different countries? How does multicultural education inform the theory and practice of youth participation in local environmental action?

Opportunities for integrating science and civic education differ significantly in countries throughout the world. For example, with its tradition of local participatory decision-making, Denmark has been at the forefront of research and practice focusing on the value and efficacy of 'action competence' and related civic action approaches to science education (see Svedbom 1994). Further afield, educators from many countries are currently developing concerted efforts under the banner of the UN *Decade of Education for Sustainable Development* (2005–2014), part of which focuses on environmental sciences education within a social action and equity context. Collaboration with international colleagues can help address the research questions outlined above, fostering learning with and from each other, and the creation of youth programmes that cross national borders. Such international collaboration will provide us with a broader perspective on the principles of youth environmental action, how to improve the practice of engaging young people in action to enhance their environment, and how participation in local environmental action contributes to science and civic learning that enhances one's ability to participate in democratic society.

References

Battistoni, R. (2002) *Civic Engagement Across the Curriculum: A Resource Book for Service-learning Faculty in All Disciplines*. Campus Compact. Providence, RI: Brown University.

Bishop, K. and Scott, W. (1998) Deconstructing action competence: developing a case for a more scientifically attentive environmental education. *Public Understanding of Science*, 7, 3: 225–236.

Boyte, H.C., Kari, N.N., Lewis, J., Skelton, N., and O'Donoghue, J. (1999) *Creating the Commonwealth*. Dayton, OH: Kettering Foundation.

Camino, L. and Zeldin, S. (2002) From periphery to centre: pathways for youth civic engagement in the day-to-day life of communities. *Applied Developmental Science*, 6, 4: 213–220.

de Winter, M. (1997) *Children and Fellow Citizens: Participation and Commitment*. Oxford: Radcliffe Medical Press.

Dewey, J. (1916) *Democracy and Education: An Introduction to the Philosophy of Education*. Toronto, ON: Collier-Macmillan.

Doyle, R. and Krasny. M. (2003) Participatory rural appraisal as an approach to environmental education in urban community gardens. *Environmental Education Research*, 9, 1: 91–115.

Dryzek, J.S. (1997) *The Politics of the Earth: Environmental Discourses*. Cambridge: Oxford University Press.

Eames-Sheavly, M.L. (1999) *Children and the Environmental Aspects of Community Development*. Project report. Ithaca, NY: Cornell University.

Eccles, J. and Gootman, J.A. (eds) (2002) *Community Programs to Promote Youth Development*. Committee on Community-level Programs for Youth, Board on Children, Youth, and Families, Division of Behavioral and Social Sciences and Education, National Research Council and Institute of Medicine. Washington, DC: National Academy Press.

Emmons, K.M. (1997) Perspectives on environmental action: reflection and revision through practical experience. *Journal of Environmental Education*, 29, 1: 34–44.

Fien, J. (1993) *Education for the Environment: Critical Curriculum Theorising and Environmental Education*. Geelong, Victoria: Deakin University Press.

Fischer, F. (2000) *Citizens, Experts, and the Environment: The Politics of Local Knowledge*. Durham, NC: Duke University Press.

Freire, P. (1973) *Education for Critical Consciousness*. New York: Continuum.

Fusco, D. and Barton, A. (2001) Creating relevant science through urban planning and gardening. *Journal of Research in Science Teaching*, 38, 8: 860–877.

Gibson, C. 2001. *From Inspiration to Participation: A Review of Perspectives on Youth Civic Engagement*. Berkeley, CA: The Grantmaker Forum on Community and National Service.

Hart, R.A. (1997) *Children's Participation: The Theory and Practice of Involving Young Citizens in Community Development and Environmental Care*. London: Earthscan.

Jensen, B.B. and Schnack, K. (1997) The action competence approach in environmental education. *Environmental Education Research*, 3, 2: 163–178.

Kirlin, M. (2003) *The Role of Civic Skills in Fostering Civic Engagement*. College Park, MD: CIRCLE (Center for Information and Research on Civic Learning and Engagement), University of Maryland School of Public Affairs.

Lederman, N.G. (1998) The state of science education: subject matter without context. *Electronic Journal of Science Education*, 3, 2. Reno, NV: University of Nevada. http://unr.edu/homepage/jcannon/ejse/lederman.html.

Melchior, A. and Bailis, L.N. (2003) *2001–2002 Earth Force Evaluation: Programme Implementation and Impacts*. Waltham, MA: Center for Youth and Communities, Heller Graduate School, Brandeis University.

Mordock, K. and Krasny, M.E. (2001) Participatory action research: a theoretical and practical framework for environmental education. *Journal of Environmental Education*, 32, 3: 15–20.

NAAEE (North American Association for Environmental Education) (2004) *Excellence in Environmental Education Guidelines for Learning (K-12)*. Washington, DC: NAAEE. http://naaee.org/pages/npeee/learner_guidelines.html.

NRC (National Research Council) (1996) *National Science Education Standards*. Washington, DC: National Academy Press.

Newmann, F.M. (1975) *Education for Citizen Action: Challenge for Secondary Curriculum*. Berkeley, CA: McCutchan Publishing.

P/PV (Public/Private Ventures) (2000) *Youth Development: Issues, Challenges and Directions*. Philadelphia, PA: P/PV.

Ross, L. and Coleman, M. (2000) Urban youth as community planners and leaders: exploring their potential with Urban Community Action Planning for Teenagers (UCAPT). *PLA Notes*, 38: 25–28.

Roth, W.M. and Désautels, J. (2004) Educating for citizenship: reappraising the role of science education. *Canadian Journal for Science, Mathematics and Technology Education*, 4, 2: 149–168.

Roth, W.M. and Lee, S. (2004) Science education as/for participation in the community. *Science Education*, 88: 263–291.

Saldivar-Tanaka, L. and Krasny, M.E. (2004) The role of NYC Latino community gardens in community development, open space, and civic agriculture. *Agriculture and Human Values*, 21: 399–412.

Schnack, K. (1994) Some further comments on the action competence debate. In: Jensen B.B. and Schnack K. (eds), *Action and Action Competence as Key Concepts in Critical Pedagogy*. Didaktiske studier, Studies in Educational Theory and Curriculum, Volume 12, Copenhagen: Royal Danish School of Educational Studies, pp. 185–190.

Skelton, N., Boyte, H.C., and Leonard, L.S. (2002) *Youth Civic Engagement: Reflections on an Emerging Public Idea*. Minneapolis, MN: Center for Democracy and Citizenship, Hubert H. Humphrey Institute of Public Affairs, University of Minnesota.

Simovska, V. (2000) Exploring student participation within health education and health promoting schools. In: Jensen, B.B., Schnack, K., and Simovska V. (eds), *Critical Environmental and Health Education: Research Issues and Challenges*. Research Centre for Environmental and Health Education, The Danish University of Education, pp. 29–43.

Svedbom, J. (1994) Science knowledge – an important part of action competence. In: Jensen B.B. and Schnack K. (eds), *Action and Action Competence as Key Concepts in Critical Pedagogy*. Didaktiske studier, Studies in Educational Theory and Curriculum, Volume 12, Copenhagen: Royal Danish School of Educational Studies, pp. 59–72.

Tompkins, L.J. (2005) A case for community-based education: students form partnerships to tackle local environmental issues. *The Science Teacher*, 72, 4: 34–36.

Torney-Purta, J., Lehmann, R., Oswald, H., and Schulz, W. (2001) *Executive Summary: Citizenship and Education in Twenty-eight Countries: Civic Knowledge and Engagement at Age Fourteen*. Amsterdam: International Association for the Evaluation of Educational Achievement (IEA).

Youngman, F. (2000) *The Political Economy of Adult Education and Development*. London: Zed Books.

Youniss, J., Bales, S., Christmas-Best, V., Diversi, M., McLaughlin, M., Silbereisen, R. (2002) Youth civic engagement in the twenty-first century. *Journal of Research on Adolescence*, 12, 1: 121–148.

Zeldin, S. *et al.* (2000) *Youth in Decision-Making: A Study on the Impacts of Youth on Adults and Organizations*. National 4-H Council. Madison, WI: University of Wisconsin.

Chapter 18
A Clash of Worlds: Children Talking About their Community Experience in Relation to the School Curriculum

Robert Barratt[1] **and Elisabeth Barratt Hacking**[2]

Keywords children's environments, participatory research, school curriculum development, local community action, urban environments

18.1 Introduction

In this chapter, we discuss our experience of developing participatory research with children. We draw on a pilot research project with 11- to 12-year-old children, studying at an urban secondary school for 11- to 18-year-olds in central England. The project was a precursor to a recent UK Economic and Social Science Research Council (ESRC) project 'Listening to Children – Environmental Perspectives and the School Curriculum'.

The chapter starts by outlining the wider theoretical basis of the project in order to ground its purpose, methodology and conceptualisation. We then describe the research context and the design of the pilot project, and discuss the tentative findings. Our approach is sympathetic to the ethnographic work of Van Maanen (1979) and the work of Chawla (Chapter 6, this volume) in attempting to understand the meanings and actions children express in their social settings, and to this end, we have developed a series of propositions to explicate the relationship between the child, their school curriculum experience, and their local community. Finally, in reflecting on the propositions, the idea of school identity is used to explore how children's participation in curriculum and community development can be promoted through a school forum. Here, the drivers (and barriers) in establishing a closer relationship between the school and the local community, as perceived by children, are identified and discussed. The chapter concludes by discussing children's prerequisites for an effective school-based participatory research project.

---

[1]School of Education, Bath Spa University, Newton Park, Newton St Loe, Bath, BA2 9BN, UK, r.barratt@bathspa.ac.uk

[2]Centre for Research in Education and the Environment (CREE), Department of Education, University of Bath, Bath, BA2 7AY, UK, edsecbh@bath.ac.uk

A. Reid et al. (eds.), *Participation and Learning*, 285–298.

Throughout the chapter, we critique the child's experience of participation and attempt to explain this experience in terms of the research context. The child's voice is used to authenticate and validate aspects of the discussion. All child-related positions have been verified by the children.

Three simple questions were posed to children at the start of the pilot project – *would you like to participate*, *why would you like to participate*, and *how can we support you to make your participation meaningful?* The pilot project focused on children's thinking about an opportunity to participate in a curriculum development project, and about their local community, environment, and school experience. In the context of this project, 'local' refers to the urban housing estate where all the children attending the school lived. Essentially we were concerned with how children's everyday experience could be defined, understood, and shaped into a dimension of a school curriculum concerned with local community action-taking. The pilot project therefore placed children's thinking at the centre of the process of curriculum development. In England, it is unusual for children to take an active role in curriculum development; indeed, given England's centralised National Curriculum, teachers no longer have a significant role in this process (Barratt Hacking *et al.* 2006).[1]

It is our belief that in order to derive meanings that can be used to develop appropriate educational experiences (as in the work of Baacke 1985), the starting point for local community participation should be an exploration of the relationship between the child and their environment. The challenge for this project was to develop methodologies that were sensitive to the personal experience of young people and ones that endeavour to engage in research with young people (Garbarino *et al.* 1989; Barratt and Barratt Hacking 2000). The pilot project attempted to develop a framework for children's participation in curriculum and community development. This framework aimed to define a way of working with children inside a school so that children could share their curriculum and community experience through a forum involving both children and adults. From here we explored how children might be given the opportunity for their curriculum and community thinking to be translated into action, with children taking the lead in the action-taking. Ultimately, the aim was to integrate this framework into school decision-making processes and into the structure of the school day.

18.2 Researching a Child's Capacity to Participate: Theoretical Matters

In our focus on the child's experience of their environment and community we have adopted an ecological perspective (Baacke 1985). Here, 'ecological' refers to the context in which the child exists – the space, the community, and the environment. In Baacke's model, the child is seen as existing not just in its 'space' but also

[1] The 'Listening to Children – Environmental Perspectives and the School Curriculum' Project was funded from May 2004–May 2005 by the UK Economic and Social Research Council (ESRC): Environment and Human Behaviour Programme [RES-221-25-0036].

as influencing and giving meaning to it. In this way children are seen to 'actively [give] meaning to their environment' (Jans 2004:35). Baacke's child-development research identifies four ecological zones each of which is occupied successively over time (Baacke 1985; Jans 2004). These include: first, the home – the 'ecological centre'; second, the neighbourhood surrounding the home – the 'ecological proximity'; third, other places that children move to for specific functions, for example, sports or shopping – the 'ecological sectors'; and fourth, the more distant places that children move to occasionally, such as on holiday – the 'ecological periphery'.

While the 11- to 12-year-old children in the project occupy all of Baacke's 'ecological zones', our primary interest is on the first three. It is here that the children are beginning to feel confident and move autonomously, as they make the transition from child to teenager and develop independence from the home and family. Nevertheless, the children are still home-centred in many ways and most see the home as a safe haven within the community.

Given that the starting point of this ecological model is that children simultaneously appropriate and influence their environment (Jans 2004), by implication participation in community development and environmental improvement has most meaning for the child in the context of 'the ecological zones', or the child's everyday world. In this project we were trying to understand and build on children's everyday participation in the local community and environment. Sauvé (2002:2) notes that,

> the local context is the first crucible for the development of environmental responsibility, in which we learn to become guardians, responsible users and builders of *Oïkos*, our common "home of life".

However, Sauvé's model of the dimensions of our environment differs from that of Baacke (1985) by focusing on the interlocking spheres of interaction in the development of a person, that is, through the interactions of oneself with others, and with the biophysical environment or 'home of life'. Nevertheless, in the context of this project we recognise that it is the local urban environment that brings together these spheres through the interaction of children with the community and environment of an economically deprived housing estate.

In developing the project methodology, we reflected on the underlying theoretical tenets of two paradigms of children's participation, as proposed by Neale (2004). Here, children are viewed as either 'welfare dependants' or as 'young citizens'. Neale (2004) developed these paradigms through his analysis of recent debates in public and policy forums about the place of children in our society, arguing that (p. 7):

> [A]t the present time, there is little consensus over these issues and they continue to be hotly debated.... It is possible to see in these debates two distinct ways of viewing children, which have important consequences for policy and professional practice.

Viewing children as 'welfare dependants' suggests that adults are in control and that children are vulnerable, incapable, and in need of guidance and protection. Viewing children as 'young citizens' offers an alternative perspective; children are viewed as having strengths and competencies and are seen as having 'an entitlement to recognition, respect and participation' (Neale 2004:7). It was the latter paradigm that informed our project methodology, alongside models such as the 'ladder of participation' (Arnstein 1969; Hart 1997 – see Chapter 1, the Introduction, and Chapter 2 by Roger Hart, this volume) in order to differentiate levels of participation and exemplify what constitutes genuine participation. In the 'ladder' model, it is at the higher rungs that the form of participation most closely fulfils the 'young citizen' paradigm; in practice, this informed our project design by, for example, giving children a significant voice in project decision-making and an equal role with teachers in the curriculum development process.

We would characterise our methodological stance as 'researching collaboratively with children' (Garbarino *et al.* 1989; Barratt and Barratt Hacking 2000), in that children were partners in making decisions about the research direction and process rather than being the objects of the research. Given that the project represented participatory research with children, we also argue that it was valid to adopt a grounded theory approach to theory generation (Strauss and Corbin 1990). Thus our research-based perspectives on participation draw directly upon the voices of children and others who have a stake in the educative process and in the local community and its future. Given this, participation is understood in relation to the dynamic that seems to exist between the school and its community. Practically, this methodological approach can be explored with children if provided with an appropriate discussion forum, in particular, by asking children to reflect upon and use their thinking to inform discussion and direct the research process. It is apparent to children that there is great value in exploring the nature of their relationship with their school and the local community. Children argue for a closer relationship between their home, school, and the local community. It is this relationship that children attempt to make sense of as they move between (and occupy) each place. The project team became aware of this personal challenge to children and attempted to understand, with children, how it affected the child's capacity to participate.

18.3 The Research Project and its Context

Many urban neighbourhoods in England are characterised by economic deprivation, socio-spatial inequality, and a lack of sustainable urban development (DEFRA 2004). Recent studies have highlighted that 'urban renewal policies in England are failing to adequately address the local environment concerns of people living in poor neighbourhoods…(such) as…litter, abandoned cars, poorly maintained public spaces and graffiti' (Lucas *et al.* 2004:1). Through their research across Britain, Spencer and Woolley (2000) found that children's concerns commonly included 'pollution, traffic, crime, and incivilities (and that)…these concerns (are) manifestly grounded in the child's own local experiences' (Spencer and Woolley 2000:189).

The project school is surrounded by a local authority housing estate that is viewed by children as their 'local environment and community' – all of the children who attend the school live in the housing estate and are able to walk, cycle, or skateboard to school. This urban environment and the concerns of its child and adult residents mirror the findings of Lucas *et al.* (2004) and Spencer and Woolley (2000). It is an environment experiencing a cycle of social, economic, and environmental deprivation with associated deleterious health issues. Its wide range of environmental issues include 'contaminated derelict sites; poor street lighting; the need for enhanced community facilities, such as parks and playgrounds; issues to do with the cleanliness of the streets and an inadequate local transport network' (Barratt and Barratt Hacking 2003:30). There are also examples of young people involved in environmental damage such as arson, and ecological and physical damage to public and private property. Thus although many of the 11- to 12-year-old children involved in this project care about their local urban environment, they often feel disenfranchised. As we have noted previously (Barratt and Barratt Hacking 2003:31) it is apparent that some of the children:

> [E]xperience high levels of frustration about the state of the local environment and their powerlessness to effect any change in it.

There is also significant local concern about children's quality of life, for example, educational achievement and opportunities, the quality of the urban environment in which children are growing up, and associated health issues. A number of children in the school live close to or in poverty, some also care for siblings and/or adults.

The project began by exploring how an urban school can support children more effectively in such deprived circumstances. It is widely recognised that urban schools in England are struggling to maintain national educational standards (Bell 2003), for example, the project school faced low literacy levels, high exclusion rates, and poor stay-on rates post-16 years. Typically, schools see themselves as having a significant role and identity in the local community. In reality it seems that some urban schools are not as valued by their communities as they might hope. As a consequence urban schools are less able to support children with the difficulty of living in a deprived space. We were therefore interested to explore the children's perceptions of the purpose of their school and how they see the school's relationship with the local community. We use the term 'school identity' to describe these perceptions later in the chapter.

We also note that there are requirements in the National Curriculum (England) documentation for Education for Sustainable Development (ESD) and Citizenship, and thus opportunities for schools to develop a curriculum that is sensitive to the local community and environmental experiences of children; this includes deprived urban settings (QCA 1998; DfES/QCA 2001). Citizenship requires schools to develop 'skills of participation and responsible action' (DfEE/QCA 1999:14). ESD 'enables pupils...to participate in decisions about the way we do things individually and collectively, both locally and globally, that will improve the quality of life now without damaging the planet for the future' (DfEE 1999:25). However, such initiatives depend on the school involving children in curriculum development or on having detailed knowledge of the local community and environment. Neither position is common in schools in England. The project team recognised that children possess expert knowledge about the particulars of urban environments that educators know little about, and this was worthy of inclusion in the project. Indeed, children participating in this project openly described 'a relevance gap' between the school curriculum and the knowledge and understanding required to live in the community. They recognised the value of contextualising learning in relation to their community experience, and hence welcomed the opportunity to reflect upon and bring the nature of their every day lives into school.

Joining the project was voluntary for the children and they were given individual support in making an informed decision whether or not to participate. We developed a framework for participation in the project school through a *school environment curriculum council* approach. While this is similar to the school council found in most schools in England, it is rare for school councils to spend time on matters to do with the local environment, the local community, learning, or the curriculum, as school councils tend to focus on social and behavioural concerns (see, e.g. Alderson 2000; Taylor and Johnson 2002).

The initial council meetings form the basis of our discussion here. The council was led by twelve 11- to 12-year-old children in collaboration with three teachers and one researcher. Initially, the children were encouraged to debate why they might undertake such a project, what the project might be about, how the project might develop into an opportunity for children to bring their school and community experience together, and how the project methodology might be developed by the children themselves. The ensuing meetings focused on exploring the children's curriculum and community experience as a basis for developing a community-based and environmentally focused curriculum whilst evaluating existing curricular provision.

In this chapter, we refer only to the meanings children attributed to their everyday experience of the school curriculum, local community, and local environment; these are encapsulated in a series of propositions. The propositions were derived through a process of data gathering and analysis. All school council meetings were audio recorded and transcribed. By reading and rereading the transcriptions, we identified patterns and themes within the discussions that we took back to the children for review. The patterns and themes were also explored further in a day at the researchers' university where the project children, together with a sample of 30 children from four other schools local to the pilot school, shared their community and school experience. Children worked in mixed-school discussion groups constructing 3D models and creating 2D multimedia visual representations of their local environments. It was through this process of developing themes, testing out themes, and through further analysis of children's evidence that the propositions emerged. These were later corroborated with other children and teachers in the research school.

18.4 What has Emerged from the Research? Understanding the Relationship Between Home, School, and the Community

We now present the five propositions that emerged from our analysis of the project data (Table 18.1) to frame our discussion of the children's conceptions of their *school identity* in relation to themselves, their home, and their community. We discuss the range and extent of each proposition, using children's (and teachers') voices to exemplify their experience.

Table 18.1 Five propositions from the analysis of the research data

1. The extent of children's local knowledge and attachment is currently unknown by adults
2. Children see little relationship between the school curriculum and the local community and its environment
3. Children's voices are not systematically represented in local community decision-making
4. There is a need for schools to promote children's participation in the learning endeavour and in community-focused curriculum development
5. Schools can provide a forum for young people's active participation in local community/ environment issues

Proposition 1. The extent of children's local knowledge and attachment is currently unknown by adults

Children in this project held expert local knowledge that was grounded in previous and current experience and shaped by what they would like to do in the future. This local knowledge carried its own language register and protocols in terms of what was fashionable and desirable, and how children worked with this knowledge and used it to interact with their community. The knowledge was culturally and socially formed and was not necessarily known by adults. The children recognised that adults did not legitimise this knowledge and that it was a powerful resource to help them operate both inside and outside the gaze of adults.

The children acknowledged that it is not always the visible that dictates what life is like in a local community. They cited the missed opportunities, untapped talent, dominant families and individuals, and poor resources as factors generating a cycle of decline. They reported how on the street, groups of young people organised and controlled or damaged what seemed to be the best places to be: 'Older children spoil the play equipment,' 'We have to pay them to use the half pipe.'[2] This threatened some children. Others accepted this dubious practice as part of growing up, forming a self-identity and shaping who and what you will become when you are an adult. The children questioned why they had to live in these conditions and why other people's local environments were better: 'We want to know why we haven't got those things in our street?.... Does there need to be places like ours for everyone else to live in the best places?'

The children conveyed a real sense of emotional attachment to, and physical engagement with, the local environment. The human qualities children bring to bear in these connections are significant in understanding the relationship between a child and their environment. Children, like adults, wish to care and be cared for and desire safety and security. We argue that human qualities such as these are also relevant to children's relationship with the local environment and their wish to care for it. If they feel distanced from the local environment, 'the environment is over there', or fearful of their safety locally, we argue that this natural caring response will be modified or inhibited, potentially leading to a lack of concern.

[2] The 'half pipe' is in the local park and a place where they can play with their skateboards or bikes.

Proposition 2. Children see little relationship between the school curriculum and the local community and its environment

> Our every day life is not understood in school.
> What we do in our spare time is not talked about in school.
> School does not have a role in improving the local environment; the school is not interested.
> School is a place where I feel safe from the local environment.

The children viewed school as a positive place that could influence the community, but in practice the children felt that the school did not play a significant role in community development. Their reasoning varied from 'the school have decided not to', to 'the school does not think about it'.

The children claimed that adults, including teachers, did not acknowledge or legitimise their local knowledge yet children argued that this knowledge was essential in order to survive in the environment outside of the school and the adults' purview. The children said they had little opportunity to discuss their ideas for the local environment. They felt that they were only invited to share ideas in relation to the teacher's agenda or interests, or the curriculum subject matter. Wider than the school context, the children claimed that their ideas about the environment did not appear in children's texts, nor in the majority of literature about the environment. Given this, the children felt that they were always working with adult ideas – some they subscribed to and gave legitimacy to, others they did not understand and took little notice of.

The project data suggested that children did not have access to current information and different views about the local environment and community in school. Equally, children had access to a filtered version of a global and national environmental agenda. The knowledge that was received did not seem relevant to children's lives and was considered biased and dated. So, understandably, the children expressed views like: 'What's it [the environment] got to do with me?' and, 'The environment is over there.' Teachers in the project acknowledged that the subject matter presented in the National Curriculum (England) in terms of Citizenship, ESD and, in some cases, their own subjects, required refinement in relation to the everyday lives of children.

Refining subject matter in relation to children's lived experiences involves departing from a national agenda and moving towards a local and child's agenda. This would recognise the importance of the local and the particular in terms of the curriculum content in order to make it relevant to children's everyday life. In this way, children would learn *through* their community. For example, the English curriculum might explore cultural texts represented in the local urban community, such as poetry, dialects, graffiti, and lyrics. Given that the urban community is such a complex space, we argue that this is fundamental to a child's developing identity and understanding of their relationship with the community and environment.

This fundamental reorientation of the purpose of schools towards the local community and environment is not only essential for children to understand themselves and their place in the local community, but is also an essential foundation for understanding others. A child's confidence and security at the local scale enables them to develop broader understandings of, for example, their place, role, and contribution in the wider community, now and in the future.

Proposition 3. Children's voices are not systematically represented in local community decision-making

> One day it's (the police station) there the next day it's gone.
> Now it's gone it's not safe, we stay at home after dark.
> [A local police station was cited as playing a significant role in moderating the behaviour of many teenagers in the community.]

Indicative accounts of many of the children's experience of community change and development include, 'We have nothing to do with it,' and, 'We haven't got a voice, our views don't count, we aren't in the know.' Children felt strongly that they were being disenfranchised from the hidden process[3] of community development, that they were rarely consulted on local issues, and they were not aware of any forum for discussion. They suggested that their community experience was somehow invisible to the decision-makers and questioned whether their experiences were too difficult to hear. Further, they recognised that 'not even the teachers know about it, unless they live here' (teachers tended to live further away from the school than the children). The children considered that issues such as litter in school were tackled because they are less sensitive.

The children were able to generate many suggestions for how the community could be organised to meet their needs, for example, 'no go areas for teenagers', 'special places for young children to play', and 'areas patrolled by adults'. The children wanted their community experience to be used to create community-based action that would make a real difference on the streets and be sustainable.

It is our view that if the child feels that they 'have a voice' and a stake in the future of the local community then they are more likely to connect with the community and get involved in the urban environment. The data showed that many children had a strong desire to improve their local area yet, in the main, neither education nor local decision-making seemed to capitalise on this effectively. This is at odds with national and international policies which have promoted residents' involvement in local development initiatives, for example, the Local Agenda 21 Strategy (derived from the 1992 World Summit in Rio de Janeiro), the community strategy in England and Wales 2000 (Local Government Association), and more recent government advice:

> Communities…need the chance to have a voice, to have a say in local decisions which affect them and so help shape their own future…and allow local people to take greater control over their local environment. (DEFRA 2004:25.)

Children's participation in everyday matters of relevance to them is an established national and international principle (UK Children's Act, DfEE 1989; UN 1990 Convention on the Rights of the Child[3]). The recent UK Children's Bill (DfES 2004) requires local authorities to recognise the contribution made by children to society. So why do the children in the project experience so little involvement in local development? The children reported a lack of access to information as one of the barriers to their local involvement. The children argued that local information was not accessible to them in any coherent or systematic way, be it based either on the 'worlds of children' or material (research evidence, national guidance) that has been transformed by

[3] The UK Government ratified the UN Convention on the Rights of the Child in 1991.

children so that it is accessible to other children. Thomas and Thompson (2004:15) had a similar response from their research with 10- to 11-year olds and found that 'very few children are aware of local authority or other public agency responsibility for public space, and do not know how to complain or access information'. If this is the case, concerns must exist over local decisions that are based on information or an evidence base that is not sensitive to community experience, needs, behaviours, and aspirations.

Proposition 4. There is a need for schools to promote children's participation in the learning endeavour and in community-focused curriculum development
and
Proposition 5. Schools can provide a forum for young people's active participation in local community/environment issues
We contend that the challenges facing the urban environment impact on their schools and young people. The project findings suggest that the school could have supported children more effectively with the challenge of living in an urban environment. There was little evidence of school-based discussion of children's ideas for the local environment ('my local experience is not talked about in school'), or of children's involvement in local community development.

The teachers conveyed a view that the school, in some way, had a role to play in local community regeneration. They were also aware of the children's desire for their school to play a role in local community change so shaping a more harmonious future. This raised significant questions for the teachers in terms of curriculum interpretation and implementation: 'can schools develop this kind of relationship with their communities' and 'how do schools position themselves inside such a community?' For teachers, their school space was a territory surrounded by another with very different value systems and learning processes.

Schools in England have some freedom to choose how to develop aspects of their curriculum, including ESD and Citizenship. Together these aspects suggest many possibilities for 'local environmental citizenship' as noted earlier. However, we found little evidence of children participating in decisions about how any of these initiatives will develop in practice. The national study of the implementation of the Citzenship curriculum orders found that 'delivery plans for citizenship education have largely been drawn up by school leaders with little or no consultation with teachers and students' (Kerr *et al.* 2002:iv). Similarly, we found that the children were passive recipients of the curriculum with few genuine opportunities to participate in curriculum development, decisions about teaching and learning, or in local community action.

As noted earlier, the teachers acknowledged that the subject matter presented in the Citizenship curriculum 'requires refinement in relation to the everyday lives of children'. They suggested that knowledge should be negotiated and directly related to the developmental process, maturation, learning, and aspirations of the child. Yet, if a position is adopted that places the child right at the centre of the curriculum planning process, as with the child's prior and current experience and future aspirations, then the teachers were concerned that the integrity of the curriculum would be brought into question.

If the school curriculum is to take more account of children's experience and involve them in participating in curriculum development and local action, fundamental changes to school practices are needed. Such changes would include placing the child's personal interests at the heart of the learning process and requiring children to think about their learning (e.g. in meta-cognitive activities). It would encompass the way the teaching and learning process accounts for the relationship between the individual and curriculum content and process (see, e.g. Minden's (1997) discussion of Goethe's *Bildung*, individualism, and socialisation).

Teaching and learning approaches offer a multitude of opportunities for developing participatory approaches. Currently, it is seen to be good educational practice to engage pupils in self-assessment, problem-solving, and thinking skills. We contend, however, that although this may lead students towards new ideas, viewpoints, or decisions, this does not necessarily result in any action-taking in terms of their everyday life (Barratt and Barratt Hacking 2004).

The experience of the school environment curriculum council suggests that schools can provide a forum for young people's active participation in curriculum and community development. It was evident that the children involved appreciated that, for the first time, someone in school was listening to their experiences, views, concerns, and aspirations. Indeed, involvement in the council appeared to have a positive effect on children's attitudes and behaviour. What was seen by the school as a group of challenging children to work with became a highly motivated and dynamic group. Equally the teachers began to question the appropriateness of the school curriculum and the way in which this is constituted and delivered. That is, there is much to do if schools are to aspire to developing community relevant curricula concerned with preparing young people for the(ir) future(s).

It became apparent that wider community representation would support the development of community-based curricular and local action. The *school environment curriculum council* argued that this could remove some of the barriers to children's participation in local decision-making. This was acknowledged and addressed in the ensuing *Listening to Children* project (Barratt Hacking *et al.* 2006).

18.5 Reflections on the Propositions and School Identity

In the main, the children held a positive image of school and wanted this resource to contribute to wider community development. They considered themselves as having a stake in the future of the community but without a forum to achieve this. The school would seem to be an appropriate forum for children, given that they participate collectively here; however, they view the school as separate from the local community with little or no current role in its development. Rather, the children considered the school's purpose to be to do with learning. They appreciated that the curriculum supports the development of their knowledge, understanding, skills, and values, but that it does not contribute to what children want for their community.

For many schools in England, their existing community role is about providing education and facilities. The project findings contribute to a rethinking of the purposes of schools in relation to their local community so that they can be seen to play a more significant role in community development. We argue that schools have a key role to play in developing the relationship between themselves and the community. They can become research-focused learning centres with the expertise to contribute to community development. Robinson and Shallcross (1998) argue for a community-focused environmental education; we have demonstrated that the context for this can be through a school forum. This involves placing the school and its children at the heart of local environmental and community development, acknowledging that children have extensive local expertise, and recognising their potential and desire to participate in local development. What interests us is the apparent extent of discord between the child's local environment knowledge, aspirations for action, and their actual participation in local issues.

Involving children and other local stakeholders in curriculum development is pivotal if schools are to ensure that their curriculum has community relevance. In this way, schools can endeavour to make the curriculum sensitive to children's everyday local experience and knowledge, develop new knowledge and understanding relevant to the local community, and provide opportunities for children to make a contribution to community development. Extending the community role would see schools with:

- An extensive knowledge-base about the local community and its environment, and how children experience this
- Specialist education and training for young people, for example, in action research, project planning, community research, developing local knowledge, sustainable solutions to environmental problems, and safety education (which also develops students' local expertise and 'employability')
- School-based initiatives to improve the local environment
- Involvement in local community initiatives in which young people work on long-term projects alongside other stakeholders from the community
- School environment curriculum councils that involve young people and other local stakeholders in curriculum development

The extension of the school's role in this way would provide children with a framework for knowledge, understanding, and skills from which to contribute to the local community beyond their school days, including through local employment and participation in local activity. We further suggest that this promotes what we would term *environmentally conscious citizenship*, whereby children are able to develop their own ideas about being a good citizen, enhance their future opportunities, and improve the quality of their local community and environment. (Developing the idea of school identity in practice also became a focus of the ensuing *Listening to Children* project, Barratt Hacking *et al.* 2006).

18.6 Final Thoughts

Three questions were posed to children at the start of the project: Would you like to participate, why would you like to participate, and how can we make your participation meaningful? We have found reflecting on these questions in relation to the propositions illuminating, particularly in terms of understanding what makes a project meaningful for children and in considering how to build on this in developing a participatory approach for the *Listening to Children* project. The children argued for a research design that considered a number of fundamental prerequisites, which circumscribe their personal needs. They are to:

- Create a local democracy (structures and processes) within school, which recognises children's contributions and supports children's decision-making
- Establish a dedicated group and a 'listening forum', which explores the child's experience of living between home, school, and community and the nature of their interconnectedness
- Appreciate that children want to bring about community change
- Recognise the apparent discord between children's personal aspirations and their opportunity to effect local action
- Consider the ways in which children's knowledge about the local environment is generated and to protect its authenticity and integrity
- Make use of children's local knowledge in curriculum and community development
- Make knowledge about other children's local community experience available to children
- Make research evidence available to children, in an accessible format, about how other schools and communities are working together

The project experience suggests that schools have the potential to provide a forum through which children can play a key role in shaping their future urban environment. We have argued that there are distinct opportunities through the National Curriculum (England) for schools to involve their students in curriculum development through Citizenship and ESD. Through this process, children can bring their everyday experience to their curriculum and learning experience.

In this project, children were willing to share their very sensitive local community knowledge and consider this in relation to the school curriculum. We believe that this has only been possible because the project acknowledged that it was important to talk with children about the research design and process. This starting point established children's trust and belief in the intentions of the project, and most importantly its relevance to them as 'young citizens' (Neale 2004). This experience has demonstrated that if children can be provided with the right conditions to participate, then the opportunity for the development of a school curriculum that is more relevant to children's everyday lives may be possible. In so doing, the apparent distance between the child's every day life and the school curriculum may be reconciled.

References

Alderson, P. (2000) School students' views on school councils and daily life at school. *Children and Society*, 14, 2: 121–134.

Arnstein, S. (1969) A ladder of citizen participation in the USA. *Journal of the American Institute of Planners*, 35, 4: 216–24.

Baacke, D. (1985) *Die 13-bis 18jährigen*: Einführung in Probleme des Jugendalters.

Barratt, R. and Barratt Hacking, E. (2000) Changing my locality: conceptions of the future. *Teaching Geography*, 25, 1: 17–21.

Barratt, R. and Barratt Hacking, E. (2003) Rethinking the geography national curriculum: a case for community relevance. *Teaching Geography*, 28, 1: 29–34.

Barratt, R. and Barratt Hacking, E. (2004) Geography's role in environmental citizenship. *Teaching Geography*, 29, 2: 70–73.

Barratt Hacking, E.C. *et al.* (2006) Education for sustainability: schools and their communities. In: Chi-Lee, J. and Williams, M. (eds), *Environmental and Geographical Education for Sustainability: Cultural Contexts*. New York: Nova Science.

Bell, D. (2003) *Access and Achievement in Urban Education: Ten Years on*. Speech to the Fabian Society [online]. (http://www.ofsted.gov.uk/news/index.cfm?fuseaction=news.details+id=1506) (Accessed 5 February 2004).

Department for Environment, Food and Rural Affairs (DEFRA) (2004) *Taking It on: Developing UK Sustainable Development Strategy Together*. London: DEFRA.

Department for Education and Employment (DfEE) (1989) *Children's Act*. London: HMSO.

DfEE/Qualifications and Curriculum Authority (QCA) (1999) *Citizenship: The National Curriculum for England, Key Stages 3–4*. London: HMSO.

DfEE (1999) *The National Curriculum: Handbook for Secondary Teachers in England*. London: HMSO.

Department for Education and Skills (DfES)/QCA (2001) *Citizenship at Key Stage 3*. London: HMSO.

DfES (2004) *Children's Bill*. London: TSO.

Garbarino, J., Stott, F. and Associates (1989) *What Children Can Tell Us*. San Francisco, CA: Jossey-Bass.

Hart, R.A. (1997) *Children's Participation: The Theory and Practice of Involving Young Citizens in Community Development and Environmental Care*. London: Earthscan & UNICEF.

Jans, M. (2004) Children as citizens: towards a contemporary notion of child participation. *Childhood*, 11: 27–44.

Kerr, D. *et al.* (2002) *Citizenship Education Longitudinal Study First Cross Sectional Survey 2001–2002, NfER/DfES Research Report RR416*. London: HMSO.

Local Government Association [online]. (http://www.lga.gov.uk/index.htm) [Accessed November 2004].

Lucas, K., *et al.* (2004) *Prioritising Local Environment Concerns: Where There's a Will There's a Way*. York, UK: Joseph Rowntree Foundation.

Minden, M. (1997) *The German Bildungsroman*. Cambridge: Cambridge University Press.

Neale, B. (ed) (2004) *Young Children's Citizenship Ideas into Practice*. York: Joseph Rowntree Foundation.

QCA (1998) *The Crick Report, Education for Citizenship and the Teaching of Democracy in Schools: Final Report of the Advisory Group on Citizenship*. London: QCA.

Robinson, J. and Shallcross, T. (1998) Social Change and Education for Sustainable Living. *Curriculum Studies*, 6, 1: 69–84.

Sauvé, L. (2002) Environmental education: possibilities and constraints. *Connect*, 27, 1–2: 1–4.

Spencer, C. and Woolley, H. (2000) Children and the city: a summary of recent environmental psychology research. *Child: Care, Health and Environment*, 28, 3: 181–198.

Strauss, A.L. and Corbin, J. (1990) *Basics of Qualitative Research: Grounded Theory Procedures and Techniques*. Newbury Park, CA: Sage.

Taylor, M.J. and Johnson, R. (2002) *School Councils: Their Role in Citizenship and Personal and Social Education*. Slough: NfER.

Thomas, G. and Thompson, G. (2004) *A Child's Place: Why Environment Matters to Children*. London: Green Alliance/DEMOS.

United Nations (1990) *Convention on the Rights of the Child*. New York: United Nations.

Van Maanen, J. (1979) The fact of fiction in organizational ethnography. *Administrative Science Quarterly*, 24: 539–550.

Willow, C. (2002) *Participation in Practice: Children and Young People as Partners in Change*. London: The Children's Society.

Chapter 19
Sustainability Education, Whole School Approaches, and Communities of Action

Tony Shallcross[1] and John Robinson[2]

Keywords sustainability education, whole school approaches, communities of practice, communities of action, participation

19.1 Introduction

Despite differences in terminology, environmental education and educations that address sustainability share a common aspiration of changing lifestyles. In this chapter we argue that a situated, whole school action-focused learning rather than behaviourist or cognitive/constructivist approaches to learning offers a better, though not the only, prospect for education to contribute to the development of more sustainable lifestyles/actions. While environmental awareness and values education approaches can generate more sustainable actions, these approaches seem incapable of contributing to sustainable lifestyles at a societal scale. Whole school approaches are not the only approach to sustainability education, but they appear to have significant benefits in promoting sustainable actions congruent with, for example, the UK Government's *Every Child Matters Strategy* in England, which promotes the principles: be safe, be healthy, achieve and enjoy, make a contribution, and be economically active. These approaches underpinned by the theoretical framework of communities of practice offer a vision for a sustainability education that has this contributory action focus.

19.2 Background

> This fundamental democratization of children is the most important aspect of their participation in the environment of their communities, more than the particular impact of any of their projects. (Hart 1997:8.)

---

[1]Institute of Education, Manchester Metropolitan University, Crewe Green Road, Crewe, CW1 5DU, UK, A.G.Shallcross@mmu.ac.uk

[2]The Education and Social Research Institute, Manchester Metropolitan University, 799 Wilmslow Road, Didsbury, Manchester, M20 2RR, UK, j.f.robinson@mmu.ac.uk

A. Reid et al. (eds.), *Participation and Learning*, 299–320.

In examining the value of whole school approaches and their intrinsic association with participatory education and communities of action in sustainability education, we need to start by clarifying some key terms. Terms linking sustainability and education are inherently problematic but we prefer sustainability education to other more prevalent terms, such as education for sustainable development (ESD), as sustainable development is anthropocentric in its ethical base, while sustainability can be ecocentric (Dobson 1998).

Sustainable development is reform environmentalism (Shiva 1992) associated with quantitative notions of maintaining yield, rather than the more qualitative conception of maintaining biodiversity. Furthermore, sustainable development is often rooted in a set of Western values that are predicated on the belief that globalisation is good because by developing a 'planetary citizenry' it emancipates individuals from their culture. Others see globalised sustainable development as a cultural homogeneity, a modernist form of 'enclosure' rather than liberation (Shiva 1992; Bowers 2004). Sustainability education, based on an eco-justice pedagogy (Bowers 2002) that revitalises and connects communities, mediates against the 'single way of knowing' that is privileged in a modernist, enclosing 'big brother' form of sustainable development (Wals and Jickling 2002) and driven by science, technology, corporations, and experts (Bowers 2004). To us, the qualitative regard for the natural environment that contemplates ecocentricity integrally within the social and economic is key to the development of a situated, participatory, education as sustainability, (Foster 2001).

Whole school approaches mean practising what is taught by striving to minimise the gap between espoused values and values in action (Posch 1993), through the integration of formal and non-formal curricula. Whole school approaches can shape 'our interaction with the environment in an intellectual, material, spatial, social and emotional sense to achieve a lasting/sustainable quality of life for all' (Posch 1999:341–342) by integrating learning with the social/organisational and technical/economic aspects of school life (see Figure 19.1). This is education *as* sustainability, a way of life (Rudduck 1999) that transforms the substance and processes of the formal curriculum, the purposes of learning, and how educational institutions and buildings work (Orr 1994). However these descriptions omit evaluation (Rauch 2000), which, when added to the other four strands in Figure 19.1, completes the picture of a cycle of plan, do and review.

In whole school approaches to sustainability education, action is seen as a stream of 'actual or contemplated causal interventions' (Giddens 1979:55) not to be confused with behavioural change whose direction is externally determined (Uzzell *et al.* 1994). Rather, research (conducted mainly with adults) suggests that activism is rooted in efficacy, engaging with moral values, based on a profound personal sense of honesty, care, and integrity, and a sense of self and communities of support as well as identity, expressed through a sense of meaning and place in a world that transcends the self (Berman 1997).

This focus on community, action identity, and context resonates with Wenger's (1999) notion of communities of practice. A community of practice privileges learning over teaching. Learning takes place through a variety of modes that are ontological rather than epistemological in nature, such as observation, replication

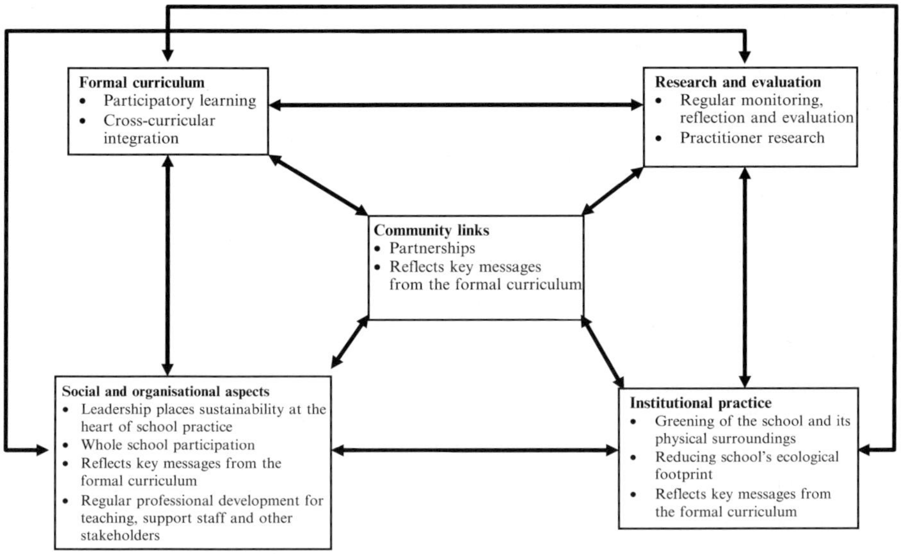

Figure 19.1 A model of a whole school approach with some characteristic features. (Adapted from Shallcross 2003; Henderson and Tilbury 2004.)

of tasks, and learning by doing. In such communities, according to Wenger, new learners are active, legitimate peripheral participants who can influence practice. They are not seen as incapable apprentices or proto-citizens in a hierarchical relationship with an expert teacher. Yet because of the dyadic relationships that exist between teachers and the taught in the majority of modern schools, and with schools tending to privilege academic rather than practically expressed knowledge; Lave and Wenger (1991) exclude schools from their considerations as viable communities of practice.

The notion of practice in relation to sustainability is troublesome because of its association with predictability, replication, and routine. What sustainability education needs is situated learning communities that identify and implement flexible, sustainable responses to new ideas and agendas for change. Because of the uncertainties surrounding the concept of sustainability and the fact that much of the exemplification of the situated learning associated with communities of practice has focused on the acquisition of manual or vocational skills (such as midwifery or farming) rather than a pantheon of social and environmental actions, we prefer to describe this contextually sensitive, situated learning approach to whole school approaches as a community of action. What becomes crucial in seeing whole school approaches in these terms is process; how actions change more than their outcomes or practice; and what these actions constitute. It is this engagement with practice/action that 'may well be a *condition* for the effectiveness of learning' (Lave and Wenger 1991:93, emphasis in the original) in communities of action (see also Chapter 8, by Vare, this volume).

19.3 Evidence About Whole School Approaches

There are many anecdotal accounts of the impact of whole school approaches (Elliot 1999; Posch 1999; Apple and Beane 1999; Rauch 2000; Farrer with Hawkes 2000) (see also Text Boxes 19.1–19.4[1]). Much of this anecdotal evidence points to a link between whole school approaches and pupil attainment. For example, Smallwood School (Text Box 19.1) recently topped its local authority league tables for its performance in national tests for 11-year-olds in mathematics, science, and language. The head teacher shunned publicity associated with this 'success' because she did not want the school to be seen as an examination factory. However, there was no doubt in her mind that this 'success' resulted from high pupil self-esteem spawning the self-efficacy and motivation that leads to improved attainment because of the school's focus on high-level pupil participation (Hart 1997). Put achievement first and attainment follows! Interestingly when the pupils were interviewed (Robinson *et al.* 2006) they felt that they had less influence on the formal curriculum of the school than they had on institutional practice. However the very question and its answer may suggest a division that is arbitrary in a whole school community of action.

Text Box 19.1 Community Links

Smallwood School is a small rural primary school in North-West England. There is an active School Council consisting of two children from each year group, all elected by their peers. Although a significant minority of pupils is from less economically advantaged backgrounds and a quarter of its pupils have special educational needs (SEN), Smallwood has recently been the top primary school in its local education authority for its pupils' performance in national examinations for 11-year-olds.

The school's *Supporting Somaliland* project began after the school heard a story about the state of education in Somaliland, one of the poorest countries in Africa. Since 2001, the school has raised over £8,000 for the Horn of Africa Learning Trust (HALT), a charity established to support the development of education in Somaliland. Much of this fund-raising has taken place out of school time. Fund-raising activities have included children singing in shopping malls, washing cars, selling cakes at local factories, and organising concerts in various venues. The school has sent two container loads to Somaliland containing 50 school boxes – each one of which could be used to establish a new

(continued)

[1] The name of this school and the others in these text boxes have not been anonymised with the agreement of the schools on the grounds that the participants and schools want their work to be recognised and celebrated.

Text Box 19.1 (continued)

'bush' school – in addition to photocopiers, fax machines, science equipment, and manual typewriters, all donated by local businesses as a result of pupils' requests. The school is now helping to build a new school in Somaliland.

The project was initiated by the head teacher, and within a few months it began to permeate the curriculum in addition to the fund-raising. After the Prime Minister of Somaliland visited Smallwood in 2003 additional links have been made by Smallwood parents who work in the health sector with a hospital in Hargeisa, the local university has established a link with a university in Somaliland, one parent has established a link with a business in Somaliland exploring the possibility of exporting ornamental goods for sale in the UK, and a local professional football club may set up a football coaching scheme in Somaliland. The head teacher and a school governor made an official visit to Somaliland in 2003.

The pupils recognise that they have participated at a high level (Hart 1997) in the *Supporting Somaliland* and other projects by:

- Writing letters to local businesses requesting donations to the project, in the process they have learned to cope with refusals
- Developing performance skills by acting and singing at fund-raising events including a joint Smallwood–Somali concert programme
- Negotiating the price for sending a container to Somaliland
- Calculating the budget for a 'school box' and becoming accustomed to bargain hunting
- Visiting a local Somali community in England, sharing Somali food and dancing Somali dances. As a result they have developed an awareness of refugee life

The major difficulties this project has encountered are time management and some initial resistance from parents. Children are called upon to sing/talk on a regular basis, often at the weekend. In addition some parents thought that the school should be raising funds for itself and for needs nearer to home. Most parents now support the project because of the video evidence they have seen, the talks given by Somalis in school, and the support of others from the local community.

Text Box 19.2 Formal Curriculum and Pedagogy

Nummenpää School is a small primary school located in a rural area in Paimio, South-West Finland. It ran an international COMENIUS project involving more than 3,000 people, including pupils, teachers, parents, and local authority officials from six different countries. The head teacher has another large

(continued)

Text Box 19.2 (continued)

school to manage; consequently Nummenpää is a very independent school when it comes to curriculum development, everyday school life and pedagogical matters. The main aims of the COMENIUS *Three Es [Environmental Education (EE) in Europe]* project were to promote:

- The ideology of children as active, being citizens deciding on, and doing their own work for their own environment, this is EE not just as knowing but as learning by doing
- Whole school approaches

Networking took place between the seven partner schools and within each school's local community. The project involved schools from Cyprus, the Czech Republic, Finland, Italy, Lithuania, and Wales. All the schools wanted to use the project to develop English language, information and communication technology (ICT) skills, and environmental work. It also developed international cooperation and understanding through visits and the use of ICT. The project outcomes were many and varied, for example, more sustainable ways to run institutional practices; new teaching methods; the production of videos, CDs, articles, booklets; and two public exhibitions of children's art work on the theme of nature. Highlight events were a balloon happening, a day without electricity, and a day without cars. Student teachers assisted the teachers with the project.

In this whole school development project adults and the schools themselves were seen as role models in the moral education of children. This is not just sustainability education being talked about in classrooms. The *Three Es project* was based on the belief that values are caught rather than taught by schools and by teachers practising what they teach. Thus the links between the formal and the non-formal curriculum became central to whole school development in sustainability education. For example, if schools teach about energy efficiency they should try to practise it. If they advocate active citizenship, children and young people should be encouraged to participate in decision-making about actions in their schools. This is education that not only makes children and young people aware of their responsibilities, but also improves children and young people's action competence – their capacity to effect their environment. The *Three Es project* tried to empower children by illustrating that what they do now as active citizens really matters for their future.

The main difficulties of the project were time pressures on teachers and bureaucracy, partly because the project became larger and more complex than originally expected. Well-documented records had to be kept because the project was funded by COMENIUS. Many local authorities only took a great interest in the project after it had secured funding and received a lot of publicity.

Text Box 19.3 Institutional Practice

The second primary school of Sykies, Thesaloniki, Greece, has 340 pupils. It is situated in the rather poor, densely built urban area of Sykies, a municipality in the west of Thesaloniki. The school consists of three-storey buildings set in very restricted, steep grounds. This makes connections dangerous between indoors and outdoors and within the building. The school is open to extra-curricular activities and has a rich school life. Local authorities are very supportive in funding extra-curricular activities, mainly environmental education projects. Most members of the teaching staff live in the neighbourhood and are very supportive of the transformation of the school into a cultural centre, open to the community with rich and varied functions and an attractive physical environment.

The overall goal of the project was to create an attractive and open learning environment in both indoor and outdoor school settings. Some aims of this project were to:

- Encourage cooperative learning
- Redevelop classrooms and the school grounds as a teaching resource
- Improve the quality of classrooms and the school grounds and make them safer
- Promote children's participation in the redevelopment process.

The school totally lacked vegetation or any other educational or play facilities, and the town also lacked open spaces for safe outdoor play. Through participatory forms of environmental education, two of the teachers collaborated with their pupils to improve the school's indoor and outdoor environments in order to use them as an educational resource. Children worked under the guidance of the teachers using action research methods with a combination of methodological tools from psychology, architecture, education, social psychology, and new technologies in order to produce proposals for the redevelopment of their classrooms and school grounds. They audited the characteristics of the classrooms and the school grounds. They also asked pupils, parents, and other stakeholders about how they thought the school grounds should be developed. All the data were used to prepare a proposed transformation plan for the school's physical environment that was presented to the whole school.

The main difficulty was that although the majority of the teachers were in favour of the redevelopment only about one fifth of the school's teaching staff became actively engaged. In addition, some of the children who participated will have left the school before the project is completed.

Text Box 19.4 Culture and Ethos

St Theresa Girls' Junior Lyceum, Malta, is a secondary school with 1,050 students. Access to the school is controlled by national examinations and students are expected to achieve good results in their studies. The School's

(continued)

Text Box 19.4 (continued)

Green Club activities are mainly carried out during morning assemblies and lunch breaks and are not related to specific curriculum subjects. Although the club only has 20 student volunteers, its activities target all the students, school staff, and parents, and these activities have become part of normal school life. The club meets regularly to discuss environmental issues identified by the students and coordinating teacher that are relevant to the school, and how these can be addressed. Activities can be grouped under two major categories:

Raising the environmental awareness of the general school population

- Club members address their peers during assemblies to encourage participation in club activities and to provide information about environmental issues through mimes and plays, and also by sitting on discussion panels in television/radio programmes. Competitions and exhibitions are organised regularly.

Improving the school environment. The club has created and manages:

- A 'nature patch' by planting indigenous trees in the grounds
- A waste recycling bank
- A compost heap
- Craft sessions that utilise waste
- A yearly 'Green Your Classroom Award' for classes adopting sustainable practices

While funding and time constraints are difficulties, the highly academic focus of the school has also made it hard for the Green Club to have a direct impact on the curriculum. Reference to academic subjects during activities is purely incidental. Nevertheless, the skills learned by club members, particularly communication skills, have had a positive effect on academic abilities. The club enjoys a very democratic structure and the ideas of every member are given their due attention. This allows the club to cater for a variety of student needs and learning styles. The underlying emphasis is on the development of a pro-environmental ethic that is manifested in sustainable actions.

Whole school approaches can lead to changes in actions (Rauch 2000), such as reducing vandalism, increasing waste separation (see Text Box 19.4), saving energy, and reducing aggression and destruction. An unforeseen benefit of a European-funded initiative to teach pupils how to play traditional children's games in the grounds of a primary school in Celle in Germany was an improved social climate as children's aggression at break times declined (see Case Study 7.3.4, Unit 7, SEEPS 2004). A similar reduction in aggressive playground behaviour was reported by Farrer with Hawkes (2000) as a consequence of a whole school approach rooted in values education. In some cases sustainability education has been targeted at integrating inattentive or difficult pupils because it affords the opportunity of conveying a feeling of self-worth and meaning to pupils (Rauch 2000). Children participated with more fervour when painting classrooms and planting their school grounds to their own designs (Rauch 2000, see Text Boxes 19.3 and 19.4). By thinking, investigating, and writing about their community, children reconfirm their own and their families worth and gained knowledge about the problems that they and their society must confront (Peterson 1999) (see Text Box 19.3). While Breiting and Mogensen (1999) report that almost three quarters of pupils may consider that concrete action towards the solving of an environmental problem is an essential part of education, research also shows that when practices do not cohere with espoused values, inauthentic participation promotes 'learned helplessness', as young people become aware that their minority status inhibits their influence as actors (Uzzell *et al.* 1994; Freeman 1999).

A strong but rarely aired argument for fostering legitimate pupil participation in whole school approaches is the emergent influence that pupils can have on actions in their local community. This notion of children as mediators of social change within a community of action has received limited attention within the social sciences (Uzzell *et al.* 1994; Rickinson 2001). Suggestions about the positive impact made on child–adult relationships by the *Ecologisation of Schools* project in Austria were all based on adult and not pupil reports (Eder 1999). However this notion of mediation can create problems if it is only cognitive and not situative in approach, because it can result in children's concerns becoming global rather than local, while their understanding of causality and responsibility remain at the concrete and personal level (Uzzell *et al.* 1994). The combination of these two trends can have a debilitating effect on action competence and mediative capacity, as found in the following example (Uzzell *et al.* 1994:200):

> The children who participated in one of the environmental education classes were three times more likely than the 'control' children to state their minority status as a barrier to future action.

Text Boxes 19.1–19.3 all identify an impact on communities. Parents of pupils at Smallwood School (Text Box 19.1) are actively involved in establishing education, health, and trade links with Somaliland. Nummenpää School (Text Box 19.2) won the Paimio Local Agenda 21 Committee's prize for linking with local and regional environmental authorities and their workers. The Second Primary School of Sykies (Text Box 19.3) got parents and other stakeholders actively involved in a project to redesign the school's grounds. The participatory action research approach used in the Sykies project subsequently became a model for redesigning school grounds in the municipality.

The success of national level school sustainability programmes, including those using whole school approaches, depends on their perceived relevance to national educational priorities, opportunities for implementation, and the flexibility of the programme to adapt to changing geographical and temporal circumstances (Henderson and Tilbury 2004:16). Henderson and Tilbury (2004) also report benefits of a number of national whole school approach programmes but observe that the limited research and evaluation base and youth of these programmes makes it difficult to identify their full impact. The main benefits of whole school approach programmes which Henderson and Tilbury could identify are: reduction of schools' ecological footprint, linking sustainability education to social skills, raising awareness of local and indigenous knowledge, questioning current practices, and curriculum integration.

The limited research evidence available about whole school approaches also indicates some limitations in their implementation. Whole school approaches seem to impact most noticeably on small schools, perhaps explaining why there seem to be far more primary than secondary schools involved in national whole school approach award schemes (Henderson and Tilbury 2004). If awards become too complex, as in the case of the Swedish Green School Award, this can limit the number of schools that submit for an award (Henderson and Tilbury 2004).

Most national whole school approach programmes demonstrate examples of awareness raising and environmental auditing, rather than participatory decision-making (Henderson and Tilbury 2004). Pupil participation is largely confined to committees and school councils that take decisions about institutional practice and/or the social and organisational strands of whole school approaches, rather than being involved in decision-making about curriculum content and pedagogy (Shallcross *et al.* 2006, see also Chapter 18 by Barratt and Barratt Hacking, and Chapter 20 by Carlsson and Sanders, this volume). To have any impact, whole school approach programmes need the support of school management teams and governing bodies (Henderson and Tilbury 2004, see also Text Boxes 19.1 and 19.3). Yet where leadership comes from a head teacher committed to liberal notions of participation, she can become identified as the heroic, visionary, leader who is both a culture founder and bearer (Nias *et al.* 1989), a status antithetical to the democratic, participatory notions of distributed leadership that she is trying to promote often in the most collaborative and transparent way (Robinson *et al.* 2006).

19.4 Whole School Approaches and the Foundations of Participation

Whole school approaches have implications for actions in each strand of Figure 19.1. Actions will not be fully educative without formal curricular input (Apple and Beane 1999), that is, what is learned in classrooms, and more importantly, how this is learned, needs to be reflected in day-to-day school life. The best hope for the action focus that sustainability education entails lies in process-focused whole school approaches based on democratic, participative, collaborative cultures that 'enhance student knowledge' (Niemi and Junn 1998:123). An authentic participatory ethos requires a collaborative culture of communication and decision-making based on mutual recognition and respect. The key features of collaborative or active school cultures (Smyth and Hattam 2002) are their beliefs, values, understandings, attitudes, meanings, norms, symbols, rituals, and ceremonies (Nias *et al.* 1989). Collaborative cultures constitute a subliminal pedagogy because they fuse the spiritual, political, and intellectual by fostering diversity while building trust – provoking and containing anxiety, engaging in knowledge creation, and combining connectedness with open-endedness (Fullan 1999). Thus collaborative schools become communities of action by changing from 'isolated islands' (Uzzell 1999) to developing environmental agency through community links (see Text Boxes 19.3 and 19.4).

Inclusive participation is a defining characteristic of radical interpretations of sustainable development and is implicit in United Nations, definitions of the term (Attfield 2003). Pupil participation is integral to collaborative school cultures as it can develop individual potential by moving from externally prescribed to more internally derived actions. Whole school approaches can empower children and young people as legitimate peripheral participation leads to progressively greater involvement in sociocultural actions (see Text Box 19.1). We would suggest that Lave and Wenger's (1991:51) observation that 'participation is always based on situated negotiation and renegotiation of meaning in the world' could have equally been written about sustainability education, particularly as communities of action evolve and transform.

However, despite the self-evident truth that participation is best learned through democratic experiences, most schools are remarkably undemocratic institutions (Beane and Apple 1999:13) and most proposals for educational reform omit considerations of citizenship and participation across the board (Orr 1992). Whole school approaches require adult and especially teacher participation and are more likely to develop where there is synergistic teaching. This involves the intentional intervention of teachers, to promote action-focused learning and widening teaching to reinforce the learning connections between structured and serendipitous formal and non-formal curricular experiences (see Text Box 19.3).

What we are advocating is a participative democracy that transfers power to people, and not necessarily a representative democracy. Participative democracy is about the common good, not just self-interest. Beane and Apple (1999) argue that democratic schools are concerned with collaboration and cooperation rather than competition. A survey by the Council of Europe (2000) shows that pupils believe that the qualities of democratic classrooms are: tolerance, mutual respect, valuing of individuals, active participation, listening, and fairness. In whole school approaches the values of participative democracy are manifest in the ways that informal social networks and interactions help make decisions in schools, whereas committees, councils, and meetings are often the overt symbols of a representative democracy that may be oligarchical, rather than representative of a deeply embedded participative democracy.

Action is important to democracy because it can reduce feelings of powerlessness if action is formulated within a culturally critical, and situated (as opposed to conditioned) approach to civic education (Niemi and Junn 1998). Learning, teaching, and action that proceed hand in hand provide opportunities to weaken defection from and strengthen attachment to sustainable actions (see Text Boxes 19.1–19.4). The continuity of social relationships in whole school approaches can promote the habituation that is indispensable in reducing defection because it engenders the mutual trust that leads to cooperation (Ridley 1996). But to foster attachment, schools have to engage with communities in new ways (see Text Boxes 19.1 and 19.3), to become active agents of change rather than 'passive transmitters of information or values' (Uzzell *et al.* 1994:13), in short, communities of action.

Thus whole school approaches require stamina as locally decided actions become routine interactions between children and adults in environments of 'ontological security' that can be established early in life (Giddens 1979:218). 'Routine is strongest when it is sanctioned, or sanctified, by tradition' (Giddens 1979:219). Through whole school approaches, moral education can commence as part of an early childhood education (Farrer with Hawkes 2000) (see Text Box 19.1), rooted in communities of action that involve pupils, teachers, and parents. An emphasis on learning to talk rather than learning from talk, enables pupils to construct their own situated meanings and identities through social interaction within these communities (Lave and Wenger 1991). Pupils learn at an early age to respect one another, to cooperate and work with each other, to listen and discuss – all democratic attributes that are essential to whole school approaches (Brain 2001).

Such whole school philosophies are reflected in the four school case studies drawn from England, Finland, Greece, and Malta (Shallcross *et al.* 2006) and discussed in Text Boxes 19.1–19.4. These case studies illustrate four of the five strands of whole school approaches shown on Figure 19.1 (there is no case study on self evaluation – for more detail on this see Wals 2006). All the case studies have been influenced to some degree by the *Sustainability Education in European Primary Schools* project (SEEPS 2004) of which a summary follows. For full details, see Shallcross *et al.* (2006).

19.5 Sustainability Education in European Primary Schools

SEEPS is a continuing professional development project designed to assist teachers through a 'training the trainers' model to developing sustainability education in their schools by promoting whole school approaches. The project promotes a school-focused, situated learning approach to continuing professional development, and provides support materials for teachers to help them to develop in-service education programmes in their own schools, after at least one member of staff has been trained in the use of the SEEPS materials. The project is based on the belief that the most effective continuing professional development integrates three areas of knowledge: theoretical, contextual, and personal. Contextualised professional learning is catered for through a strong focus on case studies and the facility to download exemplar activities in the project from its web site and subsequently adapt and/or translate these for local use. The SEEPS project has been developed by educationalists from 14 European countries including primary school teachers, teacher educators, representatives of ministries of education, and NGOs. The project has eight units:

0. Whole school approaches
1. Why bother with sustainability education?

2. Values and attitudes
3. Culture and sustainability
4. Leading and managing change
5. Teaching through the environment
6. Self-evaluation in sustainability education
7. Case studies and action research

The web site for *Educating for Sustainable Futures*: www.education.ed.ac.uk/esf or www.mmu.ac.uk/ioe (showcase) has adapted the SEEPS materials for use in initial teacher education.

As discussed earlier, whole school approaches as communities of action encourage legitimate peripheral participation (Lave and Wenger 1991) through links with issues that are authentic because they link school education to the reality that pupils will experience after school (Uzzell 1999:404) (see Text Boxes 19.1–19.4). Children support this link with the authentic: '[w]hat you learn in school needs to be put in perspective in the real world' (Connell *et al.* 1999:103). This authenticity empowers pupils by releasing their internal creative power (Begg 2000:114). However, legitimate participation will not be encouraged if schools only cater for participatory sustainability education by allocating it space in the formal curriculum. Conversely, it is equally important, that participation is not stifled by practices in the non-formal curriculum. Whole school ap proaches do not just involve changing from modernist to more holistic epistemologies; they extend to interconnections with day-to-day actions that form a matrix of experiences in schools (Capra 1996).

Yet, most school reform does not involve children, and that which does is often tokenistic (Hart 1997) emphasising participation for reasons of efficiency (Grant Lewis and Naidoo 2004) or policy accommodation, rather than for democratic reasons. Teachers are increasingly denied the right to participate in decisions about curriculum content, while pupils find that government guidelines, examination syllabuses and/or their teachers decide what and how they learn. In schools without participative structures such as student councils, few children conceive that there are ways of learning as a citizen within school or community (Davies 1999). Councils work best if they are part of whole school participatory practices that are embedded at classroom and school levels and in community links (Holden 1998) but often this is not the case (Grant Lewis and Naidoo 2004). Whole school approaches should expose spaces for pupil voice and influence in and between all five strands shown on Figure 19.1, by for example, pupils interviewing prospective teachers, mediating playground disputes (Holden 1998), and/or evaluating their own schools (see Text Box 19.3).

However in deciding where these spaces exist it is important that the perspectival nature of both participation and school culture is recognised (Smyth and Hattam 2002:377):

> [s]chool cultures are not the prerogative or domain of any one group – teachers, students, parents, politicians, and the business community or policy makers. Rather, school cultures emerge out of and are continually constructed and reconstructed through the ongoing struggles between and within each of these groups as they vie to have their particular view of schooling represented.

Pupils are aware of these cultural differences. Pupils do not regard 'traditional' schools as communities of action because their school cultures are inflexibly hierarchical and relationships with teachers are authoritarian. Conversely children in 'experimental' schools see school culture as more participatory and 'founded on a social consensus achieved through discussion and negotiation' (Emler 1992:76). Such legitimate participation encourages pupils to construct 'shared meanings in a combined exercise with teachers' (Woods 1996:39) (see Text Box 19.3) in both their formal and non-formal curricula. Pupils are motivated by such reforms. 'The more that regimes are changed to reflect the values that pupils call for (intellectual challenge, fairness, etc.), the stronger pupils' commitment to learning in schools is likely to be' (Rudduck and Flutter 2000:85) (see Text Box 19.1).

Whole school approaches are then not simply a reaction to the relative failure of awareness raising and values education to promote sustainable actions (Sterling 2001), they encapsulate positive reasons for the advocacy of collaboration and participative democracy. The focus on actions in whole school approaches has psychological benefit because the occupational culture of primary teachers, in particular, prefers the pragmatically and practically active to the theoretical (Alexander 1984). The essence of sustainable institutional practices in whole school approaches (see Figure 19.1) is their coherence with the cognitive and affective messages constructed in the formal curriculum. However, institutional practices, social organisation, and links with the community are often marginal to a school's educational endeavours and therefore areas in which innovation becomes easier. In turn, whole school approaches to sustainability education should contain a curriculum that would address synthesis (O'Sullivan 1999) partly through an interdisciplinary or transdisciplinary formal curriculum but also through sustainable actions that integrate the cognitive, affective, and active domains (see Text Boxes 19.1 and 19.2). This would be a curriculum that was proactively critical of modernist socio-economic and political structures but sensitive to the conservationist ethics and traditions of local cultures (Bowers 2002). It would integrate sociocultural knowledge with knowledge and experience of the natural world and teach things that we imagine the Earth would teach us such as humility, beauty, obligation, and wildness (Orr 1992) through a concern for environmental justice.

19.6 Recommendations for Analytical Practice

Participation is contextually variable, problematic, multilayered, multiperspectival, and sometimes corrupt or tyrannical. One reason for this situation is that when participation and many other educational matters are discussed, vision (recommendations for desirable practices), rationale (why particular visions are desirable), and designs (how we journey towards these visions) are often conflated. What is often missing from discussions about sustainability education, participation, and whole school approaches is any consideration of design: of a theory of action for change. Visions of change, in this case whole school approaches (Figure 19.1) are elusive, first, because as we journey towards a vision, especially one as new as sustainability, it changes: the act of engaging reality creates new possibilities (Chodorkoff 1990). Second, ideals themselves are contentious in education (Reisman 1990) and from a postmodern perspective, visions of sustainability education choreographed as outcomes are problematical, because of their universalising Utopianism.

When discussing visions it is also important to identify at what level vision is being described: ultimate values, platforms, decisions, or actions (Naess 1995). Whole school approaches are a platform level vision or set of principles to guide action. Vision at the level of ultimate values is volatile and the locus of much internecine academic warfare in debates about sustainability education. Outlining vision at the levels of decisions and/or actions is anathema in sustainability education because of its commitment to contextually decided sustainable solutions. Thus platforms, because of their focus on actions without prescribing practices, represent the most promising level of vision to support models for change that address a range of sustainable actions based on differing ultimate values. It was for this and other reasons that the SEEPS project advocates whole school approaches as a platform position that could be supported by a number of competing environmental rationales such as ecofeminism, deep ecology, or social ecology. But whatever the vision, gaps will still exist between rhetoric and reality.

Sustainability educators have grappled with this theory–action gap as a reflexive, existential problem of individual and/or organisational actions. However, when participation is analysed an action–values gap may appear to exist, not because of a failure of action competence, but because those who espouse transformatory visions of education rooted in, for example, an ecocentric ethic, acknowledge the potentially blinding and disempowering nature of such visions for others, and the pragmatic need for intermediate steps if these visions are ever to be approached. For example, many regard vegetarian diets as integral to an ecocentric vision of sustainability education. However it is difficult for the majority of people to act on this vision speedily. Thus an intermediate step might be to accept a carnivorous diet for the majority while seeking to ensure that any meat consumed relied on humane husbandry and slaughter (Roszak 1995:16):

> It is elementary psychology that those who wish to change the world for the better should not begin by vilifying the public they seek to persuade, or by confronting it with a task that appears impossible.

In a similar way teachers and young people may have to accept participation at a lower level than they desire, because internal or external circumstances render high-level participation impossible. For example, a pupil council might only be permitted to deal with marginal issues such as lunch breaks and discos while deliberations about learning and teaching remain off-limits (Fielding 2004; Shallcross *et al.* 2006). It is even rarer for pupils to participate in research in which meaning is constructed, conclusions drawn, and action proposed (see Text Box 19.3) consequent upon a dialogue between pupils and senior staff (Fielding 2004). However a teacher or teachers although wedded to legitimate participation may decide because of institutional or structural constraints that lower levels of participation

(Hart 1997) are better than no participation at all. What is crucial in these circumstances is that young people are not simply expected to comply with this lower level of participation but be aware of the contested nature of the vision with which current practice appears discrepant; they know participants in this false negative. In these circumstances, unless young people are aware of this espoused vision of high-level participation they will question the integrity of their teachers and/or schools (Titman 1994).

However, in analysing whole school approaches the possibility of false negatives must be acknowledged by considering the possibility that existing practices form part of a crafted theory of action that acknowledges the necessity to accept, that which is less than the preferable/desirable but an improvement on the normal/probable (see Text Box 19.4). The educator, committed to transformative participation in which students decide and act on their own with the teacher playing no more than a facilitating role, may currently be denied the role of the oppositional intellectual (Giroux 1996) on the shop floor of the current performatively focused education industry. To retain any sense of professional integrity such teachers will often need to be subversively complicit; providing what the system requires, while challenging, for example, fundamental performative values as the head teacher at Smallwood did when shunning publicity about the school's performance in national tests.

Examples of this pragmatic approach to design can be found in many case studies in the SEEPS Project (2004). Many of these case studies have visions of whole school approaches as future communities of action but the practices that many schools currently report fall short of the integrated vision in Figure 19.1. What many schools are reporting are the first steps on the way to establishing whole school approaches as communities of sustainable action. In whole school approaches learning itself is a process of participation in which learners are initially peripheral to community actions but with experience their participation becomes more authentic and less peripheral (Lave and Wenger 1991). The important question is how this peripheral participation is made legitimate, i.e. how and by whom peripheral steps are decided? The important design strategy is to acknowledge that schools' existing practices will already address some aspects of whole school approaches to sustainability education and to identify the next feasible step on the pilgrims' progress towards more integrated whole school approaches.

19.7 Participation, Power, and Democracy

Can whole school approaches ever be fully realised? Can education ever be quintessentially democratic for children and/or young people? Tensions exist between the emphasis on freedom of choice in strong versions of liberal democracy and the focus on regulation in many versions of sustainability (Attfield 2003) indicating that participation and sustainability education cannot be discussed without deliberating power and democracy. Pupil voice is not necessarily participatively democratic; it is often differentiated in source and motive. Pupil participation is normally associated with the representative democracy of a school council in which serious doubts can arise about representatives' abilities to reflect their peers' views. For example, middle class girls are more likely to speak out, which raises questions of validity and the extent to which these voices represent others (Fielding 2001).

Participation is usually granted to young people not enfranchised. Consenting to children's participation is usually the prerogative of adults (Kirby 2001). This gifting, is frequently motivated by an ideology that sees young people as 'becoming' rather than 'being' citizens (John 1996), children with responsibilities rather than young people with rights. Adults frequently give voice to pupils' research data because in order to be heard student voice has to be presented in a formal adult register. But this register can misrepresent by not conveying the same meaning contained in the knowledge, concerns, and linguistic expressions of the young people themselves (Fielding 2001). Even when there are appropriate rules and cultures to enable legitimate participation 'micro power processes' (Kapoor 2004) may still intimidate or silence young people's voice through adults' use of technical language, aggressive postures, or extrovert behaviour.

Granting students voice is often a reaction to external calls for accountability, driven by a school effectiveness agenda rather than a response to calls for democratic, participative student agency (Fielding 2001). Equally those engaged in participatory activities in schools need to give themselves permission to take risks and avoid a sort or Foucauldian 'technology of the self' by trying to second guess what everybody else might want. Discussions of participation, democratic education, and student voice are incomplete without deliberations over concepts and loci of power and the adjustment of programmes of participation to such discussions. Governments often prefer top-down expert driven 'power over' (Begg 2000) versions of participation based on implementation after restricted consultation (Wals and Jickling 2002) whereas radical interpretations of participation are more inclusive and grassroots in their focus (Attfield 2003) and situated in conceptions of power to people. Voice alone is insufficient, it has to be heard not just listened to and it has to be powerful.

There is evidence to suggest that in early adolescence, boys are more inclined towards justice and girls towards care as motives for action (Berman 1997). However, activism requires that both these voices are held in balance (Berman 1997). But caution must be omnipresent in whole school approaches, as it is too easy to unveil spaces for children to speak on behalf of adults, a form of ventriloquising that, rather than erasing the silence of children re-silences them; re-otherises rather than de-otherises them through appropriation of their voice – a form of 'double dumbing'. Initiatives that promote student voice as a data source for shaping approaches to school effectiveness in a commodified, consumerised, individualised, and vocationalised model of education have all the democracy and power of the supermarket loyalty card. On the other hand, participation should be voluntary, young people have the right to reject inclusion and participation, as in the case of the blind boy who chose to leave mainstream school and return to a school for the blind because he could only play football at the school for the blind.

Addressing democracy through political literacy alone is not sustainable and it distances schools from the real world. By making pupils aware of the deficiencies of political systems and how the schooling system denies them voice, political literacy may instil feelings of apathy and powerlessness rather than provide a solid epistemological base for active citizenship. It is participation that involves the devolution of power to pupils, schools, and their local communities to address authentic issues that many regard as the sine qua non of whole school approaches to sustainability education (Elliot 1999; Uzzell 1999), in which a crucial question may be how student voice was involved in establishing itself. These arguments raise questions about whether authentic pupil participation is possible without adopting an approach to learning, teaching, and schooling that starts with a whole school approach vision. Without such synergistic approaches to learning, teaching, and schooling the conditions for authentic pupil participation in SE may not be present. However, transforming schools through whole school approaches into communities of action that integrate educational relationships with democracy has been and always will be difficult (Fielding 2001).

19.8 Conclusion: Implications for Research

A central issue is the lack of research and evaluation into the impact and processes of whole school approaches (Henderson and Tilbury 2004). The neglect of longitudinal and community-focused research is particularly crucial to the case for the potential lifelong and lifewide impact of whole school approaches argued in this chapter. While there has been research into the specific impacts and processes of whole school approaches in individual schools, issues arise about research into genuine pupil participation in schools that adopt such approaches. These include the development of the research capacity of professionals to ask and research questions within the framework of institutional self-evaluation that is anthropological and has the capacity not only to reveal inconsistent actions but also false negatives. The clarification and subsequent classification of actions as inconsistent or as false negatives is crucial. Inconsistency between espoused values and actions calls the integrity and moral influence of adults into question while false negatives illustrate pragmatic designs for action. Although sustainability education is fundamentally concerned with changing lifestyles, there are occasions when deliberative actions such as writing to a politician or local community member (see Text Box 19.1) assume greater importance than direct actions such as recycling. Research should reveal the underlying rationales for such deliberative actions, not simply ask questions about direct actions. To be inclusive in its access to data, such research should aim to involve pupils, as well as adults, as researchers within an ethnographic, action research approach to self-evaluation of whole school approaches (Shallcross and Robinson 1999).

In order that high levels of pupil participation in sustainability education within whole school approaches can be sustained, it is important that pupils are trained as researchers. Pupils can investigate a local sustainability issues of interest to them (e.g. school grounds development, bullying or the use of school grounds, see Text Box 19.3) and teachers can investigate these same issues as well as ways of improving their own teaching (e.g. working in groups, utilising community resources, or conflict management). Although the focus of the research will vary, the process is similar in that all parties are engaged in an inquiry that identifies issues of mutual concern, constructs potential solutions and implements one or more of these, and evaluates results.

Multimodal analysis that looks at image and body language as well as text seems an obvious approach to research into whole school approaches because of the range of modalities through which learning occurs in such communities of action. However, multimodal approaches may create a problem because of their complexity. This may make them the preserve of external experts which could undermine research as an emancipatory, situated, self-evaluative enterprise (Johnston 2000). Such research, then, runs the risk of becoming the property of the expert researcher not the practitioner, and particularly not pupils.

The ethics of research into whole school approaches, especially emancipatory research, has to have a convincing story to tell about corroboration, the participation of all community members as researchers, and community ownership of solutions:

> [r]esearchers are selfish and greedy people who do not have the interest of community at heart, but are only interested in their academic advancement at the expense of community members. They often conducted their researches and never fed back the information to the community (A traditional Venda leader cited in Shallcross *et al.*, n.d., n.p.)

The situation described above raises the danger of the researcher becoming a ventriloquist who appropriates community voice. In emancipatory research corroboration of analysis and proposed action is not just an issue of epistemological and methodological validity. It is above all a central issue in whole school approaches to sustainability education as communities of action of situated ethical integrity that encapsulate the ethics of research, ethics as research, and research into ethics.

References

Alexander, R. (1984) *Primary Teaching*. London: Cassell.

Apple, M.W. and Beane, J.A. (1999) Lessons from democratic schools. In: Apple, M.W. and Beane, J.A. (eds), *Democratic Schools: Lessons from the Chalk Face*. Buckingham, UK: Open University Press, pp. 118–123.

Attfield, R. (2003) *Environmental Ethics*. Cambridge: Polity.

Beane, J.A. and Apple, M.W. (1999) The case for democratic schools. In: Apple, M.W. and Beane, J.A. (eds), *Democratic Schools: Lessons from the Chalk Face*. Buckingham, UK: Open University Press, pp. 1–29.

Begg, A. (2000) *Empowering the Earth: Strategies for Social Change*. Dartington, UK: Green Books.

Berman, S. (1997) *Children's Social Consciousness and the Development of Social Responsibility*. Albany, NY: State University of New York Press.

Bowers, C.A. (2002) Toward an eco-justice pedagogy. *Environmental Education Research*, 8, 1: 21–34.

Bowers, C.A. (2004) Revitalizing the commons or an individualised approach to planetary citizenship: the choice before us. *Educational Studies*, 36, 1: 45–58.
Brain, J. (2001) True learning. *Resurgence*, 204: 30–31.
Breiting, S. and Mogensen, F. (1999) Action competence and environmental education. *Cambridge Journal of Education*, 29, 3: 349–353.
Capra, F. (1996) *The Web of Life*. New York: Anchor Books.
Chodorkoff, D. (1990) Social ecology and community development. In: Clark, J. (ed), *Renewing The Earth: The Promise of Social Ecology*. London: Green Print, pp. 69–79.
Connell, S., Fien, J., Lee, J., Sykes, H., and Yencken, D. (1999) 'If it doesn't directly affect you, you don't think about it': a qualitative study of young people's environmental attitudes in two Australian cities. *Environmental Education Research*, 5, 1: 95–113.
Council of Europe (2000) Everyone Can Make a Difference, Council of Europe Pilot Project on Participation in and Through School, Third Training Seminar for Teachers, Report. Strasbourg: Council of Europe.
Davies, L. (1999) Researching democratic understanding in primary school. *Research in Education*, 61: 39–48.
Dobson, A. (1998) *Justice and the Environment*. Oxford: Oxford University Press.
Eder, E. (1999) From scepticism to the urge to take action, or from daily frustration to a solo initiative. *Cambridge Journal of Education*, 29, 3: 355–365.
Elliot, J. (1999) Sustainable society and environmental education: future perspectives and demands for the education system. *Cambridge Journal of Education*, 29, 3: 325–340.
Emler, N. (1992) Childhood origins of beliefs about institutional authority. *New Directions in Child Development*, 56: 65–78.
Farrer, F. with Hawkes N. (2000) *A Quiet Revolution: Encouraging Positive Values in Our Children*. London: Rider.
Fielding, M. (2001) Students as radical agents of change. *Journal of Educational Change*, 2, 2: 123–141.
Fielding, M. (2004) Transformative approaches to student voice: theoretical underpinnings, recalcitrant realities. *British Educational Research Journal*, 30, 4: 295–311.
Foster, J. (2001) Education as sustainability. *Environmental Education Research*, 7, 2: 153–65.
Freeman, C. (1999) Children's participation in environmental decision making. In: Buckingham-Hatfield, S. and Percy, S. (eds), *Constructing Local Environmental Agendas: People Places and Participation*. London: Routledge, pp. 68–80.
Fullan, M. (1999) *Change Forces: The Sequel*. London: Falmer.
Giddens, A. (1979) *Central Problems in Social Theory Action, Structure and Contradiction in Social Analysis*. London: Macmillan.
Giroux, H.A. (1996) Towards a postmodern pedagogy. In: Cahoone, L. (ed), *From Modernism to Postmodernism: An Anthology*. Oxford: Blackwell, pp. 687–697.
Grant Lewis, S. and Naidoo J. (2004) Whose theory of participation? School governance policy and practice in South Africa. *Current Issues in Comparative Education* [Online] 6, 2. Available at: http://www.tc.columbia.edu/cice/articles/sgl162.htm [11/10/04].
Greeno, J.G., Collins, A.M., and Resnick, L.B. (1996) Cognition and learning. In: Berliner, D.C. and Calfee, R.C. (eds), *Handbook of Educational Psychology*. New York: Macmillan, pp. 15–46.
Hart, R. (1997) *Children's Participation: The Theory and Practice of Involving Young Citizens in Community Development and Environmental Care*. London: UNICEF/Earthscan.
Henderson, K. and Tilbury, D. (2004) *Whole School Approaches to Sustainability: An International Review of Sustainable School Programs*. Sydney: Australian Research Institute in Education for Sustainability (ARIES) for the Department of the Environment and Heritage, Australian Government.
Holden, C. (1998) Keen at 11, cynical at 18? Encouraging pupil participation in school and community. In: Holden, C. and Clough, N. (eds), *Children as Citizens: Education for Participation*. London: Jessica Kingsley, pp. 46–62.
John, M. (ed) (1996) *Children in Charge: One Child's Right to a Fair Hearing*. London: Jessica Kingsley.

Johnston, R. (2000) Whose side, whose research, whose learning, whose outcomes? Ethics, emancipatory research and unemployment. In: Simons, H. and Usher, R. (eds), *Situated Ethics in Educational Research*. London: Routledge/Falmer, pp. 69–81.

Kapoor, I. (2004) Concluding remarks: the power of participation. *Current Issues in Comparative Education* [Online] 6, 2. Available at: http://www.tc.columbia.edu/cice/articles/ik162.htm [12/10/04].

Kirby, P. (2001) Participatory research in schools. *Forum*, 43, 2: 74–77.

Lave, J. and Wenger, E. (1991) *Situated Learning: Legitimate Peripheral Participation*. Cambridge: Cambridge University Press.

Naess, A. (1995) The apron diagram. In: Drengson, A. and Inoue, Y. (eds), *The Deep Ecology Movement: An Introductory Anthology*. Berkeley, CA: North Atlantic Books, pp. 11–12.

Nias, J., Southworth, G., and Yeomans, R. (1989) *Staff Relationships in the Primary School*. London: Cassell.

Niemi, R.G. and Junn, J. (1998) *Civic Education: What Makes Students Learn*. New Haven, CT: Yale University Press.

O'Sullivan, E. (1999) *Transformative Learning Educational Vision for the 21st Century*. Toronto, ON: University of Toronto Press.

Orr, D.W. (1992) *Ecological Literacy: Education and the Transition to a Postmodern World*. Albany, NY: State University of New York Press.

Orr, D.W. (1994) *Earth in Mind: On Education, Environment, and the Human Prospect*. Washington, DC: Island Press.

Peterson, B. (1999) La Escuela Fratney: a journey towards democracy. In: Apple, M.W. and Beane, J.A. (eds), *Democratic Schools: Lessons from the Chalk Face*. Buckingham: Open University Press, pp. 68–97.

Posch, P. (1993) Approaches to values in environmental education. In Organisation for Economic Cooperation and Development/Environment National Schools Initiative Values in Environmental Education. Conference Report. Dundee, UK: Scottish Consultative Council on the Curriculum, pp. 27–35.

Posch, P. (1999) The ecologisation of schools and its implications for educational policy. *Cambridge Journal of Education*, 29, 3: 341–348.

Rauch, F. (2000) Schools: a place of ecological learning. *Environmental Education Research*, 6, 3: 245–56.

Reisman, D. (1990) *Theories of Collective Action: Downs, Olson and Hirsch*. Basingstoke, UK: Macmillan.

Rickinson, M. (2001) Learners and learning in environmental education: a critical review of the research. *Environmental Education Research*, 7, 3: 207–320.

Ridley, M. (1996) *The Origins of Virtue*. London: Viking (Penguin).

Robinson, J., Shallcross, T., and McDonnell, P. (2006) Being sensible and social sensibility? Smallwood supporting somaliland. In: Shallcross, T., Pace, P., Robinson, J., and Wals A.E.J. (eds), *Creating Sustainable Environments in Our Schools*. Stoke-on-Trent, UK: Trentham Books, pp. 105–121.

Roszak, T. (1995) Where psyche meets Gaia. In: Roszak, T., Gomes, M.E., and Kanner, A.D. (eds), *Ecopsychology: Restoring the Earth Healing the Mind*. San Francisco, CA: Sierra Club Books, pp. 1–17.

Rudduck, J. (1999) 'Education for All', 'Achievement for All' and pupils who are 'too good to drift'. *Education Today*, 49, 2: 3–11.

Rudduck, J. and Flutter, J. (2000) Pupil participation and pupil perspective: 'carving a new order of experience'. *Cambridge Journal of Education*, 30, 1: 75–89.

SEEPS (Sustainability Education in European Primary Schools) (2004) *School Development Through Whole School Approaches to Sustainability Education*. Manchester: Manchester Metropolitan University (www.education.ed.ac.uk).

Shallcross, T. (2003) Education as Second Nature: Deep Ecology and School Development Through Whole Institution Approaches to Sustainability Education. Unpublished doctoral thesis, Manchester Metropolitan University, Manchester, UK.

Shallcross, T., Loubser C., Munzhedzi, S., and Nemutamvuni, M. (n.d.) The lake: comparative views of ideological and pedagogical tensions in a South African sustainability education project: a review of adult education practices around Lake Fundudzi. *South African Journal of Higher Education*, in review.

Shallcross, T., Pace, P., Robinson, J., and Wals, A. (eds) (2006) *Creating Sustainable Environments in Our Schools*. Stoke-on-Trent, UK: Trentham Books.

Shallcross T. and Wilkinson G. (1999) The primacy of action: reconstructing Scott's critique. *International Journal of Environmental Education*, 18, 4: 349–54.

Shallcross, T. and Robinson, J. (1999) A model of participation in continuing professional development and evaluation through action research in educating for sustainability. *Journal of In-service Education*, 25, 3: 403–422.

Shiva, V. (1992) Recovering the real meaning of sustainability. In: Cooper, D.E. and Palmer, J. (eds), *The Environment in Question Ethics and Global Issues*. London: Routledge, pp. 187–193.

Smyth, J. and Hattam, R. (2002) Early school leaving and the cultural geography of high schools. *British Educational Research Journal*, 28, 3: 375–397.

Stenhouse, L. (1983) *Authority, Emancipation and Education*. London: Heinemann.

Sterling, S. (2001) *Sustainable Education Revisioning Learning and Change*. Dartington: Green Books.

Titman, W. (1994) *Special Places; Special People*. Godalming, UK: WWFN and the Learning Through Landscapes Trust.

Uzzell, D., Davallon, J., Fontes, P.J., Gottesdiener, H., Jensen, B.B., Kofoed, J., Uhrenholdt, G., and Vognsen, C. (1994) *Children as Catalysts of Environmental Change*. Brussels: European Commission Directorate General for Science Research and Development.

Uzzell, D. (1999) Education for environmental action in the community: new roles and relationships. *Cambridge Journal of Education*, 29, 3: 397–414.

Wals, A.E.J. (2006) Reflective living practice: critically analytical case studies and school improvement. In: Shallcross, T., Pace, P., Robinson, J., and Wals A.E.J. (eds), *Creating Sustainable Environments in Our Schools*. Stoke-on-Trent, UK: Trentham Books, pp. 47–60.

Wals A.E.J. and Jickling B. (2002) 'Sustainability' in higher education: from doublespeak and newspeak to critical thinking and meaningful learning. *Higher Education Policy*, 15, 2: 121–131.

Wenger, E. (1999) *Communities of Practice: Learning, Meaning and Identity*. Cambridge: Cambridge University Press.

Woods, P. (1996) *Researching the Art of Teaching:* Ethnography for Educational Use. London: Routledge.

Chapter 20
School Councils as an Arena for Pupils' Participation in Collaborative Environmental Education Projects

Monica Carlsson[1] and Dawn Sanders[2]

Keywords school councils, collaboration, pupil participation, environmental education, everyday actor

20.1 Introduction

This chapter discusses pupils' participation in collaborative projects between schools and external actors in environmental education. The two case studies that are presented – from the Danish Eco School project, and The School Grounds Development Projects (run by *Learning Through Landscapes*) in England – are approaches to educational change in environmental education that take place within a school setting, and in both cases our interest is in the pupils' participation in school councils. The chapter focuses on the roles of pupils, teachers, and outside actors in collaborative projects, and the different, and sometimes inconsistent, understandings of the school councils as arenas for collaboration and participation.

20.2 Background

Over the last decade a range of initiatives driven by government and non-governmental organisations (NGOs) have sought to encourage and develop participatory frameworks for children and young people to engage with environmental issues. Danish and English research partners both participated in the Economic and Social Aspects of the Environment (SEER 1) project, which comprises research on possibilities in collaboration between school and local community about environmental issues. The project report, *Children as Catalysts of Environmental Change* (Uzzell *et al.* 1994), argues that partici-

[1]DPU, Research Programme for Environmental and Health Education, Danish School of Education, University of Aarhus, Tuborgvej 164, DK-2400 Copenhagen NV, Denmark, monica@dpu.dk

[2]Learning Department, The Natural History Museum Cromwell Road, London, SW7 5BD, UK, dawngarden@hotmail.com

A. Reid et al. (eds.), *Participation and Learning*, 321–337.

pation in collaborative projects can support pupils in: developing knowledge about authentic environmental issues, gaining experience with interaction and action, developing engagement and motivation, perceiving themselves as people who can influence their surroundings, and developing a sense of membership and belonging.

Within these kinds of collaboration, social relations between the social actors must be established, where the relations will demarcate connections as well as borders and differences between the actors (Katzenelson 1994). Collaboration is a process that involves the sharing of an area or field of action, and is therefore an arena *of* and *for* power. We find the arena concept useful in analysis within both the environmental and school area, since to a high extent both are characterised by instances of structural combat between agents with very different backgrounds – experts, politicians, 'souls on fire', and users. The concept covers perceptions about the social and political as staged or performed by several social actors, distributing roles to both themselves and others, and in doing so, constructing the power arena (based on Fink 1989). Within this chapter we explore participation in a specific social arena – the school council – and the roles of several social actors – pupils, teachers, and external agents.

In order to give our case studies a contextual framework we provide a brief review of environmental education change strategies in the UK and the Danish school systems, and point out potentials and challenges in collaboration between schools and external actors in relation to environmental education projects.

Both Danish and English stakeholders in the policy-making process have favoured a solution whereby central environmental education perspectives such as 'global education' and 'greening education' are integrated cross-curricularly into the curricula of different school subjects. This strategy appears to be wise considering the competition between the school subjects for resources, and the time constraints in the smaller subjects. How teachers interpret education policy into schools 'everyday' teaching is another question. The strategy of integration is based on the assumption that teachers, through activities that create room and possibilities for reflexivity (e.g. initial teacher training programmes and in-service training), have the capacity and opportunity to develop those competences that enable them to interpret and use the curricular guideline formulations of the cross-curricular perspectives. Unfortunately this does not seem to be the case since school systems tend not to offer this kind of support to their teachers. For example, cross-curricular approaches can create issues for the teaching and learning of environmental education in that by being placed across the curriculum rather than embedded within a subject area, environmental education relies heavily upon individual champions in the teaching community to develop and sustain environmental education in a cohesive fashion (Walker 1997).

Systems and agents outside the school have different values and perspectives. These can be perceived as a major benefit to schools collaborating with external agents, but only if they are made visible and are reflected on in the collaboration. The schools ability to create room and possibilities for reflexivity in curriculum development within the area of environmental education is therefore crucial.

On the one hand, the developments in the English and Danish school systems in environmental education seem to create a wider need for collaboration between schools and external agents, since schools need resources to develop teaching in environmental education, and collaboration with external actors can provide resources, such as: teaching materials, authentic learning situations, specialised knowledge, a normative support structure, underlying values, and symbols of values, such as can be found in the Eco-schools' Green Flag scheme.

On the other hand, possibilities for reflexivity in curriculum development within the area of environmental education seem to have shrunk. Existing school resources and conditions that relate to key factors such as time, teacher competences, and within-school culture, are crucial for how teachers and schools can utilise potentials in collaboration, particularly since collaboration also takes resources. The two case studies discussed in this chapter illustrate specific challenges in relation to collaboration when the arena is the school council.

20.3 Theoretical Framework

We start by discussing research on pupils' participation and school councils, drawing on educational and sociological perspectives, using this to generate perspectives on the case studies and to support reflection on them.

Action

Micheletti (2002:7) describes individualised collective action as a form of political participation in that it affords a way to excercise one's citizenship, through, for example: 'the practice of responsibility-taking through the creation of everyday settings on the part of citizens alone or together with others to deal with problems which they believe are affecting what they identify as the good life.'

Micheletti emphasises that the concept does not mean individualised political participation but is about participation in everyday politics, constructing the participants as 'everyday makers'. The *everyday maker* is more interested in making things happen, participating in local structures, and solving concrete problems, than in participating in the formal structures of representative democracy and discussing issues of political principle. The sphere is local, for example, at work, in the local community or housing area, or in schools. This can be contrasted with traditional political spheres for political actors, where participation takes place through membership, representation, and delegation of issues to relevant structural levels in the society. Examples of local actions are energy or water conservation in local settings such as schools, boycotting of problematical consumer goods or services, 'auditing', kitemarking and certification, and so forth.

The everyday maker concept thus also contrasts with the more traditional collective forms of action displayed within the context of political parties, demonstrations, and labour or trade unions. However, in environmental education theory, collective actions – or actions in the public sphere in the society – are generally described as out of reach as a social change strategy for children and youth

(Mogensen 1995; Jensen and Schnack 1997; Bishop and Scott 1998). Instead, the notion of individualised collective action seems to be useful when analysing children and youth action in environmental education. In Micheletti's conception, the first criterion of individualised collective action is citizen-initiated action – which concerns the question of whether citizens themselves decide to 'do something'. The second criterion – 'to deal with problems which they believe are affecting what they identify as the good life'– concern whether what 'is done' is dealing with problems people believe are important.

While there is not a direct correspondence, these criteria fit well with the definition of the action concept in action-oriented environmental education. On this, Jensen and Schnack (1997) put forward two defining features of such action: The first is that pupils themselves are involved in making the decision to do something. The second is that in order to be characterised as actions, they must be aimed at solving the problems that are being addressed.

School councils

In Soderberg's research on students' voices in school councils, interviews with student representatives illustrate how these students can often feel marginalised in relation to both the adults and the rest of the student community (Soderberg 1997:5). The student representatives did not feel they had the power to make significant change, and that they could find themselves in a false position of influence that gives them no more access to decisions than any other student in school. Evaluations of projects on children's participation (The Danish Children's Council 1998; Cross, 2003; Scarman Trust Can Do Project, Personal Communication) stress that it is important to create non-formal, constructive, and issue-oriented forums for children's participation, where, for example, children administrate a pool of money for their own projects along with the formal framework for participation, such as the representatives and representation in school councils.

While the school council is usually considered to be an arena where student voice can be represented, Taylor and Johnson (2002:6) have observed that, 'setting up a school council is not necessarily a guarantee of student participation, positive attitudes or progressive practices in school. As with any other learning strategy, the context in which the council operates and its processes and practices need to be supported and kept under review in order to generate positive outcomes for students.' Taylor and Johnson (2002:4) also found that there were sometimes issues within the non-participant student group, observing that 'much depended on the structures in place to support genuine participation by the student body'. These experiences with participation in school councils draw attention to the need for students to develop particular skills in order to participate in these structures for, as Kothari (2001:150) asks, 'what happens to the narratives of those who do not possess the right skills to perform as required?'

A way of understanding and constructing the roles of the 'non-participant' students in the school council is to reflect on the possibility that the students can gain control simply by not participating, because there is, as Kothari (2001:151) points out, always the possibility of subversion. A Foucauldian analysis of participation suggests non-participation can act as a subversive act or act of self-exclusion (p. 143). Kothari emphasises that such acts are often found in participatory methodologies, dividing the participants into conformist and non-conformist groups (p. 146).

Given this, a primary consideration in the nature of pupil participation in school councils is to ask if this has been seen by teachers and external actors mainly as a motivational tool to engage pupils in projects. These debates have implications for collaborative work with schools in that they question how meaningful these types of projects are in stimulating participation by students. A key question to ask, is what might be the best ways to build relationships between schools and external actors if real rather than symbolic participation by students are to be the outcomes (Hart 1992; Simovska 2004)?

From Kothari's perspective, the attempt to set up such dichotomies – for example, 'real' participation and 'symbolic' participation – is invested with notions of the morally 'bad' and 'good'. She underlines the need to disrupt the dichotomies by analysing power and social control, looking at the construction of social norms or rituals practised in society in everyday contexts (Kothari 2001:141). The division of the roles of pupils, teachers, and external actors in school councils could thus be seen as created in the council by way of norms, social practices and rituals embedded in school councils. 'Symbolic' or 'real' participation can then be seen not only as articulated in the immediate relationship between participants with different degrees of control and power, but also as historically constructed in the social practice and rituals of the school councils.

20.4 Introduction to the Case Studies

Environmental NGOs like *Learning Through Landscapes* (England) and the Danish Eco School project are relevant stakeholders in relation to environmental education, and are through collaboration with schools, able to put their issues on the school agenda. From the point of view of schools, interacting with outside actors holding different perspectives from the schools can be seen as a major potential benefit for reconceptualising or innovating existing approaches to teaching and learning in environmental education. Another potential in collaboration with external actors is the challenge they bring to the dominating school culture, for example, by representing a fundamentally different culture in relation to planning and implementation of projects. NGOs are often characteristically small flexible organisations, used to taking chances and developing ideas, and establishing an enterprising praxis, as befits their sustainability on perhaps relatively small and not necessarily stable financing circumstances. In comparison, the culture of schools is typically bound to a more traditional hierarchical organisation, with less flexibility in its planning and innovation structure (see also pp. 311–12, this volume).

In the environmental area, tasks are often solved through collaborations between governmental organisations (GO), private organisations, and NGOs, and as a consequence, many of the external actors involved in collaborative projects with schools act within complex situations at the interfaces between, for example, public financing, public policy, independent consultancy, and an NGO organisational structure. In this study, the School Grounds Development Projects were conducted by an environmental NGO, *Learning Through Landscapes* (see www.ltl.org.uk) in partnership with secondary schools in England. *Learning Through Landscapes* is a charity dedicated to supporting schools in developing school grounds as a key element of their teaching and learning environments. The Danish Eco School project is an international campaign with Danish roots supported by the EU and financed by public foundations (see www.eco-schools.org). The project concept was originally developed in 1994 through collaboration between an NGO, the Danish Outdoor Council, and the Municipality of Copenhagen.

Both cases can be described as working with an approach that focuses on the school setting, as opposed to one that works with the curriculum as the approach to educational change. The 'school council model' utilises a selection of students who collect the views of their fellow students and are seen by teachers and external actors as representing the 'student voice' within the project. In contrast, the 'within the curriculum model' focuses on situating a project in a particular curriculum area, and does not require 'student voice' as a key element as the target population tends to be teachers.

Both of the cases that we will discuss here can also be regarded as initiatives for providing authentic arenas for learning about environmental issues in schools. The Eco School project is based on the use of Eco-audit schemes, which is currently one of the primary societal strategies for dealing with environmental problems, and pupils participating in the Eco-audit schemes in schools have the opportunity to develop their own insights into the strengths and weaknesses of this particular strategy (Hoffmann and Carlsson 2003). While the School Grounds Developments Projects have the potential to bring to life the UK's national curriculum statutory orders for Citizenship and Education for Sustainable Development (ESD) owing to their participatory and cross-curricular nature (Rickinson and Sanders 2003:6).

20.5 Case 1: The School Grounds Development Projects

Our first case study examines pupil perspectives on their participation in school grounds projects that have entailed collaboration between schools and external actors (e.g. a landscape architect). The case study draws on narrative data collected from students' and teachers' perceptions of the projects including their experiences of the external collaborators (who were not interviewed) (Rickinson and Sanders 2003). The data were generated through semi-structured focus group interviews. Although the evaluation study focused on six secondary schools, this chapter considers the experiences of two of the schools, both of which utilised a school council model. These two schools subscribed to a common form of school council consisting of pupils representing their peers and several teachers facilitating the council's decision-making process. It will be argued that even though this approach might appear to be a participatory model there are intrinsic tensions to be found in the narratives of 'participant' and 'non-participant' student communities utilising this approach.

In schools that actively utilised the school council model, some of the participating students commented that this culture enabled them to be, amongst other things, 'really good at working together and sharing ideas, there wasn't any conflict or anything'. Some were also of the opinion that other students 'feel like the councillors are listening to them'. This was not always mirrored by non-participating students who felt excluded from this opportunity: '[I]t was kept between them and we weren't really told.' Clearly, there are differing student interpretations of the participation process involved in this model. For participating students, a key element of the model was to nurture students' notions of 'working as a community' and 'making decisions on what things are going to be more acceptable to people'. As one student observed: 'I think I learnt that planning is essential and that when you are choosing who you are working with and you are fortunate enough to choose, then you should analyse it and plan it and will be able to work with this person.' However, for non-participants there were issues regarding how meetings were run and the dissemination of information, typified by such statements as: 'We never really knew what was going on in the meetings.' As these extracts demonstrate, not only were there issues about involvement in the project from the viewpoint of non-participants, there was also differences in perceptions of the decision-making processes and how well these functioned.

In one of the schools evidence of a divergence of views depended upon the nature of student involvement in the project. Students who had participated at an active level were more positive in their responses to both the process and the product outcomes of the project. The responses of students that had participated at a less active level were more negative, including this expression of frustration: 'I am just more annoyed because you don't really get what you want.' Key differences here were not only dependent upon the extent of student involvement, but also on the nature of that involvement. In the case of the student who was 'more annoyed' the student's expected outcome from the project had not been achieved, causing him to disengage from the project outcomes.

Balancing individual with communal aspirations is an important challenge for schools utilising a participatory approach to this type of project. But there is more to this. How the school council process is experienced can vary, and in that variation lies the problematic of how students engage with a participatory discourse from the population at large in the school, the vast majority of whom are non-school council members. Here, the majority view can be to see participatory discourse as symbolic, and the minority, as felt by school council members, as real.

Questions regarding the involvement of the external actors drew out many issues from both teachers and students. The primary ones were a concern with the differences in the language and goals of the school members (teachers and students) and the landscape architects, and an apparent lack of empathy on the part of the architects for the way young people think and communicate. For example, some teachers considered that: 'At the end of the day they've done what they wanted, they have not taken on board what the young people wanted,' while one of the students commented, 'I don't think he was very good, he didn't really talk to us.' A teacher from one school argued that the landscape architects needed to take account of – even fit in with – young people's ideas, and the need to be clear in the discussion and presentation of ideas and plans. Another student's focus was on which of the architects might be most concerned with their opinion and that, for them, and in retrospect, being 'concerned with our opinions' was a greater priority than making the 'best impression'.

For these students and teachers the nature of the relationship with the landscape architect was crucial to how students feel their ideas were represented in the final product. Emerging from the evidence collected in this evaluation, both students and teachers were concerned with the ability of the landscape architects to engage with students. As one teacher commented, engaging pupils in this process is about building 'connections between what pupils say and what happens as part of a growing recognition of being listened to'. Ensuring that this is actually occurring within the school community and that external collaborators are also committed to this approach is a major challenge for this type of project (see also Chapter 18 by Barratt and Barratt Hacking, this volume).

Student and teacher perceptions of the way different processes for engaging students function, can best be demonstrated by their responses to one of the questions used in the evaluation, namely: 'What advice would you give to a colleague/ students in another secondary school about doing this kind of school grounds project?' A rich set of responses emerged from both teacher and student groups in all six secondary schools taking part in the study. Significantly, a key element of teacher responses was to consider who was 'in control' and how the project was enacted throughout the teacher and student community:

> I would certainly think about what level of person has control over this.
> Architects need to understand the social context of what they're trying to do. They need to understand the culture of schools.
> Don't get the landscape architect in too early because they'll "corrupt" the kids' minds too much. The school and the students need to have a vision and the landscape architect is the tool to implement it (not the other way round).
> You must have somebody there that motivates the students into thinking this can be done.

In contrast, students were particularly concerned with the participation processes and how their contributions were acknowledged or engaged with:

> Don't let them say, "OK you can't do this because we want this." It is your school - the teachers just teach there, we have to come here and learn.
> Make sure you say what you want to say, don't just agree with what they say.
> If you don't agree, if you're not happy with something and you don't say, you won't get anywhere or be happy with the result, cos we didn't really say enough, did we?

Such findings suggest that using the school council model as a vehicle for collaborative projects with external agents can provide evidence of small pockets of participatory practices but these are not necessarily anchored in a wider school culture, particularly given the experiences of non-council members in this evaluation. Clearly greater consideration needs to be given to how these pupils engage with these types of projects if student participation at a deeper level is to be a primary outcome of the partnership. For us, key points from this case study revolve around the fact that student and teacher notions of process and product are not guaranteed to be in harmony, and that this influences their feelings about the outcomes of the project, and their judgements about its overall value. Furthermore, the study affirms the need for external partners to be well briefed regarding the inclusion and representation of student thoughts and ideas.

20.6 Case 2: The Danish Eco School Project

The second case study is part of an evaluation project researching children's roles and possibilities for taking action in the area of energy (Hoffmann and Carlsson 2003). The data were generated through interviews with a school principal, a municipal educational consultant, the coordinator of the Danish Eco School project, and teachers and pupils working with the project. Furthermore, data were obtained through document analysis of material from the Danish Eco School projects web page (www.groentflag.dk), the school project web pages, and from pupil and teacher produced material.

An overarching project principle is that participating schools should establish environmental councils and involve pupils in these councils. Therefore, collaboration in the councils can be viewed as an arena for pupils' participation. The pupils are also encouraged to participate as technicians in an Eco-audit scheme where they learn about the procedures that lead to environmental certification. This certification is symbolised as a Green Flag, which can be obtained by the school if the following four project-criteria are met:

1. Pupils have to make investigations in order to 'get to know' energy
2. The school has to save at least 10% on electricity consumption
3. Pupils have to make a visible model, for example, of a windmill
4. Pupils have to communicate their results to the media, to the local community, and at home

The fourth criterion, informing others about the project and its findings (e.g. advice on how to save energy at school and at home), can be viewed as the main action dimension of the Eco School project. However, in one of the schools it was the teacher who took the initiative to set up a poster exhibition about the project in the public library. As she recounted in an interview, she made the exhibition herself because she was not sure that the pupils could do it, and she wanted the school to meet the criteria to get the green flag. In other interviews, several pupils said that they did not know that there had been an exhibition. One of the pupils who had seen it in the library expressed scepticism because it was situated in the children's library

– she did not think that the children could read and understand the exhibition and the adults would not see it there. The teacher's reason for taking the course of action herself, as opposed to letting the students take it, brings up an issue about whose judgement and criteria is of importance for the teachers – the outside actors, or the pupils? In this case the teacher found the outside actors' criteria more important, even if the project criteria emphasises that pupils have to take action and communicate results to the surrounding communities.

The overall demand for the schools to establish an environmental council does constitute a possibility for real participation by the pupils. Within these councils there are, in principle, conditions for framing a genuine dialogue between the different council members – pupils, teachers, the school principal and outside actors (including parents), technical departments in the municipality, NGOs, and so on. The coordinator of the Danish Eco School project commented (Hoffmann and Carlsson 2003:157):

> It is interesting to see if the environmental councils can motivate the children who also are sitting in the council, since it is a practical tool for planning and co-ordinating the work.

As the quotation illustrates, from the beginning there was a concern about how pupils would experience sitting on the council. In one school the principal described the environmental council as having a core function involving the different partners in environmental debate and work: the parent representatives, the after-school activity centre, the teachers, the school principal, and the pupils. Nevertheless, he admitted that the participation of pupils has demanded a lot of 'motivational work' by the teachers in both the council and in the classes. Two teachers at the school were co-ordinating the project, and the task for each class was written down in the minutes from the meetings in the council. In other words, the environmental council in this school functioned as a planning tool in the project, emphasising practical details.

A teacher from another school offered an example of an environmental council that provided possibilities for a higher degree of pupil participation. Here, the school had formulated curriculum guidelines for environmental education for all classes in the school with their starting point located within the Eco School project. In this school, the pupils in Grade 7 act as the facilitators of the project and choose tasks from the Eco School material. On the basis of their suggestions, the council works out specific plans to be carried out at the whole-school level.

Although this last example illustrates how pupils can be involved in making plans for the whole school, the Eco School project coordinator, the educational consultant from the municipality, and many of the teachers involved highlighted the lack of pupil engagement and ownership in the councils. Consequently, the project coordinator – an external actor – suggested having two different councils: one for pupils and another for the adults, because, 'It is simply too boring for the pupils to participate in planning meetings.'

Regarding the environmental council, the main conclusion was that under certain circumstances, it provided a setting for pupil participation in planning and decision-making related to environmental problems at their school. Whether or not this potential was utilised depended on how the adults looked at their own roles and the role of the council in the project. There is, for instance, a big difference between the council being used as a planning tool only or as a medium for creating ownership among all stakeholders. The role of the teacher and collaborating partners as facilitators – asking provocative questions, coming up with suggestions and ideas for action strategies, putting barriers in perspective and pointing out possible collaboration partners, and so on – can be seen as crucial in enabling pupil engagement in the process and for pupils' learning in the process. This can obviously be a challenge in a school culture that is dominated by the notion that teaching is a process where knowledge is passively transferred from the teacher to the pupil.

One could argue that the Eco School project opens up the notion of viewing the school as a political agent, since schools and pupils inform the surrounding community about environmental issues, and can gain political influence on environmental issues and strategies in the school, especially if they are related to saving energy. It is doubtful if the Eco School approach promotes the role of the school as a political agent, in the sense that teachers and pupils can be shown to be influencing fundamental political issues and decisions made in society. But as the following story illustrates, it does open up possibilities for schools and pupils within the structure of municipality and school politics to act as political consumers.

To illustrate, consider the following description by a teacher from a third school on how the Eco School project in his school is hooked up with the municipality Eco-audit scheme, and the use of economic incentives (Hoffmann and Carlsson 2003:164):

> We got self-administration of expenses for water, heating and electricity, and the savings that we could make we were offered to share with the municipality. A part of this was put aside for pupils' self-formulated projects administrated by the pupil council. That is pure motivational education. We were not sure if it was the right thing to do: It was motivating in a way that you could explain to the children that we received the 10000 Kroner, because they did so and so. But there is a danger there, because then "thinking green" doesn't have to originate from an inner conviction, but from an economic incentive. But we got the results.

Here, while the teacher questions that pupils might be more motivated by self-interest (money) than by so-called green values, the notion of the everyday maker seems to fit well with the experiences of these pupils' participation and the role of the pupils in the environmental councils. The use of economic incentives provides an opportunity to discuss and explore broader societal 'economy versus environment' dilemmas in authentic settings.

How then can schools use projects to illustrate such fundamental conflicts of interests, which are of central importance for the development of environmental education, without it leading to a situation where educational decisions are being directed by economical factors? Collaboration with external actors' demands that each partner puts forward their expectations of what the project is about and in which direction it should go.

20.7 Reflections on Challenges in Collaborative Projects

Reflecting on the themes that emerged from contrasting the two case studies, we will now discuss two further issues for schools and for future research on participation in collaborative projects between school and outside actors:

1. The roles of pupils, teachers, and outside actors in the school councils
2. The different, and sometimes inconsistent, understandings of the school councils

School councils – the roles of pupils, teachers, and outside actors

Pupils emphasise that evidence of the concern of external actors with pupil opinion is a greater priority than making the 'best impression' (Case 1). Teachers emphasise the need for landscape architects to follow up on what pupils say and what happens. Here, we see a shared expectation on the part of pupils and teachers that pupils are able to gain some control on the decision-making process in such projects.

In both case studies, teachers see themselves in the role of mediators and facilitators in the school councils, which can be seen as a strategy in strengthening pupil participation. The critical question is: *as a consequence of this strategy, which roles are available to the pupils, and how does this position them?* For example, the dominant understanding of the role of the teacher as a mediator and facilitator is that power is situated in people – there is the powerful (external actors) and the powerless (the pupils) – rather than power being embedded in social relations and situations. The solution is to empower those who are powerless – in this case the pupils in relation to the external actors – by using a 'buffer' – the teacher, whose role is to mediate between interests, languages, and perspectives, and facilitate social interaction between pupils and external actors.

This perspective highlights the need to consider the features of using a 'power buffer' in collaborative projects, for example:

- Which different perspectives and interests are represented in the collaboration?
- Which arenas for action and forms of social actions are suggested by different participants in the collaboration – pupils, external actors, and teachers?

The construction of such roles in collaborative projects is often complex. As the second case indicates, the understanding of the school council as an arena where teachers mediate between pupils and external actors is not satisfactory: one of the external actors suggested a division of pupil and adult councils in schools – the underlying assumption is that there is a need to mediate between pupils and teachers in the councils. Furthermore, the role of the teacher as a social actor with his/her own agenda is made clear in the second case were the outside actors' criteria proved to be more important for the teacher than her pupils' judgement.

Authenticity is one of the main arguments in educational theory for bringing outside reality into closer contact with the school, in order to promote motivation and engagement in school learning projects in the field of environmental education. Both case studies indicate that it does not seem to be sufficiently motivating in and of itself in order to ensure authentic strategies for dealing with environmental issues, through teachers and external actors emphasising the need to motivate pupils to participate in settings like the school councils.

As the first case study suggests, pupils are not always clear about the possible and likely outcomes of their participation in the school council, and as one of the teachers commented: 'You must have somebody there that motivates them, motivates the students into thinking this can be done.' Indeed, contingencies affect situations and social relationships, like education and school council work, in ways that mean we can never be sure of the outcome of situations. But we can have expectations, and we can communicate about these expectations and reflect on differences in expectations and perspectives related to those situations. Evidence of undertaking such reflexivity about collaborating partners' interests and perspectives can also work as a supplementary strategy in motivating students. As mentioned earlier, accessing the perspectives of outside actors can be perceived as a major benefit for schools collaborating with external agents, but it is only a potential if perspectives are made explicit and are reflected on within the collaborations in the projects.

There also seems to be an element of self-interest playing a crucial factor in the pupils' engagement in the school councils. In both pupil and teacher expectations, it is commonly understood that pupils should be able to put their ideas on the collaboration agenda (Case 1), and experience concrete gains from a 'you scratch my back, I'll scratch yours' collaboration (e.g. the pupils save energy and get to share the savings with the municipality, as described in the second case). Green values inspired by the ideal of sustainable development are based on reflections on what is in the interest of the local and the global community, as well as in the interest of the present and future generations (Stables and Scott 2002). This formulation indicates that green values relate to public interests to a higher degree than they do to self-interests. Micheletti (2002) points out, however, that it might not be fruitful to draw a sharp line between self-interest and public interest, since motivation based on fulfilment of self-interests can lead to development of motivation and ownership in relation to fundamental political issues. Thus, while the fulfilment of self-interest through the use of economic incentives is an essential part of environmental certification projects like the Eco School project, it provides an alternative route to fulfilling the ideals of sustainable development, typically framed as public or collective interests.

The school council – an arena for participation or for exercising control?

In the first case study, students were particularly concerned with how their contributions were acknowledged or engaged with, and that there was a risk of them conforming to the agenda of the external actors, for example: 'Make sure you say what you want to say, don't just agree with what they say,' indicating an awareness of different interests in the collaboration. The case also underlines the tensions to be found in the narratives of 'participant' and 'non-participant' student communities in utilising the school council model, where the first group felt that the adults were listening to them, and were included in the decision-making process, while the second group did not on both counts.

One could say that while the 'participants' in the case are showing commitment to the norms and rituals of the student council, the 'non-participants' are showing non-commitment by portraying themselves as excluded from the decisions made in the council. As Kothari points out, arenas of participation are also arenas for exercising social control, and this control or dominance is often not directly articulated (Kothari 2001:143). The school council can be seen as a ritual or social practice that is historically constructed as an arena for representative democracy. One of the functions of the school council in the first case study was to represent student voice within the project at the same time as it was a vehicle for collaborative projects with external agents. In the second case study, a core function of the school environmental council was 'involving the different partners in environmental debate and work', at the same time as functioning as a planning tool in the project, with an emphasis on practical details. Here we can find references to communitarian notions of democracy, emphasising values such as the right of pupils to be represented (Case 1) and dialogue (Case 2), alongside references to more liberalistic notions of democracy, emphasising efficiency, expressed in the pictures of the council as a vehicle (Case 1) and as a tool for planning (Case 2) (see Carlsson and Jensen 2004).

These can be very diverse and possibly contradictive functions and understandings of the school council. The question is, are the school councils meaningful arenas for pupil participation? The first case underlines the need for students to develop particular skills in order to participate in formal structures of a representative form of democracy, such as in the school councils. In the second case, the notion of the everyday maker is introduced as an alternative role for pupils, a role that seems to fit well with the pupils' experiences of participating in solving concrete environmental problems, and being rewarded economically. However, in neither case do the school councils deliberately deal with the lack or absence of pupil participation by non-council members, i.e. the power of the councils lies in their participants not their relations to the wider student body, and this might be equally regarded as their key weakness.

20.8 Concluding Remarks

In this chapter we have linked participation to a social arena – the school council – and to social actors – pupils, teachers, and external agents. We have discussed examples of an emerging form of political participation – individualised collective action, and a new role for the pupils – the everyday maker. We have also highlighted some of the complexities and contingencies involved in the construction of the various roles people play in collaborative environmental education projects, where the deliberate and unintended division of such roles raise yet more issues about power and its exercise and effects.

Our case studies indicate that school councils as arenas for participation and collaboration can be constructed in a variety of ways, and that they do not always automatically promote the development of pupil motivation, engagement, and ownership. We have also identified a number of educational challenges that arise when stakeholders with different degrees of social power are involved in environmental education projects in schools, particularly as they concern the recognition and acknowledgment of pupil voice. As we have shown, and as the earlier research in the SEER 1 project (Uzzell *et al.* 1994) suggests, the underlying assumptions about the roles and possibilities of pupils as participants (and political actors) have implications for the way collaborative projects with external actors are constructed. However, as we have pointed out in this chapter, it is the practice of those different understandings of the school councils as arenas for collaboration and participation that concretises their significance. Examples of the different purposes school councils have been assigned in the two case studies are as:

- A planning tool, emphasising practical details (Case 2)
- A medium for creating ownership (Case 2)
- Learning ground for representative democracy (Cases 1 and 2)
- A vehicle for collaborative processes (Case 1), e.g. generating experience of a different organisational culture
- An arena where teachers can mediate between pupils and external actors (Case 2)

In terms of the roles of pupils in the school councils, they illustrate examples of pupils:

- As participants in decision-making processes (Case 1)
- As non-participants – excluded or excluding themselves from decisions made in the council (Case 1)
- As 'everyday makers' (Cases 1 and 2)
- As engaged, driven by self-interest, e.g. economic incitements (Case 2)
- As powerless in relation to the external actors (Cases 1 and 2)

Researching such examples of pupil participation needs to reflect the possible changes in understandings of participation in these arenas, and of the roles and actions of the various actors therein. Questions need to be asked that consider which arenas for action and participation are constructed in environmental education within the school setting, and how are the roles between the social actors in projects that take this approach distributed, and which types of action and participation can be identified? Crucially, for both school communities and their external partners, how might research into these collaborative projects assist greater reflexivity and changes in practice to enable more participatory approaches?

References

Bishop, K. and Scott, W. (1998) Deconstructing action competence: developing a case for a more scientifically-attentive environmental education. *Public Understanding of Science*, 7, 3: 225–236.

Carlsson, M. and Hoffmann, B. (2004) Udfordringer i samarbejdet mellem skole og eksterne aktører i energi- og miljøundervisningen. (Challenges in collaboration between school and external actors in energy- and environmental education). In: Carlsson, M. and Hoffmann, B. (eds), *Samarbejde om bæredygtig udvikling*. (Collaboration about sustainable development). Copenhagen: Danish University of Education Press, pp. 145–164.

Carlsson, M. and Jensen, B.B. (2004) Encouraging environmental citizenship: the roles and challenges for schools. In: Bell, D. and Dobson, A. (eds), *Environmental Citizenship: Getting from Here to There?* Cambridge, MA: MIT Press, pp. 237–261.

Cross, B. (2003) *Scarman Trust Can Do Project*, Personal Communication.

The Danish Children's Council (1998) *Børn på banen - inddragelse af børn i lokalebeslutningsprocesser* (*Involvement of Children in Local Decision-Making Processes*). Copenhagen: Statens Information.

The Danish Eco-school Project (no date) (http://www.groentflag.dk) [Accessed 7 April 2004].

The Danish Ministry of Education (2003) *The Act of the Folkeskole*. (http://www.uvm.dk) [Accessed 7 April 2004].

The Eco-school Project (no date) (http://www.eco-schools.org) [Accessed 7 April 2004].

Fink, H. (1989) *Arenaer – om politik og iscenesættelse*. Copenhagen: Universitetsforlaget.

Hart, R. (1992) *Children's Participation: From Tokenism to Citizenship* (Innocenti Essay No. 4). Florence: UNICEF ICDC.

Hoffmann, B. and Carlsson, M. (2003) *Undervisning om energi. Børns roller og handlemuligheder på energiområdet (Children's Roles and Possibilities for Taking Action in the Energy-area).* Unpublished report, Copenhagen: Danish University of Education and The Danish Technical University.

Jensen, B.B. and Schnack, K. (1997) The action competence approach in environmental education. *Environmental Education Research*, 3, 2: 163–178.

Katzenelson, B (1994) *Homo Socius*. Socialpsykologisk grundbog, Copenhagen: Gyldendal.

Kothari, U. (2001) Power, knowledge and social control in participatory development. In: Cooke, B. and Kothari, U. (eds), *Participation: The New Tyranny?* London: Zed Books, pp. 139–152.

Micheletti, M. (2002) Individualized Collective Action. Paper presented at the Nordic Political Science Association's Meeting, Aalborg, Denmark, 14–17 August.

Mogensen, F. (1995) School initiatives related to environmental change – development of action competence. In: B.B. Jensen (ed), *Research in Environmental and Health Education*. Copenhagen: Forskningscenter for Miljø- og Sundhedsundervisning, Danmarks Lærerhøjskole, pp. 49–73.

Rickinson, M. and Sanders, D. (2003) *Grounds for Improvement. Secondary Action Research Programme Year 2 Report.* Unpublished report.

Simovska, V. (2004) Student participation: a democratic education perspective – experience form the health promoting schools in Macedonia. *Health Education Research*, 19, 2: 198–207.

Soderberg, M.B. (1997) *Student Leadership and Participation in Independent School Activities: Culture Created in Schools*. New York, NY: Columbia University, Esther A. and Joseph Klingenstein Center for Independent School Education.

Stables, A. and Scott, W. (2002) The quest for holism in education for sustainable development. *Environmental Education Research*, 8, 1: 53–60.

Taylor, M.J. and Johnson, R. (2002) *School Councils: Their Role in Citizenship and Personal and Social Education* (Research Summary). Slough: NFER. (http://www.nfer.ac.uk/htmldocs/Outcome_SCS.doc) [Accessed 1 April 2004].

Uzzell, D., Davallon, J., Bruun Jensen, B., Gottesdiener, H., Fontes, J., Kofoed, J., Uhrenholdt, G., and Vognsen, C. (1994) *Children as Catalysts of Environmental Change.* Report to DGXII/D-5

Research on Economic and Social Aspects of the Environment (SEER). Brussels: European Commission. Final Report, Contract No. EV5V-CT92-0157.

Walker, K. (1997) Environmental education and the school curriculum: the need for a coherent curriculum theory. *International Research in Geographical and Environmental Education*, 6, 3: 252–255.

Index

F

G